Developments in Animal and Veterinary Sciences, 13

Prostaglandins in animal reproduction

OTHER TITLES IN THIS SERIES

1. Guide Dogs for the Blind: their Selection, Development and Training
   by Clarence J. Pfaffenberger et al.
   1976 xii + 225 pp.

2. Ethology of Free-Ranging Domestic Animals
   by G.W. Arnold and M.L. Dudziński
   1978 xii + 198 pp.

3. Bacterial Infection and Immunity in Domestic Animals
   by J.B. Woolcock
   1979 xii + 254 pp.

4. Social Structure in Farm Animals
   by G.J. Syme and L.A. Syme
   1979 xii + 196 pp.

5. Physiology and Control of Parturition in Domestic Animals
   by F. Ellendorff, M. Taverne and D. Schmidt (Editors)
   1979 xii + 348 pp.

6. Trends in Veterinary Pharmacology and Toxicology
   Proceedings of the First European Congress on Veterinary Pharmacology and Toxicology, held in Zeist, 25–28 September, 1979
   edited by A.S.J.P.A.M. van Miert
   1980 x + 363 pp.

7. Veterinary Toxicology
   edited by M. Bartik and A. Piskac
   1981 346 pp.

8. Livestock Production in Europe
   edited by R. Politiek
   1982 viii + 336 pp.

9. Advances in Veterinary Immunology 1981
   edited by F. Kristensen and D.F. Antczak
   1982 vi + 282 pp.

10. Genetics and Animal Breeding
    Part A: Biological and Genetical Foundations of Animal Breeding
    Part B Stock Improvement Methods
    by J. Maciejowski and J. Zieba (Co-edition with PNW – Polish Scientific Publishers, Warszawa)
    1982 viii + 284 pp.
    1982 viii + 206 pp.

11. Constitutional Disorders and Hereditary Diseases in Domestic Animals
    by D. Hámori (Co-edited with Akadémiai Kiadó, Budapest)
    1983 730 pp.

12. Advances in Veterinary Immunology 1982
    edited by F. Kristensen and D.F. Antczak
    1983 v + 312 pp.

Developments in Animal and Veterinary Sciences, 13

Prostaglandins in animal reproduction

Proceedings from a symposium held at The Swedish University of Agricultural Sciences Uppsala, Sweden, May 31st–June 2nd, 1983

edited by

LARS-ERIC EDQVIST

Department of Clinical Chemistry, College of Veterinary Medicine, Swedish University of Agricultural Sciences, S-750 07, Uppsala, Sweden

and

HANS KINDAHL

Department of Obstetrics & Gynaecology, College of Veterinary Medicine, Swedish University of Agricultural Sciences, S-750 07, Uppsala, Sweden

ELSEVIER
Amsterdam – Oxford – New York – Tokyo 1984

ELSEVIER SCIENCE PUBLISHERS B.V.
Molenwerf 1
P.O. Box 211, 1000 AE Amsterdam, The Netherlands

Distributors for the United States and Canada:

ELSEVIER SCIENCE PUBLISHING COMPANY INC.
52, Vanderbilt Avenue
New York, NY 10017

Reprinted from the journal "Animal Reproduction Science" vol. 7, nos. 1-3.

Library of Congress Cataloging in Publication Data
Main entry under title:

Prostaglandins in animal reproduction II.

(Developments in animal and veterinary science ; 13)
Bibliography: p.
1. Prostaglandins--Congresses. 2. Veterinary obstetrics--Congresses. 3. Livestock--Reproduction--Congresses. I. Edqvist, Lars-Eric. II. Kindahl, Hans. III. Sveriges lantbruksuniversitet. IV. Title: Prostaglandins in animal reproduction 2. V. Title: Prostaglandins in animal reproduction two. VI. Series.
SF887.P76 1984 636.089'82 83-27476
ISBN 0-444-42294-3

ISBN 0-444-42294-3 (Vol. 13)
ISBN 0-444-41703-6 (Series)

Printed in The Netherlands

PREFACE

In 1980 we mounted a symposium on "Prostaglandins in Animal Reproduction" in order to provide a comprehensive review on this topic for the veterinary profession in Scandinavia. The proceeding from the symposium was published as Supplementum 77 to Acta veterinaria scandinavica. Later the opportunity arose to arrange a second symposium. Since the first meeting covered nearly all aspects of prostaglandins in reproduction of domestic female animals we decided that the second symposium should be more specialized. It should preferably cover an important area in animal reproduction in which there is some controversy concerning the role of prostaglandins. We decided to mount the second symposium on the role of prostaglandins in maintaining corpus luteum function during early pregnancy. The aim was to provide a forum for different views and for a review of the present state of knowledge. We believe that the symposium more than adequately fulfilled these objectives and we are grateful to all speakers and invited delegates for their participation. We also this time found it worthwhile to collect the presented scientific reports and to prepare this proceeding from the symposium. It is our hope that the publication of the proceeding will make a useful contribution to the dissemination of the latest information on the role of prostaglandins during early pregnancy.

This symposium as well as the first one was financed by the generosity of H. Lundbeck & Co. A/S, Copenhagen, Denmark and by AB LEO, Helsingborg, Sweden. We are grateful to the Swedish University of Agricultural Sciences for providing the necessary facilities for the symposium.

We are particularly indebted to Mr. Jørgen Frederiksen and Mrs. Kirsten Mærkedahl at the Veterinary Department of Lundbeck & Co. for their kind cooperation and help in arranging the meeting and for the excellent work done in preparing the manuscripts for printing. We also want to convey our sincere thanks to Mr. Imre von Szokolay and the staff at the Veterinary Department of AB Leo for their kind help in the arrangements of the symposium.

The editors

CONTENTS

SPEAKERS

Allen, W.R.	Cambridge, UK.
Bazer, F.W.	Gainesville, Florida, USA.
Flint, A.P.F.	Cambridge, UK.
Goff, A.	Montreal, Canada.
Inskeep, K.	Morgantown, West Virginia, USA.
Kindahl, H.	Uppsala, Sweden.
McCracken, J.A.	Shrewsbury, Massachusetts, USA.
Niswender, G.D.	Fort Collins, Colorado, USA.
Poyser, N.L.	Edinburgh, UK.
Schultz, R.H.	Kansas, USA.
Sharp, D.C.	Gainesville, Florida, USA.
Shemesh, M.	Beit Dagan, Israel.
Stabenfeldt, G.	Davis, California, USA.
Thatcher, W.W.	Gainesville, Florida, USA.

Animal Reproduction Science, 7 (1984) 1—30
Elsevier Science Publishers B.V., Amsterdam — Printed in The Netherlands

PROSTAGLANDIN PRODUCTION BY THE UTERUS OF THE NON-PREGNANT AND EARLY PREGNANT GUINEA-PIG

N.L. Poyser

Department of Pharmacology, University of Edinburgh, 1, George Square, Edinburgh, EH8 9JZ, Scotland, (UK).

ABSTRACT

Poyser, N.L., 1984. Prostaglandin production by the uterus of the non-pregnant and early pregnant guinea-pig. Anim. Reprod. Sci., 7: 1-30.

Oestradiol acting on a progesterone-primed uterus is the optimum stimulus for increasing prostaglandin $F_{2\alpha}$ ($PGF_{2\alpha}$) secretion from the guinea-pig uterus. This hormone combination "switches on" uterine $PGF_{2\alpha}$ secretion probably by increasing free intracellular Ca^{2+} concentrations since microsomal phospholipase A_2 (PLA_2) in the endometrium is maximally active at 7 mM Ca^{2+}, a concentration which is higher than physiological Ca^{2+} concentrations. Endometrial PG synthetase and PLA_2 activities are higher towards the end of the cycle so that, once PLA_2 has been activated, a greater quantity of arachidonic acid is released and a greater percentage of this arachidonic acid is converted into $PGF_{2\alpha}$. Oestradiol is responsible for the increase in PG synthetase activity, with the majority of the PGH_2 synthesized being directed into the $PGF_{2\alpha}$ forming pathway. Studies reported on the rat uterus indicate that oestradiol stimulates also PLA_2 activity. However, these stimulatory actions of oestradiol are attenuated by progesterone. Arachidonic acid uptake into endometrial phospholipids, particularly phosphatidylcholine (PC) and phosphatidylethanolamine (PE) increases between days 7 and 15 of the cycle. This probably occurs to compensate for the increased release of arachidonic acid from PC and PE during increased $PGF_{2\alpha}$ synthesis by the endometrium.

In early pregnancy, the guinea-pig conceptus apparently secretes an anti-luteolytic factor, probably from about days 10 to 11, which prevents the increase in oestradiol output from the ovary after day 10 which would normally occur if the animal were non-pregnant. This lack of increased oestradiol secretion is probably the reason why endometrial $PGF_{2\alpha}$ synthesizing capacity and uterine $PGF_{2\alpha}$ secretion do not increase during early pregnancy in the guinea-pig. Reduced uterine $PGF_{2\alpha}$ secretion is probably the initial cause of luteal maintenance in the pregnant guinea-pig. However, the guinea-pig conceptus secretes also a luteotrophic factor probably from about day 16 of pregnancy which further promotes luteal maintenance and progesterone secretion. In some guinea-pigs, this luteotrophic factor may be secreted as early as day 12. The luteotrophic factor does not appear to be PGE_2. The identities of the anti-luteolytic factor and of the luteotrophic factor in early pregnant guinea-pigs await discovery.

0378-4320/84/$03.00

INTRODUCTION

Prostaglandin $F_{2\alpha}$ has been identified as the uterine luteolytic hormone in several mammalian species, including the guinea-pig (see Horton and Poyser, 1976; Poyser, 1981). $PGF_{2\alpha}$ concentrations in the utero-ovarian venous blood of guinea-pigs increase after day 11 of the oestrous cycle (Blatchley et al., 1972; Earthy et al., 1975; Antonini et al., 1976), without any increase in PGE_2 concentrations (Antonini et al., 1976). Treating guinea-pigs with intra-uterine indomethacin (Horton and Poyser, 1973; Marley, 1973; Poyser and Horton, 1975; Donovan, 1975) or immunizing guinea-pigs against $PGF_{2\alpha}$ (Horton and Poyser, 1974; Hildebrandt-Stark et al., 1975; Poyser and Horton, 1975) prevents corpus luteal regression and extends the length of the oestrous cycle. These observations clearly show the importance of uterine $PGF_{2\alpha}$ for controlling the life span of the corpus luteum which, in turn, regulates the length of the cycle. If pregnancy occurs, the luteolytic influence of the uterus has to be negated since progesterone secreted by the corpus luteum is necessary for the maintenance of pregnancy in the guinea-pig up to day 28 of pregnancy (Csapo et al., 1981) or even beyond this time (Heap and Deanesly, 1966).

It is clear that mechanisms must operate in the non-pregnant guinea-pig which lead to an increase in $PGF_{2\alpha}$ output from the uterus after day 11 of the cycle. If the guinea-pig becomes pregnant, then it is equally clear that the conceptuses have to maintain luteal function in some way. This paper presents findings from experiments reported over the last decade which have been designed to investigate the hormonal stimuli and the local biochemical mechanisms controlling PG production by the non-pregnant and early pregnant guinea-pig uterus.

A. Non-pregnant uterus

i. Physiological stimulus for $PGF_{2\alpha}$ secretion. Oestradiol or progesterone treatment has been found to hasten luteal regression in several species (see Horton and Poyser, 1976), including the guinea-pig (Bland and Donovan, 1968, 1970; Choudary and Greenwald, 1968). Oestradiol was found to cause an early release of $PGF_{2\alpha}$ from the guinea-pig uterus when administered on days 4 to 6, consistent with such treatment causing premature luteolysis (Blatchley et al., 1971, 1972).

It was apparent therefore that steroidal hormones secreted by the ovary may be the physiological stimulus for increasing $PGF_{2\alpha}$ output from the guinea-pig uterus. To further test this hypothesis, guinea-pigs were ovariectomized and were treated with progesterone and/or oestradiol. $PGF_{2\alpha}$ concentrations in the utero-ovarian venous plasma were then measured and the findings are shown in Fig. 1 (Blatchley and Poyser, 1974). Progesterone treatment resulted in a plasma $PGF_{2\alpha}$ concentration of 3.7 ± 0.5 ng/ml, which is similar to the values observed between days 3 and 10 of the cycle (Blatchley et al., 1972; Earthy et al., 1975). Oestradiol treatment resulted in a mean $PGF_{2\alpha}$ concentration of 12.0 ± 5.3 ng/ml, but in only 2 of the 6 guinea-pigs were $PGF_{2\alpha}$ concentrations elevated above those seen with progesterone treatment alone. Oestradiol acting on a progesterone-primed uterus elevated plasma $PGF_{2\alpha}$ concentrations in all 5 guinea-pigs. The mean value was 34.5 ± 16.7 ng/ml. The values ranged from 11.2 to 100 ng/ml, which is reflected in the large standard error. Consequently, oestradiol acting on a progesterone-primed uterus was the optimum combination of these 2 hormones for maximum stimulation of $PGF_{2\alpha}$ release from the uterus.

This early study measured only $PGF_{2\alpha}$ output. Since $PGF_{2\alpha}$ output from the uterus increases without an increase in PGE_2 output (Antonini et al., 1976), it was decided to investigate the effects of oestradiol and progesterone on the outputs of 3 PGs from the uterus. Guinea-pigs were again ovariectomized and treated with progesterone and/or oestradiol. The outputs of $PGF_{2\alpha}$, PGE_2 and 6-oxo-$PGF_{1\alpha}$ from the uterine horns superfused in vitro were measured by radioimmunoassay (Poyser, 1983a). The outputs of these 3 PGs from the uterus of ovariectomized (control) guinea-pigs were low (0.1 to 0.2 ng/min/uterine horn). Progesterone treatment had no effect on these release rates. Oestradiol treatment increased $PGF_{2\alpha}$ output 7-fold, but it increased PGE_2 and 6-oxo-$PGF_{1\alpha}$ outputs only 1.7-fold and 2.9-fold, respectively. Progesterone priming of the uterus before oestradiol administration caused a further 2-fold stimulation of $PGF_{2\alpha}$ output without affecting the outputs of PGE_2 and 6-oxo-$PGF_{1\alpha}$, compared to oestradiol treatment alone. This study shows that not only is oestradiol acting on a progesterone-primed uterus the optimum stimulus for increased $PGF_{2\alpha}$ secretion, this combination of hormones causes a relatively specific stimulation of uterine $PGF_{2\alpha}$ secretion.

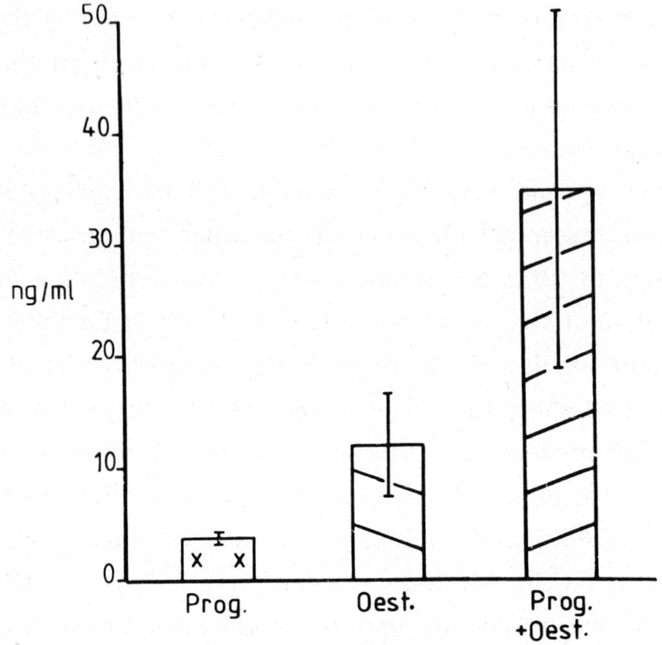

Fig. 1. Concentrations of $PGF_{2\alpha}$ in uterine venous plasma from ovariectomized guinea-pigs treated with progesterone (Prog.) and/or oestradiol (Oest.). (Blatchley and Poyser, 1974).

The importance of oestradiol for increased $PGF_{2\alpha}$ secretion by the guinea-pig uterus is apparent after hypophysectomizing guinea-pigs early in the cycle (Heap et al., 1967). Such treatment does not prevent growth of or progesterone secretion from the corpora lutea. However, luteal regression does not take place at the normal time. Heap et al. (1973) suggest that this "is probably related to the collapse of Graafian follicles and the lack of ovarian oestrogen secretion", and they infer that uterine $PGF_{2\alpha}$ secretion is not stimulated due to the lack of oestrogen. This explanation is obviously attractive, but needs testing. However, a cautionary note may be added since $PGF_{2\alpha}$ does not cause luteal regression in hysterectomized ewes in which the follicles have been destroyed by X-irradiation (Hixon et al., 1975). This suggests that follicular secretions are necessary within the ovary for $PGF_{2\alpha}$ to be luteolytic in the sheep. It is not known whether such intra-ovarian interactions are also necessary in the guinea-pig for $PGF_{2\alpha}$ to cause luteolysis.

In some species, particularly the sheep (Sheldrick et al., 1980; McCracken, 1981), oxytocin forms part of the physiological stimulus necessary for increased $PGF_{2\alpha}$ secretion from the uterus at the end of the cycle. Treatment of the day 15 guinea-pig uterus superfused *in vitro* failed to increase $PGF_{2\alpha}$ output (Poyser and Brydon, 1983). In contrast to some other species, oxytocin treatment at any stage of the cycle failed to induce premature luteal regression in the guinea-pig (Donovan, 1961). Oxytocin does not appear to be involved, therefore, in stimulating $PGF_{2\alpha}$ release from the non-pregnant guinea-pig uterus. Consequently, the evidence indicates that oestradiol, which is released from the guinea-pig ovary in increasing amounts after day 10 of the cycle (Joshi et al., 1973), acting on a uterus which has been subjected to progesterone for 7 days, is the physiological stimulus for increasing $PGF_{2\alpha}$ output from the guinea-pig uterus after day 11. This leads on to how do these ovarian hormones control uterine PG production (particularly in the endometrium, the source of the uterine luteolytic hormone) at the cellular level?

ii. Control of arachidonic acid turnover

a. Arachidonic acid release. The release of arachidonic acid from some bound source is regarded as the rate limiting step in PG synthesis by tissues. In the guinea-pig uterus, 93% of the total arachidonic acid is esterified to phospholipids, of which 80% is bound to phosphatidylcholine (PC) and phosphatidylethanolamine (PE) (Leaver and Poyser, 1981). Between days 7 and 15, there is a significant decrease ($P < 0.05$) in the amount of arachidonic acid bound as PC and an apparent decrease in the amount bound as PE (Leaver and Poyser, 1981). PC and PE may be, therefore, the source of arachidonic acid for $PGF_{2\alpha}$ synthesis by the guinea-pig uterus. Consequently, phospholipase A_2 (PLA_2) would be the enzyme involved in releasing this arachidonic acid. PLA_2 activity has been measured, therefore, in guinea-pig endometrium on days 7 and 16 of the cycle (Downing and Poyser, 1983), since these are days of low and high uterine $PGF_{2\alpha}$ outputs.

PLA_2 activity was present in both the microsomal and supernatant fractions of homogenates of guinea-pig endometrium following centrifugation at 100,000 x g for 60 min, although approximately 90% of the PLA_2 activity was present in the microsomal fraction (and was presumably membrane-bound). Both the microsomal and supernatant PLA_2 enzymes had a pH optimum of 8.0, but the former required 7 mM Ca^{2+} and the latter 2 mM Ca^{2+} for maximum activity. The activities of microsomal PLA_2 on days 7 and 16 of the cycle were 66.2 ± 7.5 and 96.0 ± 10.7 pmol/mg protein/min, respectively. The corresponding values for supernatant PLA_2 were 4.2 ± 0.8 and 7.8 ± 0.9 pmol/mg protein/min, respectively (Downing and Poyser, 1983). Endometrial PLA_2 activity increases 1.4- to 1.8-fold between days 7 and 16 of the cycle. From the activities of the microsomal PLA_2, and by knowing the total amount of protein in the guinea-pig endometrium and the percentage of the total unsaturated fatty acids in the guinea-pig uterus attributable to arachidonic acid, it can be calculated that the maximum theoretical releases rates of arachidonic acid by this enzyme in guinea-pig endometrium are 15 ng/min/uterine horn on day 7 and 22 ng/min/uterine horn on days 15 to 16 (Downing and Poyser, 1983). These values can be compared (Fig. 2) with the release rates of $PGF_{2\alpha}$ from the guinea-pig uterus superfused in vitro of 0.11

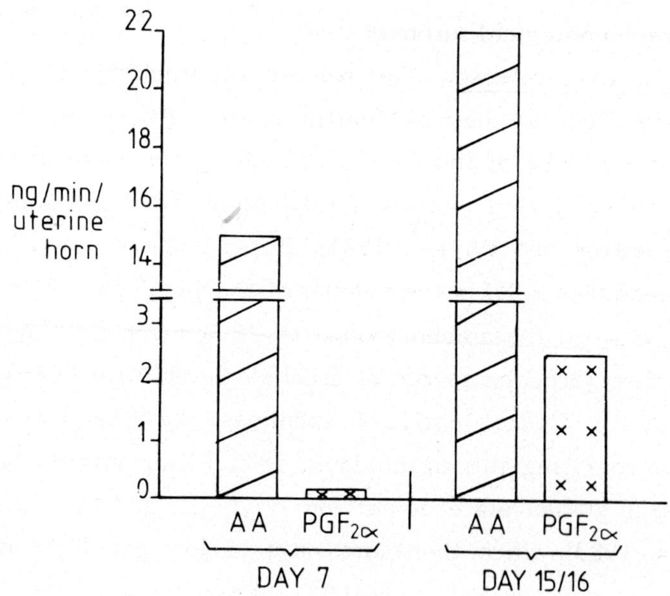

Fig. 2. A comparison of the rates of endometrial arachidonic acid (AA) release by phospholipase A_2 (theoretical maximum) and actual $PGF_{2\alpha}$ release from the uterus superfused in vitro in guinea-pigs on days 7 and 15/16 of the oestrous cycle. (Downing and Poyser, 1983; Poyser and Brydon, 1983).

ng/min/uterine horn on day 7 and 2.5 ng/min/uterine horn on day 15 (Poyser and Brydon, 1983). These $PGF_{2\alpha}$ release rates are very similar to those observed in vivo (Poyser and Maule Walker, 1979). The potential of microsomal PLA_2 for releasing arachidonic acid, particularly on day 7, greatly exceeds the actual release rate of $PGF_{2\alpha}$ from the uterus. Clearly the absolute activity of PLA_2 cannot be the factor controlling $PGF_{2\alpha}$ release from the cycle, nor can the 22-fold increase in $PGF_{2\alpha}$ release from the superfused guinea-pig uterus between days 7 and 15 of the cycle be attributable solely to a 1.5-fold increase in PLA_2 activity. Microsomal PLA_2 requires 7 mM Ca^{2+} for maximum activity, which is several fold higher than physiological concentrations. Consequently, the activation of microsomal PLA_2 by increasing intracellular free Ca^{2+} concentrations may be more important than the absolute activity of PLA_2 in controlling the supply of arachidonic acid for endometrial $PGF_{2\alpha}$ synthesis. The effect of Ca^{2+} on uterine $PGF_{2\alpha}$ release is discussed later.

b. Arachidonic acid uptake. If PC and PE are the sources of arachidonic acid for $PGF_{2\alpha}$ synthesis by guinea-pig endometrium, mechanisms must exist for the re-incorporation of arachidonic acid into these phospholiplids so that they do not become totally depleted of arachidonic acid. Studies were therefore performed to study the uptake of arachidonic acid and of oleic acid into the various lipid classes of guinea-pig endometrium on days 7 and 15 of the cycle (Ning et al., 1983). Fig. 3 shows that uptake of $[^3H]$ arachidonic acid (^3H-AA), but not of $[^3H]$ oleic acid (^3H-OA), into phospholipids (particularly PC and PE) was significantly higher ($P < 0.05$) on day 15 than on day 7, when guinea-pig endometrium was maintained in tissue culture for 24 h (Ning et al., 1983). Uptake of either acid into triglyceride was not significantly higher on day 15 than on day 7.

The incorporation of fatty acids into phospholipids during a 24 h period is considered to be due to addition to lysophospholipids rather than by de novo synthesis of phospholipids (see Trotter and Ferber, 1981). Arachidonic acid is incorporated into phospholipids predominantly by this mechanism any way (Hill and Lands, 1968; Yamashita et al., 1973). Fig. 4 outlines this process. Increased uptake of ^3H-AA into phospholipids could be due therefore to an increase in the activities of acyl-CoA:lysophosphatide acyltransferase or acyl-CoA synthetase, or to an increase in the amount of lysophospholipid present. However, if one or

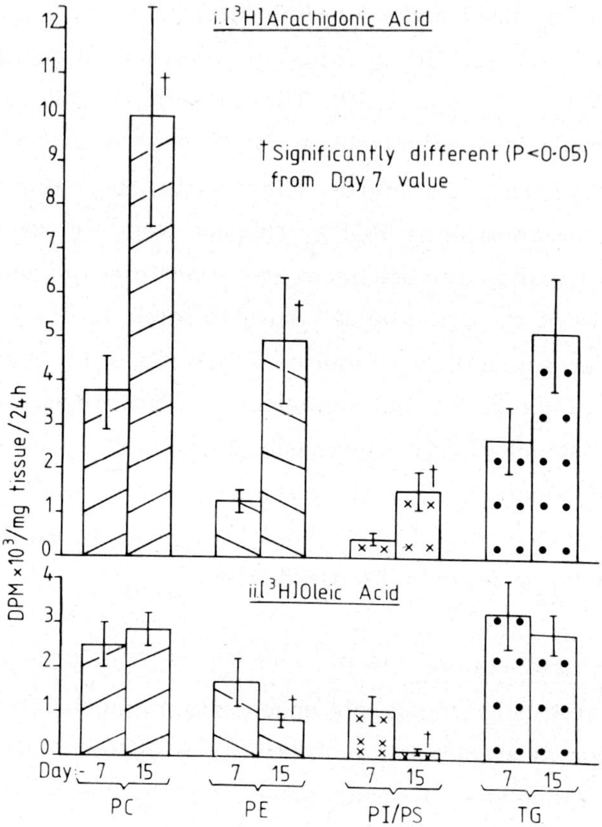

Fig. 3. Uptake of [^{3}H] arachidonic acid and [^{3}H] oleic acid into phosphatidylcholine (PC), phosphatidylethanolamine (PE), phosphatidylinositol/phosphatidylserine (PI/PS) and triglyceride (TG) of guinea-pig endometrium maintained in tissue culture on days 7 and 15 of the oestrous cycle. (Ning, Leaver and Poyser, 1983).

more of these mechanisms were operative, the uptake of ^{3}H-OA might be expected to increase also, assuming both fatty acids were good substrates for the 2 enzymes. The corresponding acyltransferase enzyme in rat liver has a specificity 8-fold higher for arachidonic acid than for oleic acid (Yamashita et al., 1973) suggesting that differences in specificity, which could change during the cycle of the fatty acids for the acyltransferase enzyme in guinea-pig endometrium, could be responsible for the differential increase in fatty acid uptake into phospholipids. Human platelets contain an acyl-CoA synthetase specific for arachidonic

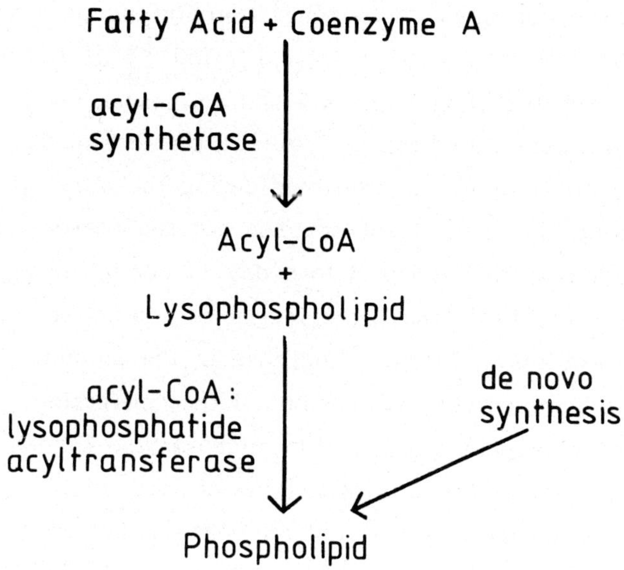

Fig. 4. An outline of the 2 possible pathways involved in incorporating fatty acids into phospholipids.

acid and a more general acyl-CoA for other fatty acids including oleic acid (Wilson et al., 1982). If 2 such enzymes exist in guinea-pig endometrium, a specific increase in the activity of arachidonyl-CoA between days 7 and 15 of the cycle could account also for the increased uptake of ^{3}H-AA, but not of ^{3}H-OA. Obviously, further investigations are required into the processes controlling arachidonic acid uptake into the phospholipids of guinea-pig endometrium. Nevertheless, it is clear that mechanisms exist in the endometrium which lead to an increased uptake of arachidonic acid into phospholiplids at the time of increased $PGF_{2\alpha}$ synthesis, probably to compensate for the increased release of arachidonic acid from PC and PE at this time.

iii. "PG synthetase" levels in the uterus. An initial study (Poyser, 1972) showed that homogenates of guinea-pig uterus, when incubated in vitro, synthesized PGs from endogenous precursors. Four to 6 times more $PGF_{2\alpha}$ than PGE_2 was formed, and the amounts of the PGs synthesized increased after day 11 of the cycle. Approximately 5 µg arachidonic acid/100 mg tissue were released from the guinea-pig uterus during this homogenization and incubation process on both days 7 and 15 (Mitchell et al., 1977). Only 1% and 3% of the arachidonic acid released is converted to $PGF_{2\alpha}$ on days 7 and 15, respectively. In addition, adding exogenous arachidonic acid to day 7 guinea-pig uterine homogenates failed to increase the amounts of PGs synthesized during the incubation process (Poyser, 1971). Consequently, lack of substrate is not the reason for the lower PG synthesizing capacity of the uterus before day 11 compared with later days. It was suggested (Poyser, 1973) that "the increase in amounts of prostaglandins formed by the uterus may reflect an increase in the amount of PG synthetase present". From studies involving estimations of the conversion of $[^{14}C]$ arachidonic acid into PGs by guinea-pig uterus microsomes, Wlodawer et al. (1976) confirmed that "the prostaglandin synthesizing activity in the guinea-pig uterus has been found to be highest at the end of the oestrous cycle".

From studies on ovariectomized guinea-pigs treated with oestradiol and/or progesterone (Naylor and Poyser, 1975), it was found that oestradiol, but not progesterone, increased the $PGF_{2\alpha}$ synthesizing capacity of the uterus. This stimulatory effect of oestradiol in the guinea-pig was confirmed by Wlodawer et al. (1976) and Thaler-Dao et al. (1982). All these studies were performed on homogenates of whole uterus whereas the endometrium is the source of the uterine luteolytic hormone. Also, since the earlier studies were performed, products synthesized from arachidonic acid by "PG synthetase" other than $PGF_{2\alpha}$ and PGE_2 have been discovered. Consequently, the production of $PGF_{2\alpha}$, PGE_2, 6-oxo-$PGF_{1\alpha}$ and TXB_2 by homogenates of endometrium, myometrium and whole uterus from guinea-pigs on selected days of the cycle, and from ovariectomized guinea-pigs treated with oestradiol and/or progesterone have been measured by radioimmunoassay (RIA), following a 90 min incubation period (Poyser, 1983b). (PGD_2 was not measured due to the lack of a RIA for this compound).

PG and TX production by the endometrium, but not by the myometrium, increased after day 11 of the cycle. The ratios of the mean endometrial productions of $PGF_{2\alpha}$, PGE_2, 6-oxo-$PGF_{1\alpha}$ and TXB_2 were 18.2 : 1.0 : 3.9 : 1.1 on day 7, 21.0 : 1.0 : 2.5 : 0.6 on day 15 and 26.8 : 1.0 : 2.0 : 0.5 at oestrus. Between days 7 and 15, the mean endometrial productions of $PGF_{2\alpha}$, PGE_2, 6-oxo-$PGF_{1\alpha}$ and TXB_2 increased 2.3-fold, 2.2-fold, 1.4-fold and 1.2-fold, respectively. Between day 7 and oestrus, the mean endometrial productions of $PGF_{2\alpha}$, PGE_2,

6-oxo-$PGF_{1\alpha}$ and TXB_2 increased 4.4-fold, 3.0-fold, 1.6-fold and 1.3-fold, respectively. It was apparent that PG and TX production by the endometrium generally increases after day 11 and that there was a preferential stimulation of $PGF_{2\alpha}$ synthesis. Furthermore, although endometrial $PGF_{2\alpha}$ synthesis is much increased on day 15, maximum $PGF_{2\alpha}$ synthesis by the endometrium occurs at oestrus when plasma progesterone concentrations are low. Little variation was seen in PG and TX production by the myometrium during the cycle, and 6-oxo-$PGF_{1\alpha}$ was the major PG synthesized. Myometrial $PGF_{2\alpha}$ synthesis was 50% lower and PGE_2 and TXB_2 synthesis were each 90% lower than the synthesis of 6-oxo-$PGF_{1\alpha}$. Changes in PG and TX production in homogenates of whole guinea-pig uterus generally reflected changes in PG and TX production by the endometrium.

Treatment of ovariectomized guinea-pigs with progesterone had no effect on endometrial PG and TX synthesis compared to control, ovariectomized guinea-pigs. Oestradiol treatment stimulated the mean endometrial productions of $PGF_{2\alpha}$, PGE_2, 6-oxo-$PGF_{1\alpha}$ and TXB_2 3.2-fold, 1.6-fold, 2.2-fold and 2.0-fold, respectively. Such treatment significantly increased the $PGF_{2\alpha}$ to PGE_2 ratio. If progesterone was given together with the oestradiol, endometrial $PGF_{2\alpha}$ synthesis was now stimulated only 1.7-fold, whereas no stimulation in the synthesis of the other 3 compounds was observed. However, the $PGF_{2\alpha}$ to PGE_2 was still significantly increased. PG and TX production by the myometrium was unaffected by steroid treatment, and changes in PG and TX production in homogenates of whole uterus again reflected PG synthesis by the endometrium. It is apparent that oestradiol causes an increase in the amount of PG synthetase in the endometrium with a preferential stimulation of $PGF_{2\alpha}$ synthesis. This stimulatory effect of oestradiol on endometrial PG synthetase levels is attenuated by progesterone. Consequently, this would explain why $PGF_{2\alpha}$ synthesis by the endometrium is highest at oestrus since plasma oestradiol levels are high and plasma progesterone levels are low at this time. This may also partly explain why $PGF_{2\alpha}$ output from the uterus increases again once plasma progesterone concentrations have fallen at the end of the cycle. The increase in endometrial $PGF_{2\alpha}$ synthesizing capacity after day 11 appears, therefore, to be the result of increased oestradiol output from the ovary after day 10 (Joshi et al., 1973).

If an increase in PG synthetase levels in the endometrium is important for luteolysis to occur, treatment of the guinea-pig uterus with a protein synthesis inhibitor should prevent this increase in enzyme level and delay luteal regression. Consequently, guinea-pigs were treated with intra-uterine actinomycin D on day 10, and uterine PG synthesizing capacity and plasma progesterone concentrations were measured on day 15 (Poyser, 1979). It was found that the local administration of actinomycin D to the uterus prevented the increase in PG synthetase levels, and plasma progesterone levels were still high indicating that luteal regression had been prevented (Fig. 5). Addition of arachidonic acid to the uterine homogenates failed to increase PG synthesis during the incubation period, indica-

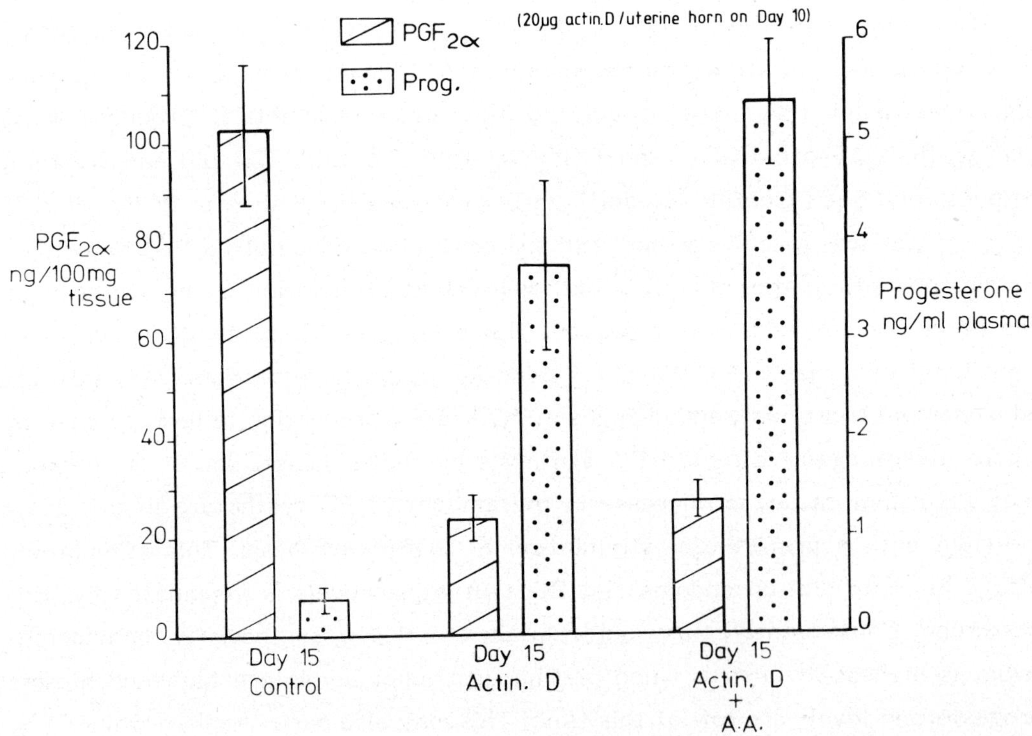

Fig. 5. Effect of intra-uterine administration of actinomycin D (Actin. D), on day 10, on uterine $PGF_{2\alpha}$ synthesizing capacity and peripheral plasma progesterone concentrations in day 15 guinea-pigs. In the right-hand column for $PGF_{2\alpha}$ synthesizing capacity, arachidonic acid (A.A., 3 µg/ml) was added to the uterine homogenates before incubation, but had no effect. (Poyser, 1979).

ting that lack of endogenous arachidonic acid was not the reason for reduced PG synthesis (Fig. 5). In a second experiment, oestradiol administration to guinea-pigs treated similarly with actinomycin D failed to stimulate uterine PG synthetase levels and plasma progesterone levels were still high (Fig. 6). Consequently, the luteolytic action of oestradiol, via increased $PGF_{2\alpha}$ production by the uterus, is blocked by inhibiting fresh protein synthesis in the uterus. Obviously, actinomycin D will not be preventing solely the increase in PG synthetase levels in the uterus, but the question arises as to whether the $PGF_{2\alpha}$ synthesizing capacity of the endometrium is the main factor regulating uterine $PGF_{2\alpha}$ output from the uterus throughout the oestrous cycle.

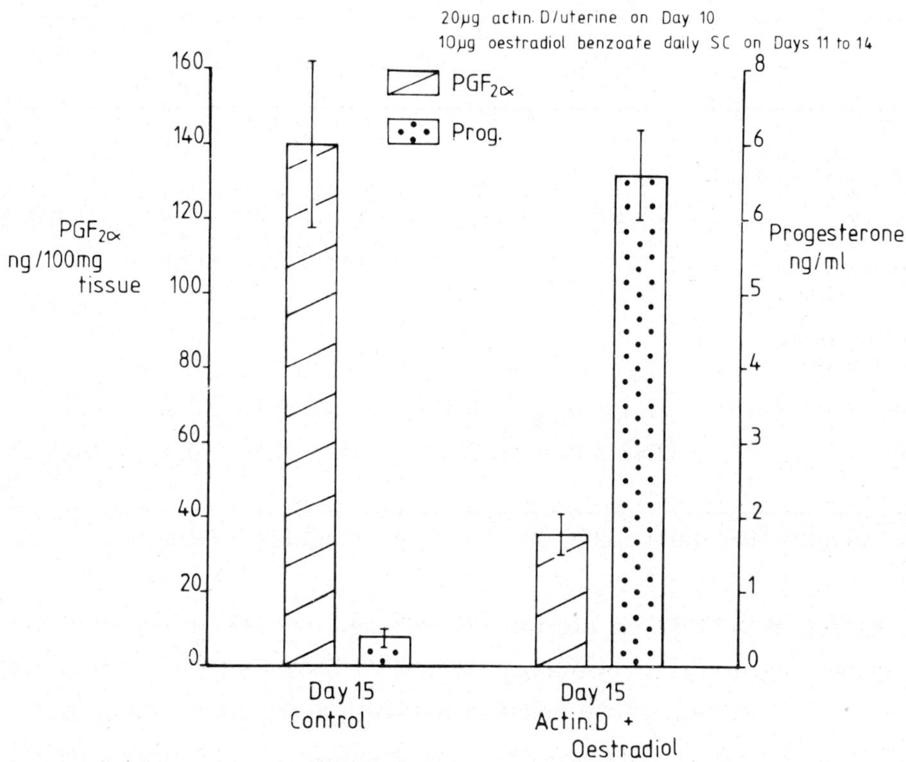

Fig. 6. The effect of administering oestradiol benzoate, on days 11 to 14 to guinea-pigs treated with intra-uterine actinomycin D (Actin. D) on day 10, on uterine $PGF_{2\alpha}$ synthesizing capacity and peripheral plasma progesterone concentrations when measured on day 15. (Poyser, 1979).

In Table I, the PG and TX synthesizing capacities of the endometrium, and the PG and TX outputs from the superfused uterus on days 7 and 15 of the cycle are compared. Although $PGF_{2\alpha}$ is the major PG which the endometrium can synthesize on day 7, $PGF_{2\alpha}$ is actually the minor PG released at this time. Also, there is a 22-fold increase in $PGF_{2\alpha}$ output from the uterus but only a 2.5-fold increase in endometrial $PGF_{2\alpha}$ synthesizing capacity between days 7 and 15 of the cycle. Clearly, the $PGF_{2\alpha}$ synthesizing capacity of the endometrium is not the main factor regulating uterine $PGF_{2\alpha}$ output during the cycle.

Table I. Comparison of the endometrial prostaglandin (PG) and thromboxane (TX) synthesizing capacities (n = 6 per day) and PG and TX outputs from the uterus superfused in vitro (n = 8 per day) from day 7 and day 15 guinea-pigs. (Poyser, 1983b; Poyser and Brydon, 1983)

	$PGF_{2\alpha}$	PGE_2	6-oxo-$PGF_{1\alpha}$	TXB_2
Synthesizing capacity (ng/100 mg tissue):-				
Day 7	63.8 ± 0.9	3.5 ± 0.2	13.5 ± 3.8	4.0 ± 0.6
Day 15	162.3 ± 16.7†	7.7 ± 1.2†	19.0 ± 2.9	4.8 ± 0.5
Output (ng/min/ uterine horn):-				
Day 7	0.11 ± 0.01	0.18 ± 0.02	0.16 ± 0.02	0.13 ± 0.01
Day 15	2.48 ± 0.28†	0.32 ± 0.05†	0.47 ± 0.05†	0.15 ± 0.02

† Significantly ($P < 0.05$) higher than corresponding day 7 value.

iv. Effect of A23187 on uterine PG release. A23187 superfused over the day 7 guinea-pig uterus significantly ($P < 0.05$) increased the outputs (ng/min/ uterine horn) of $PGF_{2\alpha}$, PGE_2 and 6-oxo-$PGF_{1\alpha}$ from 0.15 ± 0.04, 0.15 ± 0.02 and 0.05 ± 0.01 to 0.59 ± 0.06, 0.25 ± 0.03 and 0.28 ± 0.03, respectively. When superfused over the day 15 uterus, A23187 significantly ($P < 0.05$) increased the outputs (ng/min/uterine horn) of $PGF_{2\alpha}$, PGE_2 and 6-oxo-$PGF_{1\alpha}$ from 3.4 ± 0.2, 0.36 ± 0.06 and 0.18 ± 0.01 to 6.5 ± 0.05, 0.67 ± 0.13 and 0.46 ± 0.07, respectively (Poyser and Brydon, 1983). A23187 can stimulate, therefore, $PGF_{2\alpha}$, 6-oxo-

$PGF_{1\alpha}$ and, to a lesser extent, PGE_2 release from the guinea-pig uterus. It is probable that the endometrium and myometrium are the major sources of the $PGF_{2\alpha}$ and 6-oxo-$PGF_{1\alpha}$, respectively (Poyser, 1983b). Other workers have shown A23187 to stimulate $PGF_{2\alpha}$ release from the guinea-pig endometrium (Leaver and Seawright, 1982) and 6-oxo-$PGF_{1\alpha}$ from the guinea-pig myometrium (Vesin et al., 1982). Increasing intracellular Ca^{2+} concentrations stimulate, therefore, $PGF_{2\alpha}$ release from the endometrium irrespective of whether the basal output is low (day 7) or high (day 15). However, a larger release is elicited on day 15 due to the higher $PGF_{2\alpha}$ synthesizing ability of the endometrium of this time.

B. Pregnant uterus

i. Plasma $PGF_{2\alpha}$, progesterone and oestradiol concentrations. Fig. 7 compares the utero-ovarian venous plasma concentrations of progesterone, oestradiol and $PGF_{2\alpha}$ on days 12 and 15 of the oestrous cycle and of pregnancy (Blatchley et al., 1975a). In the non-pregnant guinea-pig, $PGF_{2\alpha}$ and oestradiol concentrations increased, and progesterone concentrations decreased between days 12 and 15. In the pregnant guinea-pig, $PGF_{2\alpha}$ and oestradiol concentrations remained low and progesterone concentrations remained high. These findings were confirmed by Antonini et al. (1976), who also showed that utero-ovarian venous plasma concentrations of PGE_2 did not increase in the pregnant guinea-pig during early pregnancy. The outputs of $PGF_{2\alpha}$ and PGE_2 from the day 15 pregnant uterus superfused in vitro were both low, (Poyser, unpublished results), showing that $PGF_{2\alpha}$ output from the early pregnant uterus is actually reduced and not re-directed from the venous drainage into the uterine lumen, and that there is no increase in PGE_2 output from the pregnant uterus at the time of luteal maintenance.

ii. Uterine "PG synthetase" levels. The amounts of $PGF_{2\alpha}$ and PGE_2 synthesized by uterine homogenates from day 15 non-pregnant guinea-pigs during a 90 min incubation were 113 ± 8 and 13 ± 1 ng/100 mg tissue. Corpora lutea size in these guinea-pigs were 0.18 ± 0.02 mm^3, indicating that luteal regression had taken place. The corresponding values for day 15 pregnant guinea-pigs were 13 ± 3 ng $PGF_{2\alpha}$/100 mg tissue, 9 ± 1 ng PGE_2/100 mg tissue, and 3.3 ± 0.3 mm^3 for

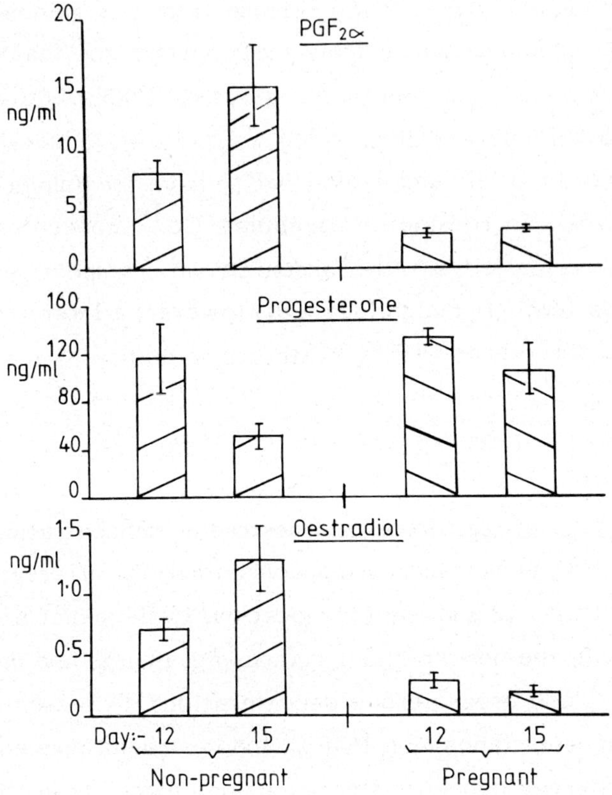

Fig. 7. Utero-ovarian venous plasma concentrations of $PGF_{2\alpha}$, progesterone and oestradiol in non-pregnant and pregnant guinea-pigs on days 12 and 15. (Blatchley, Maule Walker and Poyser, 1975a).

luteal size indicating that the corpora lutea were maintained (Maule Walker and Poyser, 1974). (The pregnant uteri had the conceptuses removed before homogenization). The reduced production of PGs by the pregnant uterus was not due to increased metabolism nor lack of endogenous arachidonic acid (Maule Walker and Poyser, 1978). The increase in uterine PG synthetase levels after day 11 in the non-pregnant guinea-pig does not take place in the pregnant guinea-pig over the equivalent period.

iii. Effect of oestradiol treatment during pregnancy. The treatment of 4 guinea-pigs with oestradiol benzoate (10 μg daily) between days 10 and 21 of pregnancy resulted in abortion (Poyser and Maule Walker, 1979). The $PGF_{2\alpha}$ synthesizing capacity of homogenates of the uteri (freed of conceptuses) from these 4 animals, during a 90 min period, on day 21 was 144.5 ± 8.9 ng/100 mg tissue. This compares with a value of 66.5 ± 6.1 ng/100 mg tissue in day 21, non-treated guinea-pigs. In 3 of the 4 guinea-pigs treated with oestradiol, corpora lutea size was 0.36 ± 0.05 mm^3 and plasma progesterone concentrations were < 0.2 ng/ml (when measured on day 21), indicating that luteal regression had occurred due probably to the increase in $PGF_{2\alpha}$ synthesizing capacity of the uterus to levels associated with causing luteolysis. The measurement of peripheral plasma progesterone concentrations in these 3 guinea-pigs between days 10 and 21 showed that luteal regression had actually occurred between days 11 and 13, which was 2 days earlier than in the oestrous cycles immediately prior to mating. Obviously, abortion would occur under these circumstances and, in these 3 guinea-pigs, the conceptuses were found in the vagina on day 21.

In the fourth guinea-pig which received oestradiol, on day 21 the corpora lutea were well maintained (4 mm^3) and plasma progesterone concentrations were high (31 ng/ml), in spite of the high $PGF_{2\alpha}$ synthesizing capacity of the uterus (126.3 ng/100 mg tissue). Plasma progesterone concentrations dropped 30% between days 10 and 12, but then recovered and increased rapidly after day 16 (Poyser and Maule Walker, 1979). Clearly abortion in this guinea-pig was not due to lack of progesterone, but abortion was more difficult since the conceptuses were found in the uterus, and not in the vagina as in the other 3 guinea-pigs. Oestradiol treatment of guinea-pigs during mid-pregnancy also causes abortion and plasma progesterone levels to fall (Evans et al., 1981). This suggests that oestradiol treatment is also having detrimental effects on the placenta, and this may be the reason why abortion occurred in the fourth guinea-pig treated with oestradiol between days 10 and 21.

iv. Effect of unilateral pregnancy. In day 15 unilaterally pregnant guinea-pigs, the concentration of $PGF_{2\alpha}$ in the utero-ovarian venous plasma on the non-pregnant side was greater than on the pregnant side (6.1 ± 1.8 v 2.7 ± 0.3 ng/ml). These values can be compared with the corresponding concentrations of 15.1 ± 3.6 and 3.1 ± 0.1 ng/ml in day 15 non-pregnant and day 15 bilaterally pregnant guinea-pigs, respectively, (Blatchley et al., 1975a, b). Oestradiol concentrations in the utero-ovarian venous plasma of day 15 unilaterally pregnant guinea-pigs were 461 ± 78 pg/ml on the pregnant side and 699 ± 248 pg/ml on the non-pregnant side. The corresponding oestradiol concentrations for day 15 non-pregnant and day 15 bilaterally pregnant guinea-pigs were 1281 ± 287 and 188 ± 36 pg/ml, respectively (Blatchley et al., 1975a, b). Progesterone output was well maintained from both ovaries in unilaterally pregnant guinea-pigs and in bilaterally pregnant guinea-pigs on day 15, but had fallen in day 15 non-pregnant guinea-pigs (Blatchley et al., 1975a, b).

The amounts of $PGF_{2\alpha}$ synthesized by homogenates of the non-pregnant uterine horn and the pregnant uterine horn (freed of conceptuses), from day 15 unilaterally pregnant guinea-pigs, when incubated for 90 min were 22 ± 2 and 45 ± 3 ng/100 mg tissue, respectively. These values can be compared with 13 ± 3 ng $PGF_{2\alpha}$/100 mg tissue synthesized by homogenates of uterine horns (freed of conceptuses) from day 15 bilaterally pregnant guinea-pigs and 113 ± 8 ng $PGF_{2\alpha}$/100 mg tissue by homogenates of uterine horns from day 15 non-pregnant guinea-pigs (Maule Walker and Poyser, 1974). The corpora lutea on the non-pregnant side tended to be smaller than on the pregnant side in day 15 unilaterally pregnant guinea-pigs (2.0 ± 0.2 mm^3 v 4.3 ± 0.2 mm^3). These values can be compared with 3.3 ± 0.3 mm^3 and 0.18 ± 0.02 mm^3 for the corpora lutea from day 15 bilaterally pregnant and day 15 non-pregnant guinea-pigs, respectively (Maule Walker and Poyser, 1974).

It is clear that pregnancy is well-maintained in unilaterally pregnant guinea-pigs. However, the $PGF_{2\alpha}$ synthesizing capacity of and the amounts of $PGF_{2\alpha}$ released from the non-pregnant uterine horn are approximately 2-fold higher than the corresponding values for the pregnant uterine horn in day 15 unilaterally pregnant guinea-pigs. Corpora lutea size is 50% less on the non-pregnant side compared to the pregnant side in such guinea-pigs. These differences may be due to the higher output of oestradiol from the ovaries of day 15 unilaterally pregnant guinea-pigs, especially on the non-pregnant side, compared to day 15 bilaterally pregnant guinea-pigs. However, the $PGF_{2\alpha}$ synthesizing capacity of and $PGF_{2\alpha}$ output from the non-pregnant uterine horn of day 15 unilaterally pregnant guinea-pigs do not reach, by a long way, the values associated with causing luteal regression (as seen in day 15 non-pregnant guinea-pigs), although some attenuation of luteal growth is seen. However, this attenuation is insufficient to affect adversely progesterone secretion.

If the non-pregnant horn in unilaterally pregnant guinea-pigs is transected at the cervical end (so that there is now no continuity between the 2 uterine horns), $PGF_{2\alpha}$ synthesizing capacity of the non-pregnant horn was 72 ± 5 ng/100 mg tissue compared to 25 ± 3 ng $PGF_{2\alpha}$/100 mg tissue by the pregnant horn (freed of conceptuses) on day 15 of pregnancy. The sizes of the corpora lutea were 1.9 ± 0.3 mm^3 on the non-pregnant side and 4.0 ± 0.4 mm^3 on the pregnant side. Peripheral plasma progesterone concentrations on day 15 in these unilaterally pregnant guinea-pigs with a transected, non-pregnant horn were 2.5 ± 0.5 ng/ml, which were approximately 60% lower than in day 15 unilaterally pregnant guinea-pigs with the non-pregnant horn intact (5.8 ± 0.6 ng/ml) and in day 15 bilaterally pregnant guinea-pigs (6.7 ± 1.7 ng/ml) (Poyser and Maule Walker, 1979). On day 20, corpora lutea size was 3.1 ± 0.2 mm^3 on the non-pregnant side with the uterine horn transected and 4.1 ± 0.2 mm^3 on the pregnant side. Plasma progesterone concentrations in these unilaterally pregnant guinea-pigs with the non-pregnant horn transected were 19.0 ± 1.2 ng/ml on day 20. These values can be compared with 4.3 ± 0.2 mm^3 on the non-pregnant side, 4.8 mm^3 on the pregnant side (for luteal size) and 29.6 ± 1 ng/ml (for plasma progesterone concentrations) in day 20 unilaterally pregnant guinea-pigs with the pregnant horn intact (Poyser and Maule Walker, 1979).

Disconnecting the non-pregnant horn from the pregnant horn in unilaterally pregnant guinea-pigs causes a further stimulation of $PGF_{2\alpha}$ synthesizing capacity in the non-pregnant horn on day 15 compared to when the 2 uterine horns are attached. Corpora luteal size and progesterone output from the ovary on the non-pregnant side are considerably reduced. However, the $PGF_{2\alpha}$ synthesizing capacity of the transected, non-pregnant horn is still insufficient to cause luteolysis, and the corpora lutea and progesterone output do recover. This finding in guinea-pigs can be contrasted with that in sheep where corpora lutea in the ovary adjacent to a non-pregnant transected uterine horn in unilaterally pregnant ewes regress completely (see Moor, 1968).

DISCUSSION

A. Non-pregnant uterus

Three biochemical processes probably associated with $PGF_{2\alpha}$ synthesis in the guinea-pig endometrium show increased activity towards the end of the cycle at the time of increased $PGF_{2\alpha}$ secretion since, between days 7 and 15/16 of the oestrous cycle, there is:-

a) Increased incorporation of arachidonic acid into phospholipids, particularly PC and PE.
b) Increased activity of PLA_2.
c) Increase in $PGF_{2\alpha}$ synthesizing capacity, due probably to an increase in the amount of the cyclo-oxygenase present, with the majority of the PGH_2 synthesized being directed in some way into the $PGF_{2\alpha}$ forming pathway.

The increase in PG synthetase levels, with direction to the $PGF_{2\alpha}$ pathway, is caused by oestradiol (Poyser, 1983b). It has been shown in rats that oestradiol stimulates uterine PLA_2 activity (Dey et al., 1982), so oestradiol may have a similar effect in guinea-pigs. Both these stimulatory actions of oestradiol are attenuated by progesterone. The increased incorporation of arachidonic acid into phospholipids may be due also to stimulation in activity of the appropriate enzyme by oestradiol. The increase in output of oestradiol from the guinea-pig uterus after day 10 (Joshi et al., 1973) is consistent with oestradiol being the physiological stimulus for these increases in enzyme activity. However, the combined increase in the activities of PLA_2 and cyclo-oxygenase alone is insufficient to account for the increased output of $PGF_{2\alpha}$ from the uterus at the end of the oestrous cycle. In addition, since A23187 causes a 2-fold stimulation of $PGF_{2\alpha}$ output from the day 15 guinea-pig uterus (presumably by stimulating arachidonic acid release), it is clear that the cyclo-oxygenase enzyme is not operating at V_{max}. Consequently, although cyclo-oxygenase levels in the uterus are important in regulating the amounts of $PGF_{2\alpha}$ synthesized the availability of free arachidonic acid is clearly the rate limiting step in this process since increasing arachidonic acid release increases the rate of $PGF_{2\alpha}$ synthesis.

An anomaly which argues also against changes in endometrial enzyme levels being responsible solely for the changes in $PGF_{2\alpha}$ synthesis and release is that oestradiol is incapable of maximally stimulating $PGF_{2\alpha}$ output from the uterus unless the uterus has been primed with progesterone, yet progesterone attenuates the stimulatory effect of oestradiol on endometrial PLA_2 (as shown in the rat) and cyclo-oxygenase levels. This is clearly demonstrated in Fig. 8 which compares endometrial $PGF_{2\alpha}$ synthesizing capacity and uterine $PGF_{2\alpha}$ release

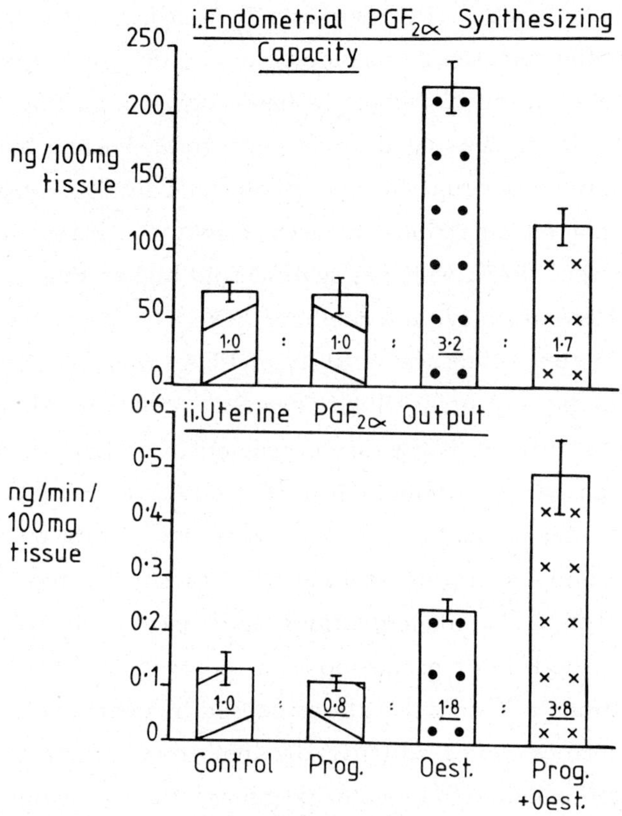

Fig. 8. A comparison of endometrial $PGF_{2\alpha}$ synthesizing capacity and $PGF_{2\alpha}$ output from the uterus superfused in vitro of ovariectomized guinea-pigs which had no treatment (Control) or were treated with progesterone (Prog.) and/or oestradiol (Oest.). (Poyser, 1983a, b).

from ovariectomized guinea-pigs treated with progesterone and/or oestradiol (Poyser, 1983a, b). Endometrial $PGF_{2\alpha}$ synthesizing capacity is greatest in those guinea-pigs treated with oestradiol alone, but uterine $PGF_{2\alpha}$ output is greatest in those guinea-pigs receiving progesterone and oestradiol. Consequently, although progesterone alone is without effect, it has a potentiating role on the stimulatory action of oestradiol on $PGF_{2\alpha}$ release which more than compensates for its inhibitory effect on the stimulation by oestradiol of the activities of the enzymes involved in PG synthesis. What is this potentiating action of progesterone?

The rate limiting step in PG synthesis by tissues (including the uterus, as outlined above) is the release of arachidonic acid from some bound source, which is usually considered to be phospholipid (see Vogt, 1978). Fig. 2 shows that the potential of PLA2 in guinea-pig endometrium for releasing arachidonic acid is more than adequate to account for the amounts of $PGF_{2\alpha}$ released. Microsomal PLA2 in guinea-pig endometrium requires 7 mM Ca^{2+} for maximum activity (Downing and Poyser, 1983), which is several fold higher than physiological Ca^{2+} concentrations. Consequently, intracellular free Ca^{2+} concentrations may be more important in controlling the activity of PLA_2 than the absolute activity of the enzyme. Up to day 11, intracellular free Ca^{2+} concentrations in the endometrial cells which synthesize $PGF_{2\alpha}$ are presumed to be low, so that little arachidonic acid is released and little $PGF_{2\alpha}$ is produced. From day 11 oestradiol, released in increasing amounts from the ovary and acting on the endometrium which has been primed with progesterone for 7 to 8 days, may cause an increase in intracellular free Ca^{2+} concentrations and thereby "switch on" endometrial $PGF_{2\alpha}$ synthesis. The role of progesteone is therefore to facilitate the increase in intracellular free Ca^{2+} concentrations caused by oestradiol. The eventual fall in plasma progesterone levels will lift the inhibitory effect of progesterone on the stimulation of endometrial cyclo-oxygenase (and possibly PLA_2 levels) by oestradiol, and this may contribute partly to the further increase in uterine $PGF_{2\alpha}$ output after luteolysis has occurred. However, the withdrawal of progesterone, and the decline in ovarian oestradiol output after oestrus will result in a decrease in arachidonic acid release, due to the sequestering of intracellular Ca^{2+}, and $PGF_{2\alpha}$ secretion becomes "switched off". Endometrial enzyme activities will also fall after oestrus due to the lack of oestradiol. Further study is necessary to investigate these suggestions but Fig. 9 outlines some of the biochemical processes believed to be involved in stimulating $PGF_{2\alpha}$ output from the guinea-pig uterus.

In the larger species (such as sheep, cow and pig) where chronic sampling of the uterine venous blood is possible, increased uterine $PGF_{2\alpha}$ secretion after day 11 is not continuous, but occurs as pulses. To obtain a pulse of $PGF_{2\alpha}$ release from the uterus, there must be a synchronous increase in $PGF_{2\alpha}$ synthesis by and release from the majority of the endometrial cells which produce $PGF_{2\alpha}$. This necessitates, therefore, some form of "pulse generator", which may be formed by

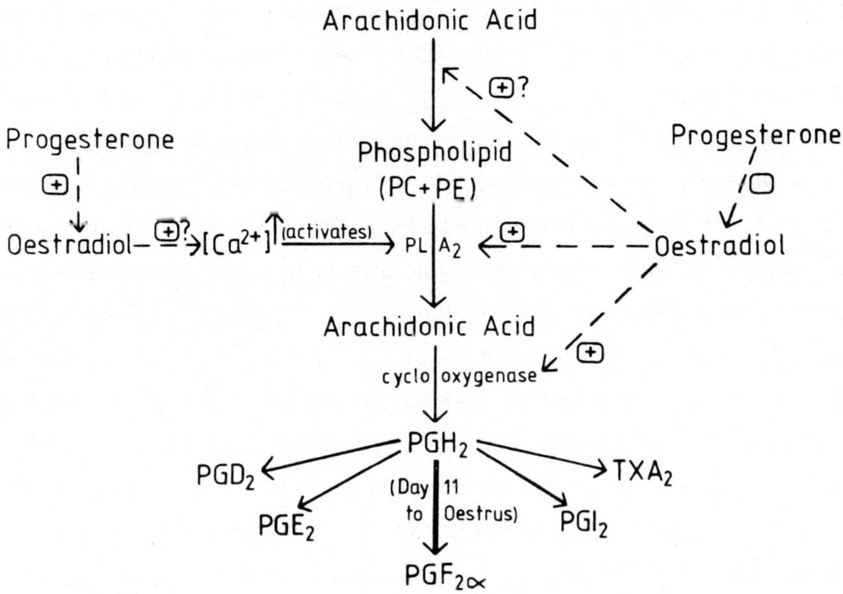

Fig. 9. Proposed mechanisms involved in stimulating $PGF_{2\alpha}$ synthesis in guinea-pig endometrium from day 11 of the oestrous cycle. The "thickened arrow" indicates that there is a preferential stimulation of endometrial $PGF_{2\alpha}$ synthesis and release between days 11 and oestrus.

the pulsatile nature of some hormone "input" to the uterus. It has been implied that the pulsatile nature of oxytocin released from the posterior pituitary gland may be responsible for the pulsatile nature of $PGF_{2\alpha}$ release from the sheep uterus (Fairclough et al., 1980). Recent evidence suggests that the oxytocin may be produced by the ovary instead of the pituitary gland (Wathes and Swann, 1982; Flint and Sheldrick, 1982). However, the release of $PGF_{2\alpha}$ from the sheep uterus occurs from day 12, which is about 2 days before the sheep uterus becomes sensitive to the $PGF_{2\alpha}$-releasing action of oxytocin (Roberts et al., 1975). Furthermore, maximum $PGF_{2\alpha}$ release from the sheep uterus occurs when plasma oxytocin concentrations during the oestrous cycle are at their lowest (Sheldrick and Flint, 1981; Webb et al., 1981; Schams et al., 1982), although this inverse relationship may be compensated for by the increase in oxytocin binding sites in the endometrium from day 15 of the cycle (Roberts et al., 1976). The treatment of ovariectomized sheep with twice daily injections of oestradiol and progesterone results in pulses of $PGF_{2\alpha}$ being released from the uterus (Louis et al.,

1977). Infusions of oestradiol or oxytocin through the sheep uterus also result in a pulse of $PGF_{2\alpha}$ being released (Barcikowski et al., 1974; Roberts et al., 1975). It is clear that the pulsatile nature of $PGF_{2\alpha}$ from the uterus is independent of the nature of the hormonal stimulus, and occurs even when the plasma concentrations of the hormones are relatively stable. Consequently, the pulsatile nature of $PGF_{2\alpha}$ release from the uterus appears to be an intrinsic property of the uterus.

A23187 superfused over the guinea-pig uterus in vitro causes also a pulse of $PGF_{2\alpha}$ to be released (Poyser and Brydon, 1983). It is proposed, therefore, that intermittent increases in intracellular free Ca^{2+} concentrations in the endometrial cells which produce the $PGF_{2\alpha}$ may be the cause of pulsatile $PGF_{2\alpha}$ release from the uterus. This infers that the "pulse generator" is actually "switched on" by oestradiol acting on a progesterone-primed uterus. If this is so, how does such a "pulse generator" cause a synchronous increase in intracellular free Ca^{2+} levels in the endometrial cells? Perhaps in another decade the answer will be known.

B. Pregnant uterus

If the guinea-pig becomes pregnant, increased $PGF_{2\alpha}$ release into the uterine venous drainage after day 11 does not occur. $PGF_{2\alpha}$ release is not re-directed into the uterine lumen, as proposed for the pig (Bazer and Thatcher, 1977). Increases in uterine PG synthetase levels and uterine $PGF_{2\alpha}$ release are prevented in the guinea-pig during early pregnancy. This is in agreement with the much lower levels of the main $PGF_{2\alpha}$ metabolite in guinea-pig urine during days 12 to 20 of pregnancy compared to days 12 to 17 of the cycle (Granström and Kindahl, 1976). The lack of increased oestradiol output from the ovary after day 10 in the pregnant guinea-pig is probably responsible for the lack of an increase in uterine $PGF_{2\alpha}$ production. It appears that the conceptuses secrete an anti-luteolytic factor which prevents an increase in oestradiol output from the ovary, thus removing the stimulus for increased $PGF_{2\alpha}$ production by the endometrium. Results from the unilaterally pregnant guinea-pigs, with the non-pregnant horn intact or transected, indicate that the anti-luteolytic factor acts predominantly in a local manner but does have a systemic effect as well. This latter observation is consistent with conceptuses grafted to the spleen being able to maintain luteal function in guinea-pigs (Bland and Donovan, 1969). The anti-luteolytic factor is probably secreted after about day 10 of pregnancy.

The administration of oestradiol to 3 guinea-pigs from day 10 of pregnancy resulted in an increase in the $PGF_{2\alpha}$ synthesizing capacity of the uterus and a decline in plasma progesterone concentrations due to regression of the corpora lutea. Progesterone concentrations fell about 2 days earlier than they would have done if the guinea-pigs were not pregnant. These findings illustrate the importance of reduced oestradiol output from the ovary for luteal maintenance during early pregnancy. In a fourth pregnant guinea-pig, oestradiol treatment from day 10 caused an initial drop in plasma progesterone concentrations, but then the progesterone concentrations recovered and increased further in spite of the increase in $PGF_{2\alpha}$ synthesizing capacity of the uterus and, presumably, an increase in uterine $PGF_{2\alpha}$ output. It appears that the conceptuses in this guinea-pig secreted also a luteotrophic factor from about day 12, which maintains and actually stimulates progesterone secretion from the corpora lutea in spite of increased $PGF_{2\alpha}$ production by the uterus. The pregnant sheep uterus also secretes a luteotrophic factor which acts locally to maintain corpora luteal function and prevent the luteolytic action of the uterus (Mapletoft et al., 1975, 1976). Since PGE_2 prevents the luteolytic action of $PGF_{2\alpha}$ in sheep (Henderson et al., 1977), it has been proposed that PGE_2 is the luteotrophic factor secreted by the pregnant sheep uterus. Evidence both in favour and against this proposal has been reported (see Poyser, 1981). However, it is clear in the guinea-pig that PGE_2 is not the luteotrophic factor, since the secretion of PGE_2 from the uterus is not increased in early pregnancy during the critical period of luteal maintenance. The identities of both the anti-luteolytic factor and the luteotrophic factor secreted during early pregnancy in the guinea-pig await elucidation.

In the study of Heap et al. (1973), progesterone production rates by the corpus luteum were compared in hyphophysectomized, hysterectomized and early pregnant guinea-pigs. This comparison showed that the guinea-pig conceptuses secreted the luteotrophic factor from about day 16 of pregnancy. However, from the limited study on the effects of oestradiol treatment on early pregnancy, it was apparent that occasionally this luteotrophic factor may be secreted as early as day 12 of pregnancy. Nevertheless, it is clear that secretion of the anti-luteolytic factor precedes the secretion of the luteotrophic factor by several days, and it is the secretion of this anti-luteolytic factor which is responsible initially for luteal maintenance.

CONCLUSION

Oestradiol acting on a progesterone-primed uterus is apparently the physiological stimulus for increasing $PGF_{2\alpha}$ output from the non-pregnant uterus after day 11 of the cycle. Oxytocin does not appear to form part of this stimulus in the guinea-pig. Luteal maintenance during early pregnancy in the guinea-pig appears to be due initially to the secretion by the conceptuses of an anti-luteolytic factor, from about day 10, which acts to prevent the increase in oestradiol output from the ovary and thereby removes the stimulus for increased $PGF_{2\alpha}$ release from the uterus. The conceptuses secrete also a luteotrophic factor from about day 16 of pregnancy, which further stimulates progesterone secretion by the corpora lutea. In some guinea-pigs this luteotrophic factor (which is not PGE_2) may be secreted from as early as day 12 of pregnancy. The identities of these factors involved in luteal maintenance await elucidation.

ACKNOWLEDGEMENTS

These investigations were supported by grants from the M.R.C. and the Wellcome Trust. Authentic prostaglandins and thromboxanes were supplied by the Upjohn Company, U.S.A. 6-oxo-$PGF_{1\alpha}$ used for raising antibodies was supplied by ICI Ltd., U.K.

REFERENCES

Antonini, R., Turner, T.T. and Pauerstein, C.J., 1976. The hormonal control of the guinea-pig corpus luteum during early pregnancy. Fert. Steril., 27: 1322-1325.

Barcikowski, B., Carlson, J.C., Wilson, L. and McCracken, J.A., 1974. The effect of endogenous and exogenous and exogenous estradiol 17β on the release of prostaglandin $F_{2\alpha}$ from the ovine uterus. Endocrinology, 95: 1340-1349.

Bazer, F.W. and Thatcher, W.W., 1977. Theory of maternal recognition of pregnancy in swine based on estrogen controlled endocrine versus exocrine secretion of prostaglandin $F_{2\alpha}$ by uterine endometrium. Prostaglandins, 14: 397-402.

Bland, K.P. and Donovan, B.T., 1968. Gonadal hormones and the regression of the corpora lutea in the guinea-pig. Nature (Lond.), 220: 179-180.

Bland, K.P. and Donovan, B.T., 1969. Control of luteal function during early pregnancy in the guinea-pig. J. Reprod. Fert., 20: 491-501.

Bland, K.P. and Donovan, B.T., 1970. Oestrogen and progesterone and the function of the corpora lutea in the guinea-pig. J. Endocr., 47: 225-230.

Blatchley, F.R. and Poyser, N.L., 1974. The effect of oestrogen and progesterone on the release of prostaglandins from the uterus of the ovariectomized guinea-pig. J. Reprod. Fert., 40: 205-209.

Blatchley, F.R., Donovan, B.T., Poyser, N.L., Horton, E.W., Thompson, C.J. and Los, M., 1971. Identification of prostaglandin $F_{2\alpha}$ in the utero-ovarian blood of guinea-pigs following oestrogen treatment. Nature (Lond.), 230: 243-244.

Blatchley, F.R., Donovan, B.T., Horton, E.W. and Poyser, N.L., 1972. The release of prostaglandins and progesterone into the utero-ovarian venous blood of guinea-pigs during the oestrous cycle and following oestrogen treatment. J. Physiol. (Lond.), 228: 69-88.

Blatchley, F.R., Maule Walker, F.M. and Poyser, N.L., 1975a. Utero-ovarian venous plasma levels of prostaglandin F_{α}, progesterone and oestradiol in non-pregnant and early pregnant guinea-pigs. J. Endocr., 64: 12P.

Blatchley, F.R., Maule Walker, F.M. and Poyser, N.L., 1975b. Progesterone, prostaglandin $F_{2\alpha}$ and oestradiol in the utero-ovarian venous plasma of non-pregnant and early unilaterally pregnant guinea-pigs. J. Endocr., 67: 225-229.

Choudary, J.B. and Greenwald, G.S., 1968. Luteolytic effect of oestrogen on the corpora lutea of the cyclic guinea-pig. J. Reprod. Fert., 16: 333-341.

Csapo, A.I., Puri, C.P. and Tarro, S., 1981. Relationship between timing of ovariectomy and maintenance of pregnancy in the guinea-pig. Prostaglandins, 22: 131-140.

Dey, S.K., Hoversland, R.C. and Johnson, D.C., 1982. Phospholipase A_2 activity in the rat uterus - modulation by steroid hormones. Prostaglandins, 23: 619-630.

Donovan, B.T., 1961. The role of the uterus in the regulation of the oestrous cycle. J. Reprod. Fert., 2: 508-510.

Donovan, B.T., 1975. Indomethacin, ketoprofen and corpus luteum regression in the guinea-pig. Br. J. Pharmac., 53: 225-227.

Downing, I. and Poyser, N.L., 1983. Estimation of phospholipase A_2 activity in guinea-pig endometrium on days 7 and 16 of the estrous cycle. Prostaglandins, Leukotrienes and Medicine, in press.

Earthy, M., Bishop, C. and Flack, J.D., 1975. Progesterone and prostaglandin F concentrations in utero-ovarian venous plasma of cyclic guinea-pigs. J. Endocr., 64: 11P-12P.

Evans, J.J., Sin, I.L., Merrick, A.Y. and Kelleher, P.R., 1981. Effect of oestrogen on pregnancy of guinea-pigs. J. Reprod. Fert., 62: 205-211.

Fairclough, R.J., Moore, L.G., McGowan, L.T., Peterson, A.J., Smith, J.F., Tervit, H.R. and Watkins, W.B., 1980. Temporal relationship between plasma concentrations of 13,14-dihydro-15-keto-prostaglandin F and neurophysin I/II around luteolysis in sheep. Prostaglandins, 20: 199-208.

Flint, A.P.F. and Sheldrick, E.L., 1982. Ovarian secretion of oxytocin is stimulated by prostaglandin. Nature (Lond.), 297: 587-589.

Granström, E. and Kindahl, H., 1976. Radioimmunoassay for urinary metabolites of prostaglandin $F_{2\alpha}$. Prostaglandins, 12: 759-784.

Heap, R.B. and Deanesly, R., 1966. Progesterone in systemic blood and placentae of intact and ovariectomized pregnant guinea-pigs. J. Endocr., 34: 417-423.

Heap, R.B., Perry, J.S. and Rowlands, I.W., 1967. Corpus luteum function in the guinea-pig: arterial and luteal progesterone levels and the effects of hysterectomy and hypophysectomy. J. Reprod. Fert., 13: 537-553.

Heap, R.B., Illingworth, D.V. and Perry, J.S., 1973. The secretory activity of the corpus luteum in the guinea-pig and its role in the estabishment and maintenance of pregnancy. In: R. Denamur and A. Netter (Editors), Le Corps Jaune, Masson, Paris. pp. 69-80.

Henderson, K.M., Scaramuzzi, R.J. and Baird, D.T., 1977. Simultaneous infusion of prostaglandin E_2 antagonizes the luteolytic action of prostaglandin $F_{2\alpha}$ in vivo. J. Endocr., 72: 379-382.
Hildebrandt-Stark, H.E., Marcus, G.J., Yoshinga, K., Behrman, H.R. and Greep, R.O., 1975. Influence of prostaglandin F autoantibodies on the estrous cycle of the guinea-pig. Prostaglandins, 9: 231-244.
Hill, E.E. and Lands, W.E.M., 1968. Incorporation of long-chain and polyunsaturated acids into phosphatidate and phosphatidylcholine. Biochim. Biophys. Acta, 152: 645-648.
Hixon, J.E., Gengenbach, D.R. and Hansel, W., 1975. Failure of prostaglandin $F_{2\alpha}$ to cause luteal regression in ewes after destruction of ovarian follicles by X-irradiation. Biol. Reprod., 13: 126-135.
Horton, E.W. and Poyser, N.L., 1973. Elongation of oestrous cycle in the guinea-pig following subcutaneous or intra-uterine administration of indomethacin. Br. J. Pharmac., 49: 98-105.
Horton, E.W. and Poyser, N.L., 1974. Elongation of oestrous cycle in the guinea-pig following active immunisation against prostaglandin $F_{2\alpha}$. Prostaglandins, 5: 349-354.
Horton, E.W. and Poyser, N.L., 1976. Uterine luteolytic hormone: A physiological role for prostaglandin $F_{2\alpha}$. Physiol. Rev., 56: 595-651.
Joshi, H.S., Watson, D.J. and Labhsetwar, A.P., 1973. Ovarian secretion of oestradiol, oestrone, 20-dihydroprogesterone and progesterone during the oestrous cycle of the guinea-pig. J. Reprod. Fert., 35: 177-182.
Leaver, H.A. and Poyser, N.L., 1981. Distribution of arachidonic acid and other fatty acids in the lipids of guinea-pig uterus and plasma in relation to uterine prostaglandin synthesis. J. Reprod. Fert., 61: 325-333.
Leaver, H.A. and Seawright, A., 1982. Control of endometrial prostaglandin output in vitro during the estrous cycle of the guinea-pig: influence of estradiol-17β, progesterone, oxytocin and calcium ionophore A23187. Prostaglandins, Leukotrienes and Medicine, 9: 657-668.
Louis, T.M., Parry, D.M., Robinson, J.S., Thorburn, G.D. and Challis, J.D.G., 1977. Effects of exogenous progesterone and oestradiol on prostaglandin F and 13,14-dihydro-15-oxo-prostaglandin $F_{2\alpha}$ concentrations in uteri and plasma of ovariectomized ewes. J. Endocr. 73: 427-439.
Mapletoft, R.J., Del Campo, M.R. and Ginther, O.J., 1975. Unilateral luteotrophic effect of uterine venous effluent of a gravid uterine horn in sheep. Proc. Soc. Exp. Biol. Med., 150: 129-137.
Mapletoft, R.L., Lapin, D.R. and Ginther, O.J., 1976. The ovarian artery as the final component of the local luteotrophic pathway between a gravid uterine horn and ovary in ewes. Biol. Reprod., 15: 414-421.
Marley, P.B., 1973. Indomethacin lengthens the oestrous cycle of the guinea-pig when given orally with oestrogen or when implanted within the uterine lumen. Prostaglandins, 4: 251-262.
Maule Walker, F.M. and Poyser, N.L., 1974. Production of prostaglandins by the early pregnant guinea-pig uterus in vitro. J. Endocr., 61: 265-271.
Maule Walker, F.M. and Poyser, N.L., 1978. The metabolism of prostaglandins by the guinea-pig uterus with particular reference to corpus luteal maintenance in early pregnancy. Br. J. Pharmac., 62: 177-183.
McCracken, J.A., 1981. The identification of prostaglandin $F_{2\alpha}$ as a uterine luteolytic hormone and the hormonal control of its synthesis. Acta Vet. Scand., Suppl. 77: 71-88.

Mitchell, S., Poyser, N.L. and Wilson, N.H., 1977. Effect of p-bromophenacyl bromide, an inhibitor of phospholipase A_2, on arachidonic acid release and prostaglandin synthesis by the guinea-pig uterus in vitro. Br. J. Pharmac., 59: 107-113.

Moor, R.M., 1968. The corpus luteum of the sheep: functional relationship between the embryo and corpus luteum. J. Anim. Sci., 27, Suppl. 1: 97-118.

Naylor, B. and Poyser, N.L., 1975. Effect of oestradiol and progesterone on the in vitro production of prostaglandin $F_{2\alpha}$ by the guinea-pig uterus. Br. J. Pharmac., 55: 229-232.

Ning, A.C.W.S. , Leaver, H.A. and Poyser, N.L., 1983. Arachidonic acid uptake into and release from guinea-pig endometrium in vitro on days 7 and 15 of the oestrous cycle. Prostaglandins, Leukotrienes and Medicine, 10: 369-380.

Poyser, N.L., 1971. Release of prostaglandins from the uterus. Ph.D. Thesis, University of Edinburgh.

Poyser, N.L., 1972. Production of prostaglandins by the guinea-pig uterus. J. Endocr., 54: 147-159.

Poyser, N.L., 1973. The physiology of prostaglandins. In: J.A. Loraine (Editor), Chapter 2 in Clinics in Endocrinology and Metabolism Vol. 2 (No. 3) "Reproductive Endocrinology and World Population", Saunders and Co., London, pp. 393-410.

Poyser, N.L., 1979. Effect of actinomycin D on uterine prostaglandin production and oestrous cycle length in guinea-pigs. J. Reprod. Fert., 56: 559-565.

Poyser, N.L., 1981. Prostaglandins in Reproduction. Research Studies Press, Division of John Wiley and Sons Ltd., Chichester, 260 pp.

Poyser, N.L., 1983a. Effect of treating ovariectomized guinea-pigs with estradiol and progesterone on basal and A23187-stimulated release of prostaglandins from the uterus superfused in vitro. Prostaglandins, Leukotrienes and Medicine, 11: 345-360.

Poyser, N.L., 1983b. Differential stimulation of prostaglandin and thromboxane synthesizing capacities in guinea-pig uterus and ovary. Prostaglandins, Leukotrienes and Medicine, 10: 167-177.

Poyser, N.L. and Horton, E.W., 1975. Plasma progesterone levels in guinea-pigs actively immunized against prostaglandin $F_{2\alpha}$, hysterectomized or treated with intra-uterine indomethacin. J. Endocr., 67: 81-88.

Poyser, N.L. and Maule Walker, F.M., 1979. Antiluteolytic effect of the embryo. In: J. Whelan (Editor), "Maternal Recognition of Pregnancy", Ciba Foundation symposium No. 64 (new series), Excerpta Medica, Amsterdam, pp. 261-292.

Poyser, N.L. and Brydon, L.J., 1983. Prostaglandin release from the guinea-pig uterus superfused in vitro. Effect of stage of estrous cycle, progesterone, estradiol, oxytocin and A23187. Prostaglandins, 25: 443-456.

Roberts, J.S., Barcikowski, B., Wilson, L., Skarnes, R.C. and McCracken, J.A., 1975. Hormonal and related factors affecting the release of prostaglandin $F_{2\alpha}$ from the uterus. J. Steroid Biochem., 6: 1091-1097.

Roberts, J.S., McCracken, J.A., Gavagan, J.E. and Soloff, M.S., 1976. Oxytocin-stimulated release of prostaglandin $F_{2\alpha}$ from ovine endometrium in vitro. Correlation with estrous cycle and oxytocin-receptor binding. Endocrinology, 99: 1107-1114.

Schams, D., Lahlou-Kassi, A. and Glatzel, P., 1982. Oxytocin concentrations in peripheral blood during the oestrous cycle and after ovariectomy in 2 breeds of sheep with low and high fecundity. J. Endocr., 92: 9-14.

Sheldrick, E.L. and Flint, A.P.F., 1981. Circulating concentrations of oxytocin during the estrous cycle and early pregnancy in sheep. Prostaglandins, 22: 631-636.

Sheldrick, E.L., Mitchell, M.D. and Flint, A.P.F., 1980. Delayed luteal regression in ewes immunized against oxytocin. J. Reprod. Fert., 59: 37-42.

Thaler-Dao, H., Ramonatxo, M., Saintot, M., Chaintreuil, J. and Crastes De Paulet, A., 1982. Effect of oestradiol-17β on prostaglandin biosynthesis by the uterus of the ovariectomized rat and guinea-pig. Prostaglandins, 24: 149-164.

Trotter, J. and Ferber, E., 1981. CoA-dependent cleavage of arachidonic acid from phosphatidylcholine and transfer to phosphatidylethanolamine in homogenates of murine thymocites. FEBs Lett., 128: 237-241.

Vesin, M.F., Leiber, D. and Harbon, S., 1982. Contribution of lipoxygenase and cyclo-oxygenase metabolites in the modulation of cyclic nucleotides in the guinea-pig myometrium. Differential effects of carbachol and ionophore A23187. Prostaglandins, 24: 851-871.

Vogt, W., 1978. Role of phospholipase A_2 in prostaglandin formation. In: C. Galli, G.Galli and G. Porcellati (Editors), Advances in Prostaglandin and Thromboxane Research Vol. 3, Raven Press, New York, pp. 89-95.

Wathes, D.C. and Swann, R.H., 1982. Is oxytocin an ovarian hormone? Nature (Lond.), 297: 225-227.

Webb, R., Mitchell, M.D., Falconer, J. and Robinson, J.S., 1981. Temporal relationships between peripheral plasma concentrations of oxytocin, progesterone and 13,14-dihydro-15-keto-prostaglandin $F_{2\alpha}$ during the oestrous cycle and early pregnancy in the ewe. Prostaglandins, 22: 443-454.

Wilson, D.B., Prescott, S.M. and Majerus, P.W., 1982. Discovery of an arachidonyl coenzyme A synthetase in human platelets. J. Biol. Chem., 257: 3510-3515.

Wlodawer, P., Kindahl, H. and Hamberg, M., 1976. Biosynthesis of prostaglandin $F_{2\alpha}$ from arachidonic acid and prostaglandin endoperoxides in the uterus. Biochim. Biophys. Acta, 431: 603-614.

Yamashita, S., Hosaka, K. and Numa, S., 1973. Acyl-donor specificities of partially purified 1-acylglycerophosphate acyltransferase, 2-acylglycerophosphate acyltransferase and 1-acylglycerophosphorylcholine acyltransferase from rat liver microsomes. Eur. J. Biochem., 38: 25-31.

Animal Reproduction Science, 7 (1984) 31—55
Elsevier Science Publishers B.V., Amsterdam — Printed in The Netherlands

HORMONE RECEPTOR CONTROL OF PULSATILE SECRETION OF $PGF_{2\alpha}$ FROM THE OVINE UTERUS DURING LUTEOLYSIS AND ITS ABROGATION IN EARLY PREGNANCY

J.A. McCracken, W. Schramm and W.C. Okulicz

Worcester Foundation for Experimental Biology, Inc., 222 Maple Avenue, Shrewsbury, Massachusetts 01545, (USA).

ABSTRACT

McCracken, J.A., Schramm, W. and Okulicz, W.C., 1984. Hormone receptor control of pulsatile secretion of $PGF_{2\alpha}$ from the ovine uterus during luteolysis and its abrogation in early pregnancy. Anim. Reprod. Sci., 7: 31-55.

Studies on the control of $PGF_{2\alpha}$ secretion by hormone receptors in the ovine uterus based on in vivo infusion experiments in the autotransplanted uterus are summarized. In addition, direct measurement of hormone receptor dynamics for oestradiol-17β, progesterone and oxytocin in the ovine uterus under experimental conditions are reported. Investigation of the antiluteolytic effect of early pregnancy in sheep was approached in two ways: (a) studies on the antiprostaglandin secreting effect of the embryo and (b) studies on a putative luteoprotective effect of the embryo. In the former study, the measurement of $PGF_{2\alpha}$ in samples of uterine venous plasma collected hourly from 5 sheep on days 12 to 17 post-mating failed to reveal the presence of any major peaks of $PGF_{2\alpha}$ secretion. In the latter approach, several putative luteoprotective substances were examined for their ability to block the luteolytic effect of $PGF_{2\alpha}$ given as 5 separate hour-long pulses over 25 h into the arterial supply of the autotransplanted ovary. These substances included: day 14-15 uterine venous plasma from pregnant sheep, PGE_2, ovine prolactin, ovine LH, ovine placental lactogen and porcine relaxin. In no case was a convincing luteoprotective effect demonstrated suggesting that the antiprostaglandin secreting effect of the embryo may be the more important of these two antiluteolytic mechanisms. Experiments were performed to investigate circulating levels of endogenous oxytocin by monitoring intramyometrial pressure on a continuous basis in conscious sheep. These studies revealed the presence of hour-long periods of intense uterine activity at intervals of 5.5 to 6.5 hours on days 14 to 16 in the cyclic animal, but not in the early pregnant animal. Based on these results and on recent reports on the production of oxytocin by the corpus luteum, a hypothesis is presented to explain the hormone receptor control of episodic pulsatile secretion of $PGF_{2\alpha}$ in the cyclic sheep, including a possible mechanism for its abrogation during early pregnancy.

INTRODUCTION

Corpus luteum (CL) regression in sheep is induced by prostaglandin $F_{2\alpha}$ ($PGF_{2\alpha}$), which is secreted by the uterus into the uterine vein at the time of luteolysis and reaches the ovary directly via a counter current transfer mechanism in the utero-ovarian pedicle (McCracken et al., 1971, 1972, 1973). Such a transfer mechanism permits small amounts of $PGF_{2\alpha}$ to reach the ovary directly without passing through the pulmonary vascular bed, where it would be rapidly degraded to its inactive metabolite (Piper et al., 1970). It was shown that $PGF_{2\alpha}$ is released from the ovine uterus as a series of episodic surges lasting about one hour with approximately five such pulses occurring each 24 hours over luteolysis on days 14 and 15 of the cycle (Thorburn et al., 1973; Barcikowski et al., 1974; Baird et al., 1976). This episodic release of $PGF_{2\alpha}$ subsequently was confirmed by the detection of a series of intermittent pulses of the primary metabolite of $PGF_{2\alpha}$, 15-keto-13,14-dihydro-$PGF_{2\alpha}$ in the peripheral blood of sheep around luteolysis (Kindahl et al., 1976a; Peterson et al., 1976).

In view of the pulsatile nature of $PGF_{2\alpha}$ secretion from the uterus at luteolysis, pulsed infusions of $PGF_{2\alpha}$ were given directly into the arterial supply of the autotransplanted ovary in the sheep (Schramm et al., 1982, 1983). It was found that four separate hour-long infusions of $PGF_{2\alpha}$ (0.04 to 0.08 µg/h) given over 19 hours caused permanent CL regression in only one out of four animals. The addition of a fifth hour-long pulse of $PGF_{2\alpha}$ (5 pulses in 25 hours, 0.04 to 1.2 µg/h), caused complete CL regression in four out of four animals. A single daily pulse (0.1 µg/h) given once per day for four consecutive days caused a temporary fall in progesterone after each daily infusion followed by recovery. Permanent CL regression did not occur with this protracted regimen, suggesting that a relatively short pulse frequency of $PGF_{2\alpha}$ over a minimal period of about 25 to 30 hours is a necessary condition for physiological regression of the CL in the sheep. The minimal effective dose of $PGF_{2\alpha}$ was found to be 1/40th of the amount of $PGF_{2\alpha}$ (2.5 µg/h) which previously was required to cause luteolysis when given as a continuous infusion for 6 to 19 hours into the ovarian artery (Goding et al., 1972; McCracken et al., 1973). However, since the secretion of $PGF_{2\alpha}$ into the uterine vein at luteolysis is in the order of 25 µg/h (McCracken et al., 1972), 2.5 µg/h would be equivalent to a counter current transfer of 10% from the uterine vein to the ovarian artery when, in fact, using ^{3}H-$PGF_{2\alpha}$, the transfer was found to be 1% or less (McCracken et al., 1972; Land et al., 1976). The lower effective dose found by pulsed infusions is now in close agreement with the quantitative aspects of the counter current transfer system.

The present paper (1) summarizes and extends our findings on the hormonal regulation of $PGF_{2\alpha}$ synthesis in the ovine endometrial cell; (2) reports on the antiluteolytic effect of the embryo via (a) an antiprostaglandin secreting effect and (b) a putative luteoprotective effect; (3) investigates the role of endogenous levels of oxytocin (OT) in the process of $PGF_{2\alpha}$ secretion; (4) describes a hypothesis for the pulsatile secretion of $PGF_{2\alpha}$ from the ovine uterus based on the regulation of receptors for OT (rOT) by oestradiol-17β (E_2-17β), progesterone (P), and their respective receptors.

EXPERIMENTAL RESULTS

(1) Hormonal regulation of $PGF_{2\alpha}$ synthesis

While our early studies indicated a role for both E_2-17β and P in the regulation of $PGF_{2\alpha}$ synthesis in the uterus (Barcikowski et al., 1974), we also showed that both mechanical stimulation of the uterus and intra-arterial infusions of OT could evoke $PGF_{2\alpha}$ secretion (Roberts et al., 1975), but only early and late in the cycle, the response to these stimuli being absent during the mid-luteal phase (see Figs. 1 and 2). Thus, the ability of mechanical stimulation of the uterus to evoke $PGF_{2\alpha}$ secretion appeared to depend on two independent, but simultaneously operating steroid hormone-dependent variables: (i) the variation in the central OT-releasing reflex during the cycle (ii) the variation in the ability of the uterus to relase $PGF_{2\alpha}$ in response to OT. Subsequently, we showed that the relative concentration of rOT in the endometrium, the principal site of $PGF_{2\alpha}$ secretion (Wilson et al., 1972), was found to vary cyclically (Roberts et al., 1976) with the highest concentration of rOT occurring late in the oestrous cycle (Fig. 3).

To further examine the role of E_2-17β and P in regulating uterine rOT, the autotransplanted uterus in the ovariectomized sheep was used as an in vivo perfusion system (McCracken, 1980). The ability of a brief (10 min) intra-arterial infusion of OT into the uterus to evoke $PGF_{2\alpha}$ secretion was used as an in vivo probe for the presence of rOT since we had previously shown that OT could evoke $PGF_{2\alpha}$ secretion only in the presence of its receptor. As shown in Fig. 4, E_2-17β given alone as a constant infusion caused the formation of rOT in about 6

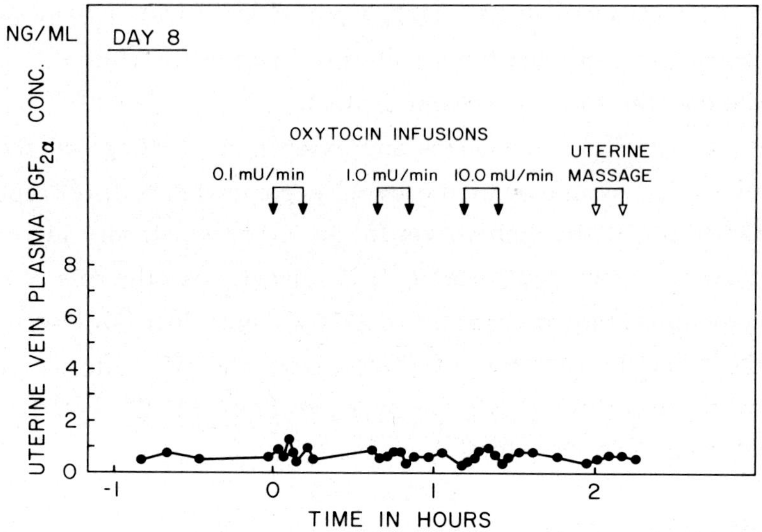

Fig. 1. The concentration of $PGF_{2\alpha}$ in uterine venous plasma after infusing oxytocin for 10 minutes into the adjacent uterine artery and after mechanical stimulation of the uterus by massage for 10 minutes on day 8 of the cycle in sheep.

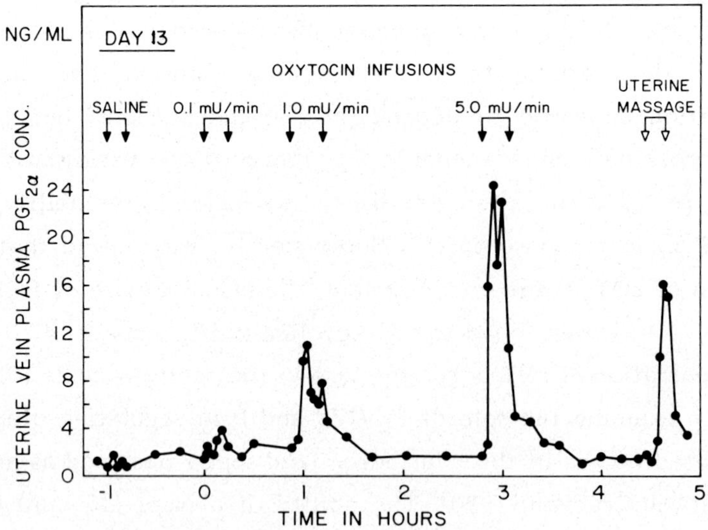

Fig. 2. The concentration of $PGF_{2\alpha}$ in uterine venous plasma after infusing oxytocin for 10 minutes into the adjacent uterine artery and after mechanical stimulation of the uterus by massage for 10 minutes on day 13 of the cycle in sheep.

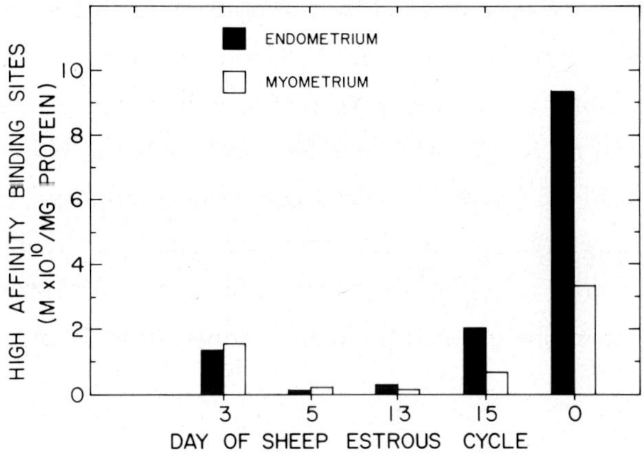

Fig. 3. Concentration of high-affinity binding sites for oxytocin in endometrium and myometrium from in situ ovine uterus on different days of the oestrous cycle.

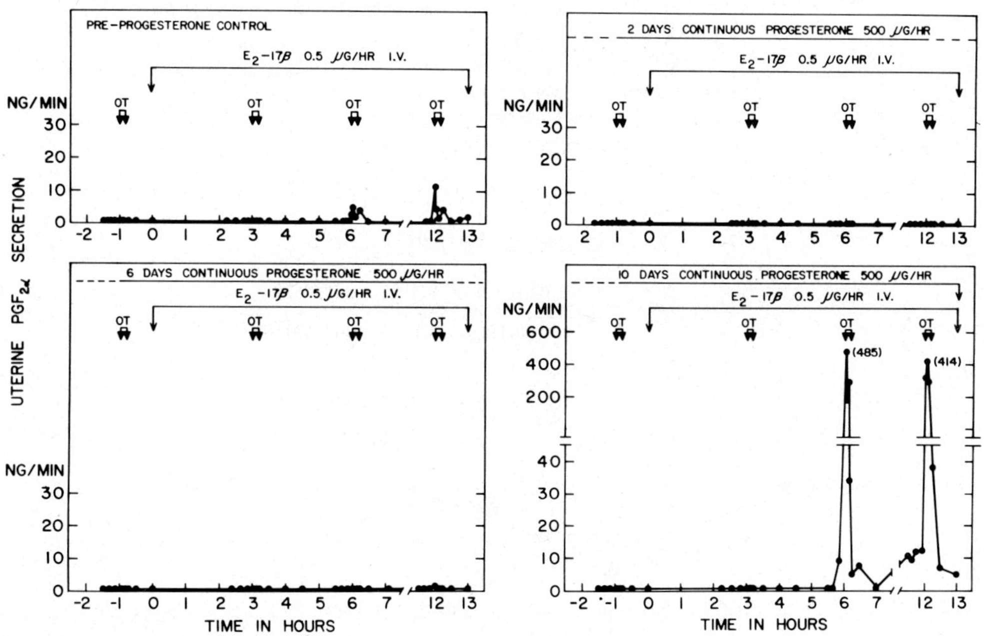

Fig. 4. Effect of 12-hour infusions of E_2-17β on days 2, 6, and 10, superimposed on the continuous 10-day infusion of progesterone on OT-induced $PGF_{2\alpha}$ secretion from the transplanted uterus in ovariectomized sheep. Although progesterone initially inhibits the previously observed stimulating effect of E_2-17β on OT-induced $PGF_{2\alpha}$ secretion, eventually, after 10 days, it greatly amplifies the response to OT in terms of $PGF_{2\alpha}$ secretion.

hours, an effect which was abolished by the continuous infusion of P for 10 days (inhibition of rE_2-17β replenishment by P). However, after 10 days of P infusion, E_2-17β again was able to induce rOT possibly because rP replenishment was impaired and P could no longer suppress oestrogen action. However, at this time OT caused a 100-fold increase in $PGF_{2\alpha}$ secretion compared to the response to OT when E_2-17β was given alone (P may increase lipid precursors for $PGF_{2\alpha}$ synthesis). These events are summarized in Fig. 5 which represents our working hypothesis on the hormone-mediated events controlling $PGF_{2\alpha}$ synthesis in the endometrial cell.

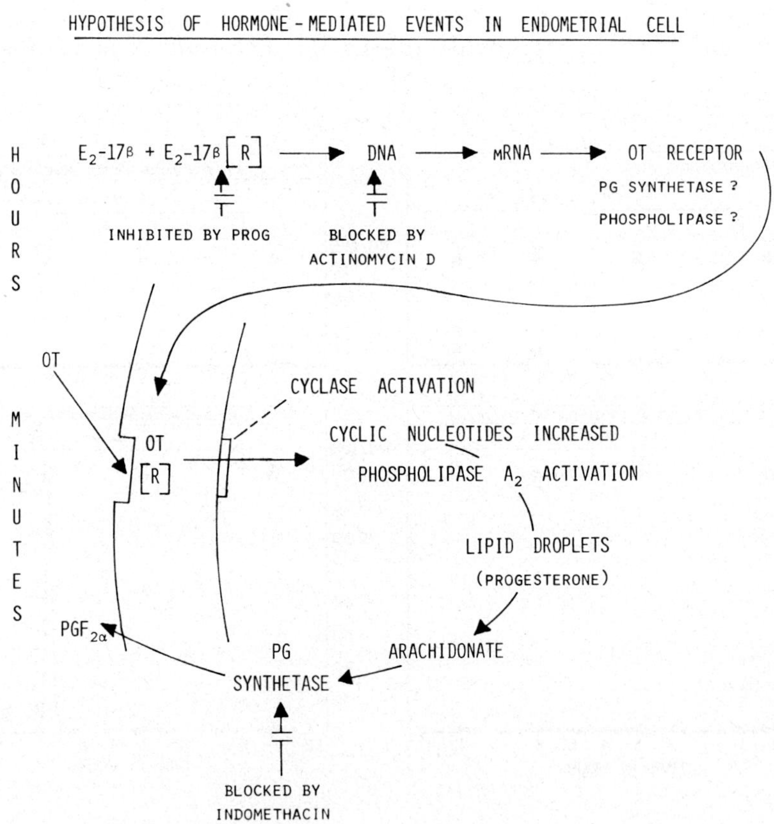

Fig. 5. Working hypothesis on the hormone receptor mediated events controlling $PGF_{2\alpha}$ secretion from the endometrial cell.

This general hypothesis is now supported by work carried out at the Worcester Foundation to directly measure receptor dynamics for E_2-17β, P and OT in uterine tissues from ovariectomized sheep sampled sequentially in vivo after E_2-17β and P treatment (McCracken et al., 1982; Okulicz et al., 1982). Endometrial and myometrial tissues were obtained from individual animals by biopsy at -1 h, +3 h, +6 h and +12 h after stopping a continuous 5-day infusion of P (500 μg/h) in the ovariectomized sheep continuously infused with E_2-17β (0.5 μg/h). By 6 h after stopping P there was a significant rise in nuclear rE_2-17β in endometrium and myometrium and a parallel rise in plasma membrane bound rOT (oestrogen-dependent event) which continued to rise between 6 h and 12 h. Control uterine tissues all sampled simultaneously prior to P withdrawal showed that regional differences in receptor populations in the uterus did not account for the sequential changes in receptors observed following the withdrawal of P.

Recently, P has been shown to rapidly and selectively down-regulate nuclear rE_2-17β in the hamster (Evans et al., 1980) and rat uterus (Okulicz et al., 1981a). This effect of P results in the preferential loss of the biologically active form of nuclear rE_2-17β, the occupied form (Okulicz et al., 1981b), and appears to involve the induction of an rE_2-17β regulatory factor (ReRF) by P (Okulicz et al., 1981c; Leavitt et al., 1983). This novel effect of P may serve as a mechanism by which P modulates E_2-17β action. We conclude, therefore, that the loss of P influence, either naturally (at the end of the luteal phase), or experimentally (by stopping an i.v. infusion of P) permits the rise of nuclear rE_2-17β in the presence of E_2-17β and, hence, the synthesis of rOT. At the end of the luteal phase when these rOT are activated by endogenous levels of OT, synthesis of $PGF_{2\alpha}$ occurs in the endometrium and causes regression of the CL in the adjacent ovary via a counter current system.

(2) The antiluteolytic effect of early pregnancy

Early studies in the sheep by Moor and Rowson (1966) demonstrated that the transfer of an embryo to the uterus of a synchronized non-pregnant recipient sheep would prevent cyclic regression of the corpus luteum in the recipient even if transferred as late as day 13 of the cycle. Later work indicated that the same effect could be achieved by infusing extracts or homogenates of embryos into the uterus (Moor, 1968; Martal et al., 1979; Ellinwood et al., 1979a). Following the identification of $PGF_{2\alpha}$ as a luteolytic hormone in the sheep (McCracken et al., 1972), it seemed likely that the capacity of the embryo to prevent corpus luteum regression might be related to either (a) suppression of $PGF_{2\alpha}$ release from the uterus by the embryo or its products (anti-prostaglandin secreting effect) or (b) the ability of a secretory product of the embryo to protect the corpus luteum against potentially luteolytic releases of $PGF_{2\alpha}$ from the uterus (luteoprotective effect). One or both of these mechanisms would allow the corpus luteum to continue to secrete the progesterone which is necessary for the maintenance of early gestation in most species. These two different mechanisms have been loosely termed "the antiluteolytic effect of the embryo", which is an unfortunate phrase since it does not indicate the nature of the effect, i.e. an indirect one via suppression of $PGF_{2\alpha}$ secretion from the uterus or a direct one on the corpus luteum by protecting it against the effects of $PGF_{2\alpha}$. The terms (a) "antiprostaglandin secreting" and (b) "luteoprotective", respectively, would seem more appropriate.

(a) Antiprostaglandin secreting effect of the embryo: Reports on the effect of the presence of an embryo on $PGF_{2\alpha}$ secretion from the uterus in early pregnancy, until recently, have been conflicting. The levels of $PGF_{2\alpha}$ at this time have been reported to be diminished or absent (Barcikowski et al., 1974; Thorburn et al., 1973), unchanged (Pexton et al., 1975a; 1975b; Cerini et al., 1976; Nett et al., 1976; Lewis et al., 1977; 1978), or higher (Wilson et al., 1972; Ellinwood et al., 1979b). Results showing unchanged or higher levels of $PGF_{2\alpha}$ in early pregnancy are difficult to equate with the low basal levels of the stable metabolite of $PGF_{2\alpha}$ seen in the peripheral blood of most cyclic species in early pregnancy compared to the elevated levels of the metabolite observed at the time of luteolysis. Such a difference, either in 15-keto-13,14-dihydro-$PGF_{2\alpha}$ or in uterine vein $PGF_{2\alpha}$, between cyclic and early pregnant animals has been reported for the sheep (Peterson et al., 1976), cow (Kindahl et al., 1976b), pig (Moeljono et al., 1977; Shille et al., 1979; Bosc et al., 1981; Kindahl et al., 1981), mare (Stabenfeldt et al., 1981; Kindahl et al., 1982) and guinea-pig (Poyser and Maule-Walker, 1979).

Recently, we have monitored changes in the concentration of $PGF_{2\alpha}$ in uterine venous plasma and the concentration of P in utero-ovarian venous plasma sampled hourly in five conscious sheep on days 12 to 17 post-mating. The average changes in concentration of PGF2α and P over this sampling period are shown in Fig. 6 and Fig. 7. The 100% (± SD) value in Fig. 6 represents the mean average concentration of PGF2α (0.95 ng/ml) in uterine venous plasma in five sheep and in Fig. 7 represents the mean average concentration of P (1.9 μg/ml) in utero-ovarian venous plasma. It can be seen that PGF2α levels are relatively constant except for a small rising trend between days 13 and 14. However, the major luteolytic pulses seen in the cyclic sheep on days, 14, 15 and 16 appear to be absent in the pregnant animal thus confirming many earlier reports (see above). This finding of a lack of PGF2α pulses in early pregnant sheep is in keeping with our observation (McCracken et al., 1979) that rOT levels are significantly lower (33 fmoles/mg protein) on day 16 of pregnancy compared to the levels seen in the cyclic animal at this time (>250 fmoles/mg protein, see Table I).

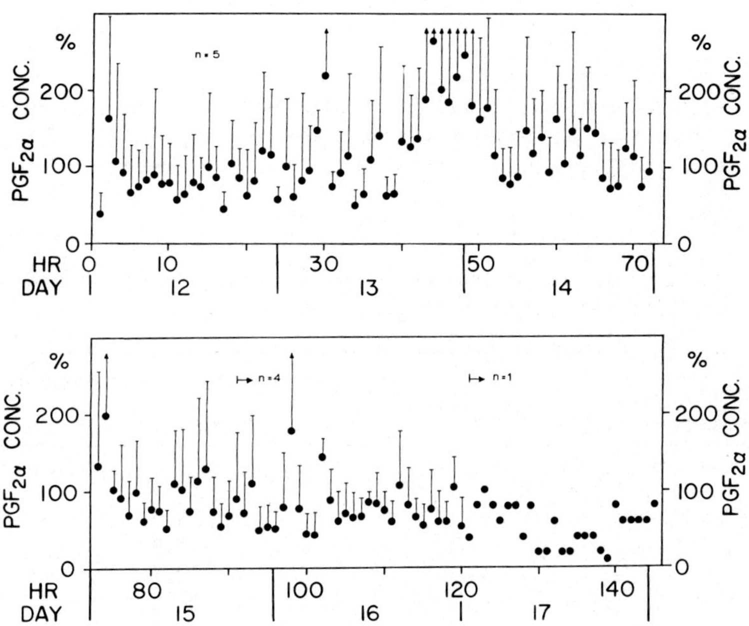

Fig. 6. Mean average hourly changes in PGF2α concentration in uterine venous plasma from five pregnant sheep sampled on days 12 to 17 post-mating. The 100% value (± SD) represents the mean average concentration of PGF2α in five sheep (0.95 ng/ml).

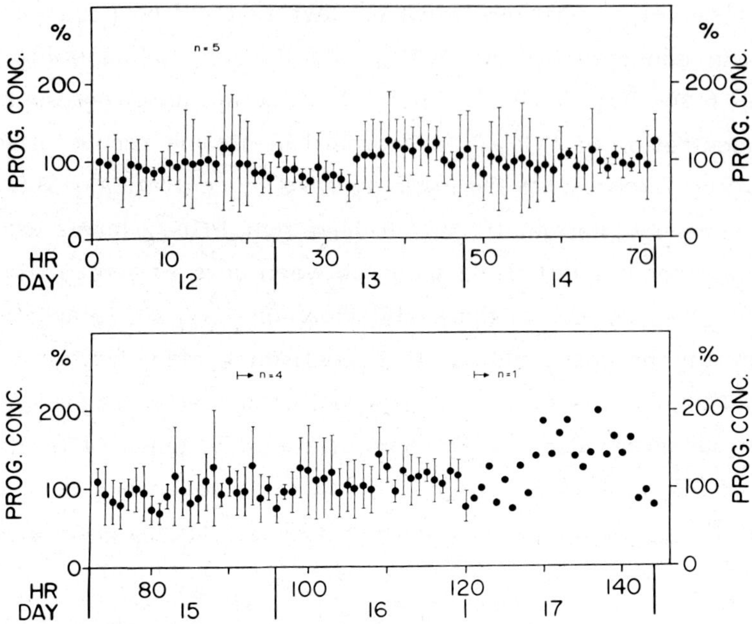

Fig. 7. Mean average hourly changes in progesterone conc. in utero-ovarian venous plasma from five pregnant sheep sampled on days 12 to 17 post-mating. The 100% value (± SD) represents the mean average concentration in five sheep (1.9 μg/ml).

Table I. Comparison of concentration of OT receptor binding sites in endometrium of pregnant and non-pregnant sheep on day 16

Day of pregnancy		Concentration of OT receptor binding sites (fmoles/mg of protein)	
16		24.3	
16		39.3	
16		33.8	
16		34.6	
	Mean	33.0	fmoles/mg
Cycle controls (Day 16/17)	Mean	>250.0	fmoles/mg

This difference in rOT between the pregnant and cyclic animal does not appear to be related to the concentration of E_2-17β in maternal blood since these were not significantly different (Table II). It is more likely that this difference is due to the persistence of P action via rP (see later) which may thus block the rOT-synthesizing action of E_2-17β in the pregnant sheep at this critical time.

Table II. OT - Stimulated $PGF_{2\alpha}$ concentration (uterine vein) and steroid concentration (utero-ovarian vein) in day 16 pregnant versus non-pregnant sheep

	Non-pregnant			Pregnant		
	OT↑ $PGF_{2\alpha}$	PROG	E_2-17β	OT↑ $PGF_{2\alpha}$	PROG	E_2-17β
	(ng/ml)	(ng/ml)	(pg/ml)	(ng/ml)	(µg/ml)	(pg/ml)
	15.5	2.2	65	< 1.0	2.6	55
	8.2	1.5	20	< 1.0	0.4	65
	30.1	5.5	150	< 1.0	1.5	40
	23.5	3.0	10	< 2.5	3.1	10
Mean	19.3	3.1	61	< 1.5	1.9	43

Recently, we have also examined the effect of embryo removal on the life span of the CL of pregnancy in sheep, 21 days post-mating. As shown in Fig. 8 peripheral levels of P persist for 20 to 40 days after embryo removal, indicating that once the embryo abrogates luteolysis by suppressing $PGF_{2\alpha}$ secretion on days 14 and 15, the CL is maintained for a considerable period of time in the absence of an embryo. In the case of a twin pregnancy, the extension of the life span of the CL appears to be more pronounced. A similar effect of embryo removal on the life span of the CL in sheep has been reported by Martal et al. (1979).

(b) Luteoprotective effect of the embryo: There are several indications that the CL in early pregnancy is refractory to the luteolytic action of prostaglandins. Ginther (1976) demonstrated in a series of experiments involving the anastomoses of the uterine vein and the ovarian artery from the pregnant horn to the non-pregnant horn and vice versa, that the inhibition of luteolysis in the

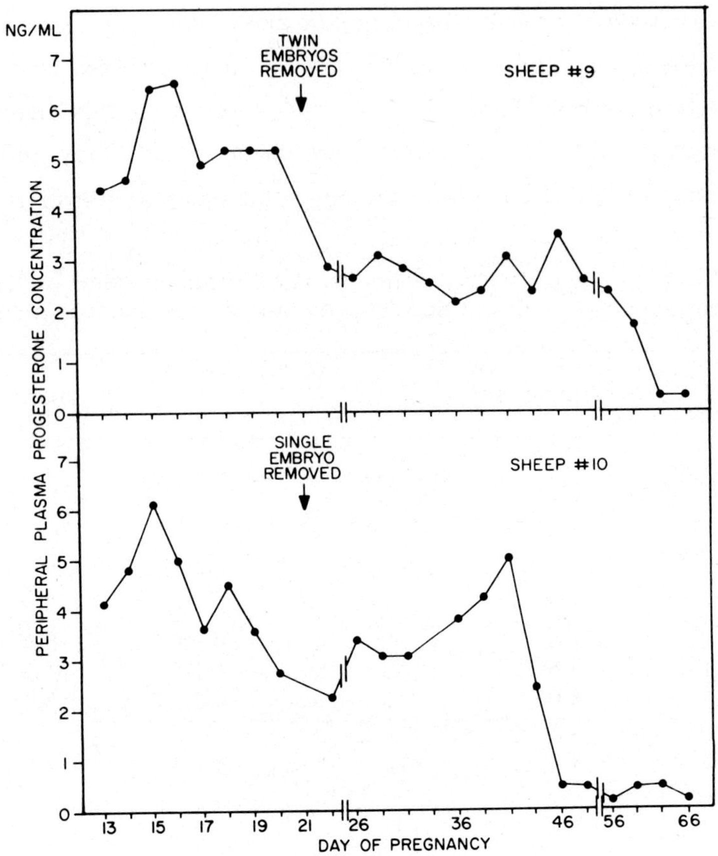

Fig. 8. Peripheral plasma progesterone concentration before and after removal of embryos on day 21 of pregnancy in sheep. Progesterone concentration shows a decline after embryo removal, but persists for a further 20 to 40 days.

sheep by the conceptus involved a local vascular pathway. Moreover, the evidence suggested that putative fetal luteotrophins may overcome the effect of luteolysins. Furthermore, infusion of a relatively large amount of $PGF_{2\alpha}$ (200 µg) into an ovarian artery of 7 pregnant sheep caused luteolysis in the adjacent ovary in only two sheep (Mapletoft et al., 1976). The same amount of $PGF_{2\alpha}$ injected into follicles of pregnant sheep resulted in CL regression in only 17% of the animals in comparison to 83% of non-pregnant sheep undergoing the same treatment (Inskeep et al., 1975; Pratt et al., 1977). Likewise, the administration of E_2-17β in a concentration which induced luteal regression in cyclic sheep is only partially effective in pregnant sheep (Kittok and Britt, 1977). Nancarrow et al. (1982) found that multiple embryos in the sheep were more protective than a single embryo against the luteolytic effect of 100 µg cloprostenol.

It has been suggested that PGE_2 counteracts the luteolytic effect of $PGF_{2\alpha}$ (Henderson and McNatty, 1975). As a stimulator of cAMP and a vasodilator, PGE_2 has indeed properties antagonistic to $PGF_{2\alpha}$. So far, however, there is insufficient evidence that PGE_2, in physiological amounts can abolish the luteolytic action of $PGF_{2\alpha}$ for other than short periods of time (Henderson et al., 1977; Pratt et al., 1977; Magness et al., 1978; Colcord et al., 1978). Moreover, an elevation in the concentration of PGE_2 in the uterine vein in early pregnancy has not so far been demonstrated (Lewis et al., 1978). Nevertheless, PGE_2 might be one of several factors prolonging the life span of the CL. Because of its low molecular weight, it could probably be transported locally from the uterus to the ovary via a countercurrent system.

We have examined several substances for their ability to counteract the luteolytic effect of $PGF_{2\alpha}$ given in physiological amount as five-hour long pulses into the arterial supply of the transplanted ovary. The following putative luteoprotective substances were infused into the ovary prior to and during an effective luteolytic regimen of $PGF_{2\alpha}$: day 14-15 uterine venous plasma from pregnant sheep (Fig. 9); PGE_2 (Fig. 10); ovine placental lactogen (Fig. 11); ovine prolactin (Fig. 12); porcine relaxin (Fig. 12); ovine LH (Fig. 12). None of these substances was able to completely protect the CL against a physiological luteolytic dose of $PGF_{2\alpha}$. We conclude that, while a luteoprotective substance may exist, we have so far been unable to identify the nature of such a putative substance.

3. The role of endogenous levels of OT in the regulation of $PGF_{2\alpha}$ secretion

To examine the presence of oxytocic substances in the circulating blood, we have monitored intramyometrial pressure (IMP) in conscious sheep by means of miniature pressure-sensitive balloons chronically implanted in the myometrium of both uterine horns (Schramm and McCracken, 1982). In ovariectomized sheep, simultaneous bursts of uterine contractions (50 to 100 mm Hg) occurred in both uterine horns with a mean duration of 5.9 min ± 1.7 SD, followed by a period of relaxation lasting 14.2 min ± 3.4 SD. These bursts of IMP could be replicated in

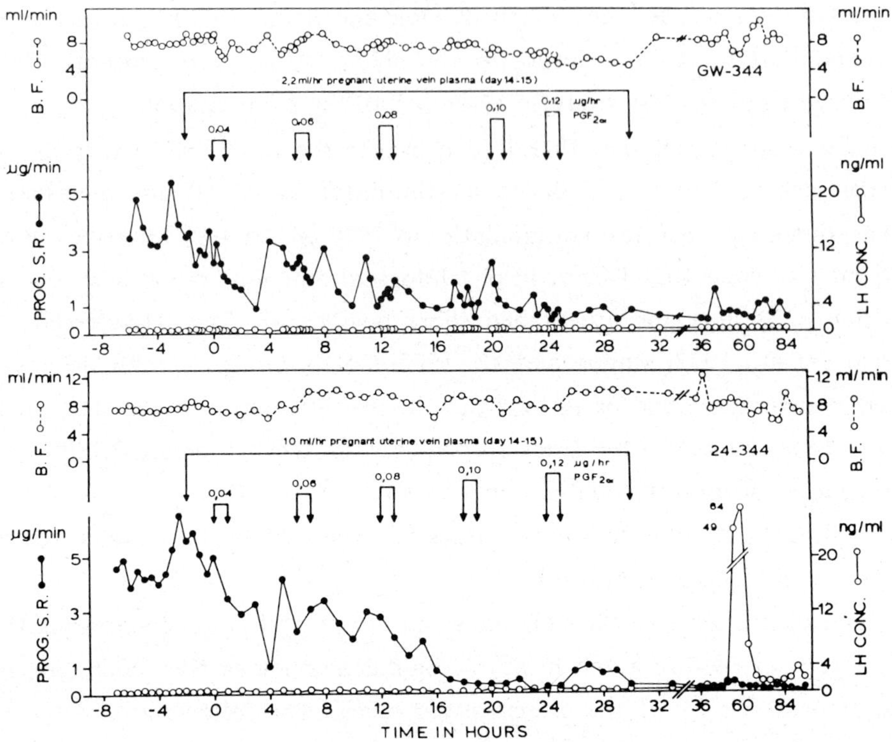

Fig. 9. The effect of infusing uterine venous plasma harvested from sheep on days 14 and 15 of pregnancy on the luteolytic action of 5 hour-long pulses of $PGF_{2\alpha}$ given over a period of 25 hours. In this and subsequent experiments of a similar type substances were infused separately into the arterial supply of the autotransplanted ovary on day 12 of a cycle induced by prior treatment with $PGF_{2\alpha}$ (two injections of Lutalyse given i.m. four hours apart on day -2).

one horn by a one-minute infusion of OT (0.01 mU) into the adjacent uterine artery or in both horns by a one minute infusion of OT (10.0 mU) into a jugular vein. The infusion of a potent antagonist of OT into one uterine horn completely blocked the bursts of IMP only locally in the infused horn. Based on the foregoing results, we conclude that these bursts of IMP are caused by episodic releases of OT into the systemic circulation. The bursts of IMP gradually diminished 10 to 20 days after ovariectomy, but could be maintained by the continuous infusion of E_2-17β (5 to 10 μg/day). An infusion of P (10 mg/day) superimposed on the infusion of E_2-17β caused these bursts to disappear after 3 days. When P was stopped, the bursts of IMP gradually returned over a period of 3 days.

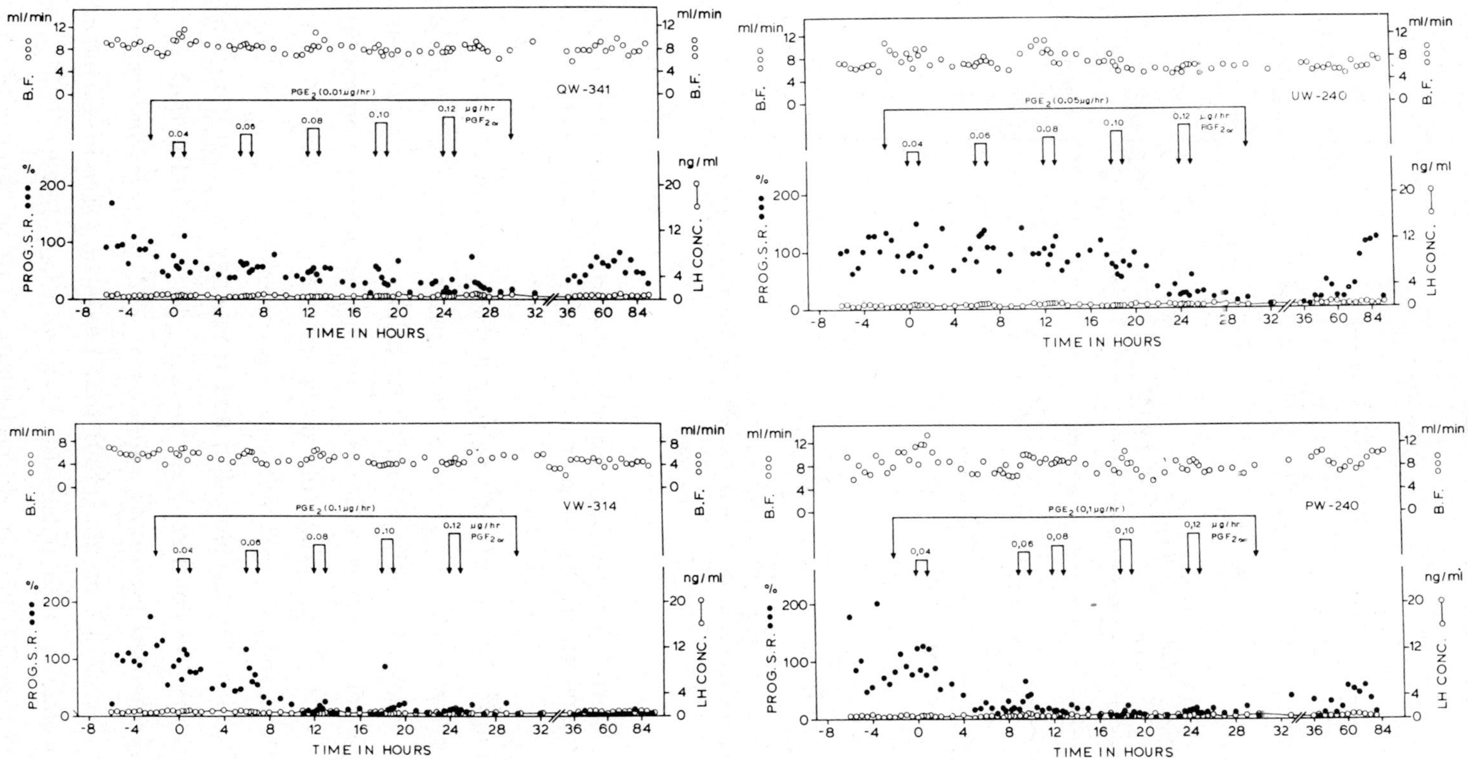

Fig. 10. The effect of infusing PGE_2 on the luteolytic action of 5 hour-long pulses of $PGF_{2\alpha}$ given over a period of 25 hours. Lower infusion rates of PGE_2 seemed to be more effective in retarding the fall in progesterone than the higher infusion rates of PGE_2.

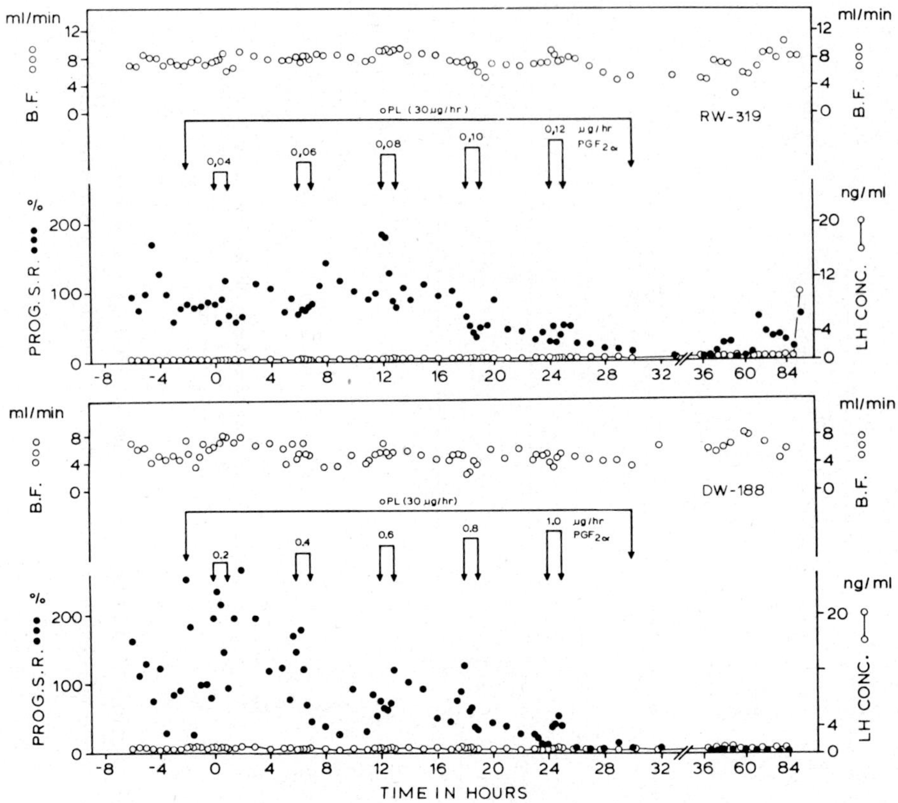

Fig. 11. The effect of infusing purified ovine placental lactogen on the luteolytic action of 5 hour-long pulses of $PGF_{2\alpha}$ given over a period of 25 hours.

In cyclic sheep, similar bursts of IMP were observed in both uterine horns from day 13 of one cycle until day 5 of the next cycle when they gradually disappeared over 2 to 3 days. In cyclic sheep, in addition to these regular episodic bursts of IMP, hour-long periods of intense continuous uterine contractions were observed at intervals of approximately six hours from days 14 to 16 of the cycle (see Fig. 13). However, no such activity was recorded in two pregnant sheep at this time. Because of the duration and frequency of these hour-long episodes of continuous uterine contractions, we considered initially that these hour-long episodes represented luteolytic releases of $PGF_{2\alpha}$ from the endometrium. The report by Fairclough et al. (1980) that peaks of OTN in peripheral blood occurred synchronously with pulses of 15-keto-13,14-dihydro-$PGF_{2\alpha}$ in cyclic sheep, suggested that OT itself might contribute to these observed periods

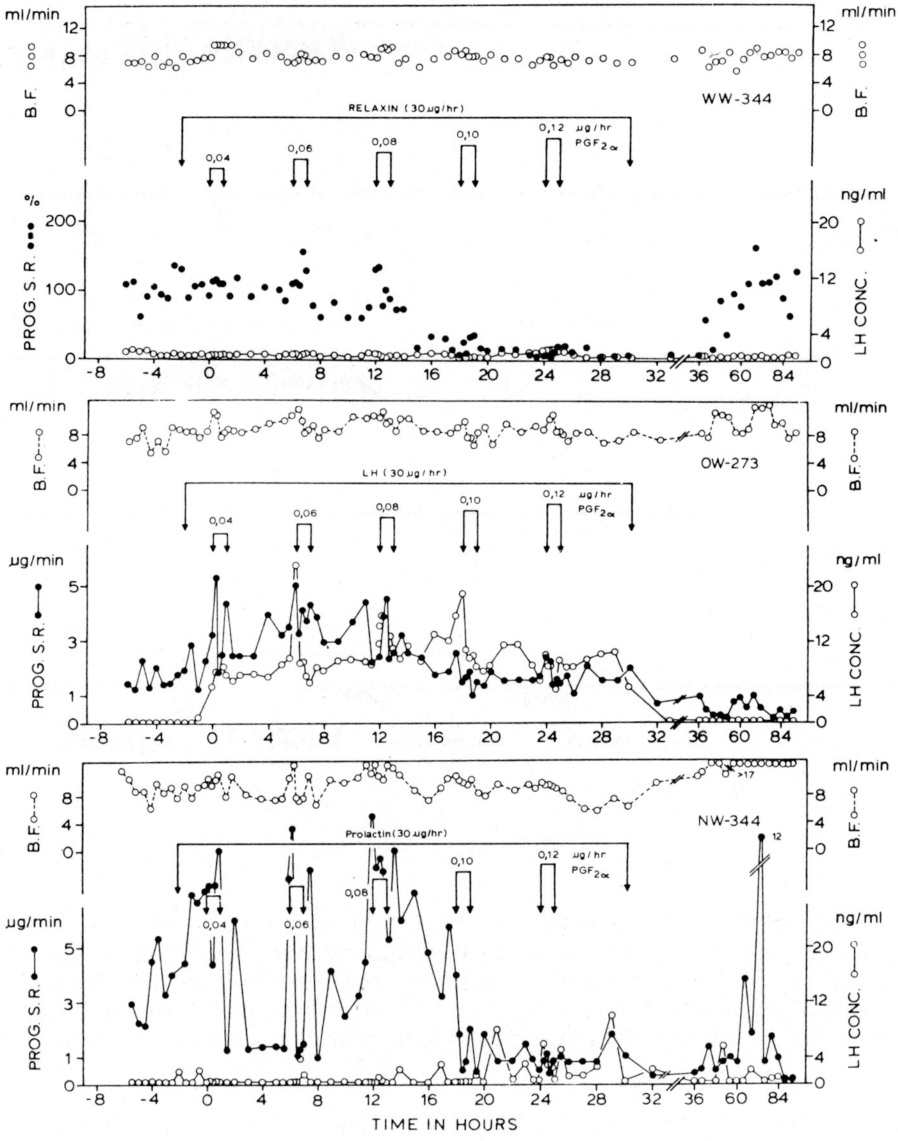

Fig. 12. The effect of infusing porcine relaxin, ovine LH, or ovine prolactin on the luteolytic action of 5 hour-long pulses of $PGF_{2\alpha}$ given over a period of 25 hours. LH infusion appeared to maintain progesterone secretion rate, but only for as long as the infusion of LH was given.

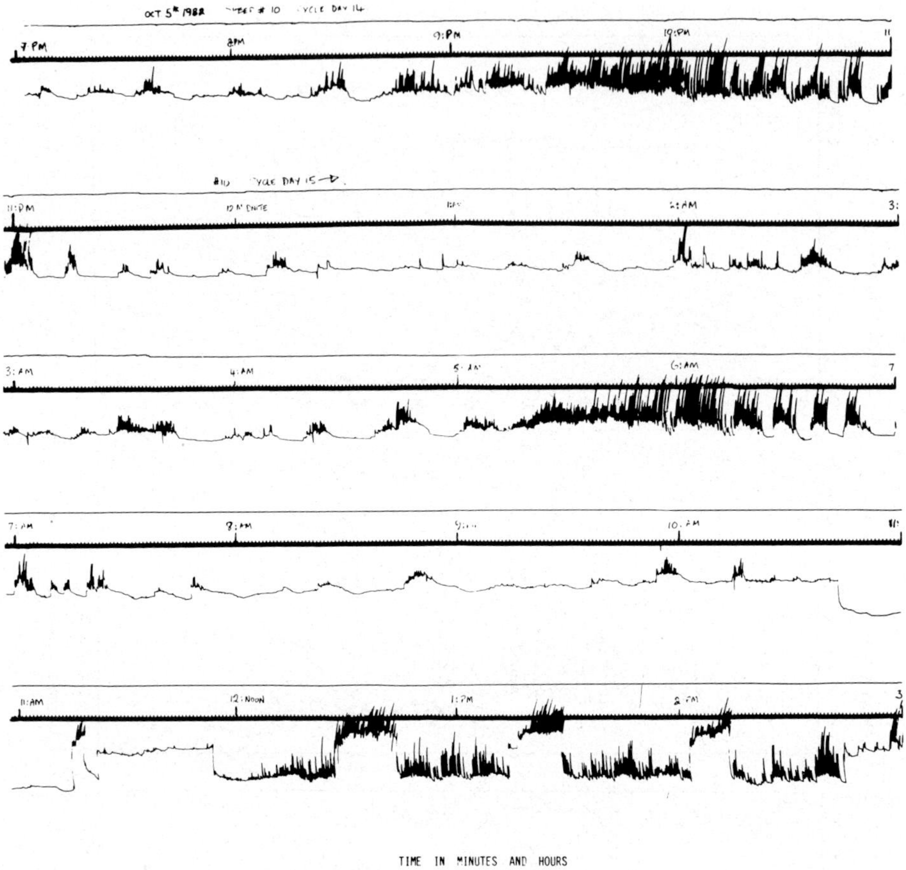

Fig. 13. Intramyometrial pressure in the uterus of a conscious sheep on days 14 to 15 of the oestrous cycle recorded continously over a period of 20 hours. In addition to small episodic bursts of activity, three periods of high intensity contractions, lasting about 2 hours, occurred at intervals of 5.5 to 6.5 hours. These episodes were not observed in pregnant sheep over the same period post-mating.

of intense contractile activity in the uterus. On the other hand, subsequent reports indicated that basal levels of OT were relatively high during the luteal phase of the cycle in sheep, with lower levels occurring around the time of luteolysis (Sheldrick and Flint, 1981; Webb et al., 1981; Schams et al., 1982). However, the rather surprising reports that the ovine CL contained OT (Flint and Sheldrick, 1982; Wathes and Swann, 1982) taken together with the parallelism of peripheral levels of OT and P in the sheep cycle, suggested that the CL contributed to the basal levels of OT seen during the cycle. Moreover, there appears to be a pulsatile component in the peripheral blood levels of OT in the sheep (Mitchell et al., 1982) which may represent a neural contribution especially since we have evidence for an episodic release of OT in the ovariectomized sheep as described above (Schramm and McCracken, 1982). Recently Flint and Sheldrick (1983) have shown that OT is discharged from the sheep ovary following treatment with cloprostenol and during endogenous surges of $PGF_{2\alpha}$ from the uterus. They suggested that ovarian OT so released may reinforce the synthesis of $PGF_{2\alpha}$ in the uterus. Moreover, they suggest that the pulses of OTN reported in the sheep during luteolysis could be of ovarian, rather than of neural origin. Thus the hour-long periods of intense uterine contractions on days 14 to 16 may represent either $PGF_{2\alpha}$ synthesis in the uterus or the release of OT from the ovary or perhaps even a combination of both.

4. Hypothesis to explain the intermittent episodes of $PGF_{2\alpha}$ release from the uterus over the time of luteolysis in sheep

Based on the above investigations and reports, we propose the following sequence of events for the regulation of episodes of secretion $PGF_{2\alpha}$ from the ovine uterus during luteolysis (Fig. 14).

1. Because of declining P action on the uterus towards the end of the luteal phase, E_2-17β is now able to stimulate the formation of endometrial rOT.
2. Endogenous levels of OT (of ovarian and/or pituitary origin) interact with newly synthesized rOT.
3. $PGF_{2\alpha}$ secretion from the uterus rapidly follows coupling of OT to its receptor.
4. 15-Keto-13,14-dihydro-$PGF_{2\alpha}$ levels increase in peripheral blood due to metabolism of secreted $PGF_{2\alpha}$ primarily in the pulmonary vascular bed.
5. P secretion from the CL begins to decline (luteolysis) as a result of the action of $PGF_{2\alpha}$ locally transferred to the ovary.
6. OT is released from the CL by $PGF_{2\alpha}$ and may reinforce the secretion of $PGF_{2\alpha}$ from the uterus.
7. OTN levels in peripheral blood increase probably because this carrier-protein is released from the CL along with OT.

Since rOT are likely to be down-regulated (desensitized) by the release of OT from the CL, subsequent hour-long episodes of $PGF_{2\alpha}$ release from the endome-

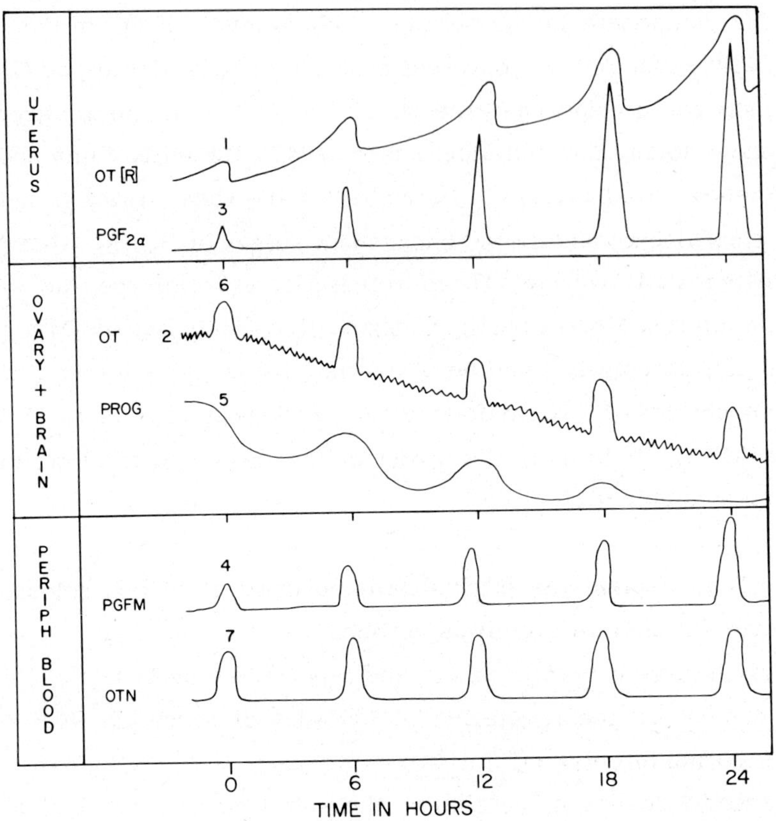

Fig. 14. Hypothesis to explain the intermittent episodes of $PGF_{2\alpha}$ release from the uterus during luteolysis in sheep. See text for explanation of numbers. (PGFM = 15-keto-13,14-dihydro-$PGF_{2\alpha}$).

trium will follow at intervals of approximately six hours, the regeneration time of rOT. The most fundamental way to abrogate this system of episodic releases of $PGF_{2\alpha}$ during early pregnancy, would be for a secretory product of the embryo to maintain P action in the endometrium. This could be achieved directly through enhanced rP replenishment, thus promoting P suppression of nuclear rE_2-17β accumulation. Alternatively, direct inhibition of nuclear rE_2-17β accumulation by an embryonic product could occur. Such effects of the embryo would inhibit oestrogen-dependent synthesis of rOT and hence block the subsequent chain of events. It is also possible that an embryonic product could directly inhibit rOT formation. These postulated effects of the embryo on endometrial cell receptor dynamics would appear to be of considerable duration as witnessed by the survival of the CL for 20 to 40 days after the removal of the embryo from the sheep uterus on day 21 post-mating. We also know that rOT concentrations are significantly lower in the early pregnant sheep compared to the cyclic sheep at this time. The nature of such a putative embryonic product which might bring about changes in receptors is presently unknown. However, as shown more than a decade ago by Moor (1968), the infusion of homogenates of ovine preimplantation blastocysts into the sheep uterus extends the life span of the CL. Such an effect of the blastocyst has been confirmed by other investigators (Martal et al., 1979; Ellinwood et al., 1979a). The active substance is thought to be proteinaceous in nature and has been termed "trophoblastin" by Martal et al. (1979).

The above hypothesis is clearly open for refinement and modification as new information on this topic unfolds. Certainly the key role postulated for rP-dependent suppression of nuclear rE_2-17β in the above hypothesis is subject to experimental test. Indeed, investigations are currently underway in this laboratory to examine in detail the sequential changes in the biological status of rP as well as other receptors in the cyclic and early pregnant sheep over this critical period of time.

ACKNOWLEDGEMENTS

We are grateful to Merrilyn E. Glew, Larry C. Bovaird, Gudrun Schramm and Arthur Copp for expert assistance in carrying out these studies. The work was supported by USPHS Grant HD 08129, by a Research Fellowship from the Deutsche Forschungsgemeinschaft (WS), and by an NIH Research Fellowship HD 06295 (WCO).

REFERENCES

Baird, D.T., Land, R.B., Scaramuzzi, R.J. and Wheeler, A.G., 1976. Endocrine changes associated with luteal regression in the ewe: The secretion of ovarian estradiol, progesterone and androstenedione and uterine prostaglandin $F_{2\alpha}$ throughout the oestrous cycle. J. Endocrin., 69: 275-286.

Barcikowski, B., Carlson, J.C., Wilson, L. and McCracken, J.A., 1974. The effect of endogenous and exogenous estradiol-17β on the release of prostaglandin $F_{2\alpha}$ from the ovine uterus. Endocrinology, 95: 1340-1349.

Bosc, M.J., Martinat-Botte, F., and Terqui, M., 1981. Practical uses of prostaglandins in pigs. Acta vet. scand., Suppl. 77: 209-226.

Cerini, J.C., Cerini, M.E., Cumming, I.A., Findlay, J.K. and Lawson, R.A.S., 1976. Prostaglandin F secretion during early pregnancy in sheep. J. Reprod. Fert., 46: 533 (abstract).

Colcord, M.L., Hoyer, C.L. and Weems, C.W., 1978. Effects of prostaglandin E_2 (PGE-2) as an antiluteolysin on estrogen-induced luteolysis in ewes. J. Anim. Sci., (Suppl. 1) 47: 352-353.

Ellinwood, W.E., Nett, T.M. and Niswender, G.D., 1979a. Maintenance of the corpus luteum of early pregnancy in the ewe. I. Luteotropic properties of embryonic homogenates. Biol. Reprod., 21: 281-288.

Ellinwood, W.E., Nett, T.M. and Niswender, G.D., 1979b. Maintenance of the corpus luteum of early pregnancy in the ewe. II. Prostaglandin secretion by the endometrium in vitro and in vivo. Biol. Reprod., 21: 845-856.

Evans, R.W. and Leavitt, W.W., 1980. Progesterone inhibition of uterine nuclear estrogen receptor: Dependence on RNA and protein synthesis. Proc. Natl. Acad. Sci., 77: 5856-5860.

Fairclough, R.J., Moore, L.G., McGowan, L.T., Peterson, A.J., Smith, J.F., Tervit, H.R. and Watkins, W.B., 1980. Temporal relationship between plasma concentrations of 13,14-dihydro-15-keto-prostaglandin F and neurophysin I/II around luteolysis in sheep. Prostaglandins, 20: 199-208.

Flint, A.P.F. and Sheldrick, E.L., 1982. Ovarian oxytocin is stimulated by prostaglandins. Nature, 297: 587-588.

Flint, A.P.F. and Sheldrick, E.L., 1983. Evidence for a systemic role for ovarian oxytocin in luteal regression in sheep. J. Reprod. Fert., 67: 215-225.

Ginther, O.J., 1976. Comparative anatomy of utero-ovarian vasculature. Veterinary Scope, (Kalamazoo) 20: 1-17.

Goding, J.R., Baird, D.T., Cumming, I.A. and McCracken, J.A., 1972. The functional assessment of autotransplanted endocrine glands. Acta Endocr., Suppl. 158: 169-199.

Henderson, K.M. and McNatty, K.P., 1975. A biochemical hypothesis to explain the mechanism of luteal regression. Prostaglandins, 9: 779-797.

Henderson, K.M., Scaramuzzi, R.J. and Baird, D.T., 1977. Simultaneous infusion of PGE_2 antagonizes the luteolytic action of $PGF_{2\alpha}$ in vivo. J. Endocrin., 73: 379-383.

Inskeep, E.K., Smutny, W.J., Butcher, R.L., and Pexton, J.E., 1975. Effects of intrafollicular injections of prostaglandins in non-pregnant and pregnant ewes. J. Anim. Sci., 41: 1098-1104.

Kindahl, H., Granström, E., Edqvist, L.-E., and Eneroth, P., 1976a. Prostaglandin levels in peripheral plasma during the reproductive cycle. Adv. Prosta. Thromb. Res., 2: 667-671.

Kindahl, H., Edqvist, L.-E., Bane, A. and Granström, E., 1976b. Blood levels of progesterone and 15-keto-13,14-dihydro-prostaglandin $F_{2\alpha}$ during the normal oestrous cycle and early pregnancy in heifers. Acta Endocr., 82: 134-149.

Kindahl, H., Lindell, J.O. and Edqvist, L.-E., 1981. Release of prostaglandin $F_{2\alpha}$ during the oestrous cycle. Acta vet. scand., Suppl. 77: 143-158.

Kindahl, H., Knudsen, O., Madej, A. and Edqvist, L.-E., 1982. Progesterone, prostaglandin $F_{2\alpha}$, PMSG and oestrone sulphate during early pregnancy in the mare. J. Reprod. Fert., Suppl. 32: 353-359.

Kittok, R.J. and Britt, J.H., 1977. Corpus luteum function in ewes given estradiol during the estrous cycle or early pregnancy. J. Anim. Sci., 45: 336-341.

Land, R.B., Baird, D.T. and Scaramuzzi, R.J., 1976. Dynamic studies of prostaglandin $F_{2\alpha}$ in the utero-ovarian circulation of the sheep. J. Reprod. Fert., 47: 209-214.

Leavitt, W.W., MacDonald, R.G. and Okulicz, W.C, 1983. Hormonal regulation of estrogen and progesterone receptor systems. In: Biochemical actions of hormones, Vol X (G. Litwack, Editor) Academic Press, New York, pp. 324-356.

Lewis, G.S., Wilson, L., Wilks, J.W., Pexton, J.E., Fogwell, R.L., Ford, S.P., Butcher, R.L., Thayne, W.V. and Inskeep, E.K., 1977. $PGF_{2\alpha}$ and its metabolites in uterine and jugular venous plasma and endometrium of ewes during early pregnancy. J. Anim. Sci., 45: 320-327.

Lewis, G.S., Jenkins, P.E., Fogwell, R.L. and Inskeep, E.K., 1978. Concentrations of prostaglandins E_2 and $F_{2\alpha}$ and their relationship to luteal function in early pregnant ewes. J. Anim. Sci., 47: 1314-1323.

Magness, R.R., Huie, J.M. and Weems, C.W., 1978. Effect of contralateral and ipsilateral intrauterine infusion of prostaglandin E_2 (PGE_2) on luteal function in the nonpregnant ewe. J. Anim. Sci., (Suppl. 1), 47: 376.

Mapletoft, R.J., Lapin, D.R. and Ginther, O.J., 1976. The ovarian artery as the final component of the local luteotropic pathway between a gravid uterine horn and ovary in ewes. Biol. Reprod., 15: 414-421.

Martal, J., Lacroix, M.C., Loudes, C., Saunier, M. and Winterberger-Torres, S., 1979. Trophoblastin, an antiluteolytic protein present in early pregnancy in sheep. J. Reprod. Fert., 56: 63-73.

McCracken, J.A., 1980. Hormone receptor control of $PGF_{2\alpha}$ secretion by the ovine uterus. Adv. Prosta. Thromb. Res., 8: 1329-1344.

McCracken, J.A., Baird, D.T. and Goding, J.R., 1971. Factors affecting the secretion of steroids from the transplanted ovary in the sheep. Rec. Prog. Horm. Res., 27: 537-582.

McCracken, J.A., Carlson, J.C., Glew, M.E., Goding, J.R., Baird, D.T., Green, K. and Samuelsson, B., 1972. Prostaglandin $F_{2\alpha}$ identified as a luteolytic hormone in sheep. Nature New Biol., 238: 129-134.

McCracken, J.A., Barcikowski, B., Carlson, J.C., Green, K. and Samuelsson, B., 1973. The physiological role of prostaglandin $F_{2\alpha}$ in corpus luteum regression. Adv. Biosci., 9: 599-624.

McCracken, J.A., Glew, M.E., Underwood, L.F., Gavagan, J.E. and Roberts, J.S., 1979. Diminshed $PGF_{2\alpha}$ secretion from the ovine uterus in early pregnancy: The role of oxytocin receptor availability. Fourth International Prostaglandin Conference, Washington, DC., Raven Press, New York; p. 78.

McCracken, J.A., Schramm, W., Okulicz, W., Robidoux, W. and Leavitt, W.W., 1982. Control of $PGF_{2\alpha}$ secretion from the ovine uterus: Dynamic studies on uterine receptors for estrogen, progesterone and oxytocin. Fifth International Conference on Prostaglandins, Florence, Italy, May 18-21, Abst. No. 384.

Mitchell, M.D., Kraemer, D.L., Brennecke, S.P. and Webb, R., 1982. Pulsatile release of oxytocin during the estrous cycle, pregnancy and parturition in sheep. Biol. Reprod., 27: 1169-1173.

Moeljono, M.P., Thatcher, W.W., Bazer, F.W., Frank, M., Owens, L. and Wilcox, C.J., 1977. A study of prostaglandin $F_{2\alpha}$ as the luteolysin in swine: II. Characterization and comparison of prostaglandin F, estrogens and progestin concentrations in utero-ovarian vein plasma of nonpregnant gilts. Prostaglandins, 14: 543-555.

Moor, R.M., 1968. The effect of the embryo on corpus luteum function. J. Anim. Sci., (Suppl. 1), 27: 97-118.

Moor, R.M. and Rowson, L.E., 1966. Local uterine mechanisms affecting luteal function in sheep. J. Reprod. Fert., 11: 307-310.

Nancarrow, C.D., Evison, B.M. and Connell, P.J., 1982. The effect of embryos on luteolysis and termination of early pregnancy in sheep with cloprostenol. Biol. Reprod., 26: 263-269.

Nett, T.M., Staigmiller, R.B., Akbar, A.M., Diekman, N.A., Ellinwood, W.E. and Niswender, G.D., 1976. Secretion of prostaglandin $F_{2\alpha}$ in cycling and pregnant ewes. J. Anim. Sci., 42: 876-880.

Okulicz, W.C., Evans, R.W. and Leavitt, W.W., 1981a. Progesterone regulation of estrogen receptor in the rat uterus: a primary inhibitory influence on the nuclear fraction. Steroids, 37: 463-470.

Okulicz, W.C., Evans, R.W. and Leavitt, W.W., 1981b. Progesterone regulation of the occupied form of nuclear estrogen receptor. Science, 213: 1503-1505.

Okulicz, W.C., MacDonald, R.G. and Leavitt, W.W., 1981c. Progesterone-induced estrogen receptor-regulatory factor in hamster uterine nuclei: preliminary characterization in a cell-free system. Endocrinology, 109: 2273-2275.

Okulicz, W.C., McCracken, J.A., Robidoux, W.F., Schramm, W. and Leavitt, W.W., 1982. Rapid recovery of nuclear estrogen receptors in the ovine uterus following progesterone withdrawal. 64th Annual Meeting Endocrine Society, San Francisco, June 16-18. Endocrinology, 110: A 122 Abstr. No. 618.

Peterson, A.J., Tervit, H.R., Fairclough, R.J., Havik, P.G. and Smith, J.F., 1976. Jugular levels of 13,14-dihydro-15-keto-prostaglandin F and progesterone around luteolysis and early pregnancy in the ewe. Prostaglandins, 12: 551-558.

Pexton, J.E., Weems, C.W. and Inskeep, E.K., 1975a. Prostaglandin F in uterine venous plasma, ovarian arterial and venous plasma and in ovarian and luteal tissue of pregnant and nonpregnant ewes. J. Anim. Sci., 41: 154-159.

Pexton, J.E., Weems, C.W. and Inskeep, E.K., 1975b. Prostaglandins F in uterine and ovarian venous plasma from nonpregnant ewes collected by cannulation. Prostaglandins, 9: 501-509.

Piper, P.J., Vane, J.R. and Wyllie, J.H., 1970. Inactivation of prostaglandins by the lungs. Nature, 225: 600-604.

Poyser, N.L. and Maule Walker, F.M., 1979. Maternal recognition of pregnancy. Excerpta Medica, Amsterdam. Ciba Foundation Symposium 64: 261-292.

Pratt, B.R., Butcher, R.L. and Inskeep, E.K., 1977. Antiluteolytic effect of the conceptus and of PGE_2 in ewes. J. Anim. Sci., 46: 784-791.

Roberts, J.S., Barcikowski, B., Wilson, L. Jr., Skarnes, R.C. and McCracken, J.A., 1975. Hormonal and related factors affecting the release of $PGF_{2\alpha}$ from the uterus. J. Steroid Biochem., 6: 1091-1097.

Roberts, J.S., McCracken, J.A., Gavagan, J.E. and Soloff, M.S., 1976. Oxytocin-stimulated release of prostaglandin $F_{2\alpha}$ from ovine endometrium: Correlation with estrous cycle and oxytocin-receptor binding. Endocrinology, 99: 1107-1114.

Schams, D., Lahlou-Kassi, A. and Glatzel, P., 1982. Oxytocin concentrations in two breeds of sheep with high and low fecundity. J. Endocrinology, 92: 9-13.

Schramm, W. and McCracken, J.A., 1982. Steroid modulation of episodic myometrial activity in conscious sheep. 64th Annual Meeting Endocrine Society, San Francisco, June 16-18. Endocrinology, 110: A 234 Abstr. No. 170.

Schramm, W., Bovaird, L., Glew, M.E. and McCracken, J.A., 1982. Corpus luteum regression induced by ultra-low pulses of prostaglandin $F_{2\alpha}$. Vth International Conference on Prostaglandins, Florence, Italy, May 18-21. Abstr. No. 385.

Schramm, W., Bovaird, L., Glew, M.E., Schramm, G. and McCracken, J.A., 1983. Corpus luteum regression induced by ultra-low pulses of prostaglandin $F_{2\alpha}$. Prostaglandins, 26: 347-364.

Sheldrick, E.L. and Flint, A.P., 1981. Circulating concentrations of oxytocin during the estrous cycle and early pregnancy in sheep. Prostaglandins, 22: 631-636.

Shille, V.M., Karlbom, I., Einarsson, S., Larsson, K., Kindahl, H. and Edqvist, L. E., 1979. Concentrations of progesterone and 15-keto-13,14-dihydro-prostaglandin $F_{2\alpha}$ in peripheral plasma during the estrous cycle and early pregnancy in gilts. Zbl. Veterinaermed. A, 26: 169-181.

Stabenfeldt, G.H., Kindahl, H., Hughes, J.P., Neeley, D.P., Liu, I and Pascoe, D., 1981. Control of luteolysis in the mare. Acta vet. scand., Suppl. 77: 159-170.

Thorburn, G.D., Cox, R.I., Currie, W.B., Restall, B.J. and Schneider, W., 1973. Prostaglandin F and progesterone concentrations in the utero-ovarian venous plasma of the ewe during the oestrous cycle and early pregnancy. J. Reprod. Fert., Suppl. 18: 151-158.

Wathes, D.C., and Swann, R.W., 1982. Is oxytocin an ovarian hormone. Nature, 297: 225-227.

Webb, R., Mitchell, M.D., Falconer, J. and Robinson, J.S., 1981. Temporal relationships between peripheral plasma concentrations of oxytocin, progesterone and 13,14-dihydro-15-keto-prostaglandin $F_{2\alpha}$ during the oestrous cycle and early pregnancy in the ewe. Prostaglandins, 22: 443-453.

Wilson, L., Cenedella, R.J. and Butcher, R.L., 1972. Prostaglandin $F_{2\alpha}$ in the uterus of ewes during early pregnancy. Prostaglandins, 1: 479-482.

Animal Reproduction Science, 7 (1984) 57—74
Elsevier Science Publishers B.V., Amsterdam — Printed in The Netherlands

CELLULAR AND MOLECULAR MECHANISMS INVOLVED IN LUTEOLYSIS AND MATERNAL RECOGNITION OF PREGNANCY IN THE EWE[1]

W.J. Silvia[2], T.A. Fitz[2], M.H. Mayan and G.D. Niswender

Department of Physiology and Biophysics, Colorado State University, Fort Collins, Colorado 80523, (USA).

ABSTRACT

Silvia, W.J., Fitz, T.A., Mayan, M.H. and Niswender, G.D., 1984. Cellular and molecular mechanisms involved in luteolysis and maternal recognition of pregnancy in the ewe. Anim. Reprod. Sci., 7: 57-74.

The corpus luteum is composed of two distinct steroidogenic cell types, both of which contribute significantly to the total progesterone secreted during the luteal phase of the oestrous cycle. The small luteal cells secrete very little progesterone unless stimulated by luteinizing hormone (LH) while large luteal cells appear to spontaneously secrete progesterone at a high rate.

Large luteal cells contain receptors for prostaglandin (PG)$F_{2\alpha}$. Steroid secretion and viability of these cells in vitro are suppressed by $PGF_{2\alpha}$. Prostaglandin $F_{2\alpha}$ may also reduce the ability of LH to stimulate steroidogenesis in the small cells, but this effect appears to be exerted through a signal originating from the large cells since the small cells are essentially devoid of receptors for $PGF_{2\alpha}$.

The corpus luteum of pregnancy was demonstrated to be resistent to the luteolytic effect of $PGF_{2\alpha}$ since higher doses of $PGF_{2\alpha}$ were required to cause luteolysis in pregnant than in non-pregnant ewes. In addition, more $PGF_{2\alpha}$ was required to cause luteolysis in ewes with two embryos when compared to ewes with one. Prostaglandin E_2, of embryonic and/or endometrial origin, is secreted from the uterus during early pregnancy and may be responsible for luteal resistence to the luteolytic effect of $PGF_{2\alpha}$. Prostaglandin E_2 has been demonstrated to protect luteal tissue from the actions of $PGF_{2\alpha}$ both in vivo and in vitro.

---

1 Research was supported by grants from NIH (HD-11590) and the Colorado State Experiment station.

2 Supported by NIH Training Grant HD-07031.

0378-4320/84/$03.00

INTRODUCTION

The cellular and molecular mechanisms by which luteotropins regulate synthesis and secretion of progesterone now appear to be far more complex than we had envisioned as recently as 3 years ago (Niswender et al., 1980). At that time, we proposed a model for how luteinizing hormone (LH) interacts with a "typical" ovine luteal cell. Yet, we now know that there is no "typical" luteal cell. The corpus luteum is composed of at least two distinct types of steroidogenic cells, distinguishable by differences in size and in function (Fitz et al., 1982). The larger cells (22-40 μm in diameter) secrete progesterone at a relatively high rate in the absence of LH. They have few receptors for LH and do not respond to treatment with LH in vitro. The smaller cells (12-22 μm in diameter) have receptors for LH and secrete very little progesterone in vitro unless exposed to at least 100 pg LH/ml. During the luteal phase of the oestrous cycle the concentration of LH in serum varies from 200 pg to 5 ng/ml (Niswender et al., 1969; Baird et al., 1976; Yuthasastrkosol, 1977); therefore, progesterone secreted at this time is derived from both cell types. The relative contribution of each cell type to the total amount of progesterone secreted by the corpus luteum is thus determined not only by the concentration of LH in serum but the number of cells of each type found in the gland at that time.

Fitz and Sawyer (1982) have quantified the numbers of large and small luteal cells throughout the oestrous cycle and pregnancy. There are very few large cells present early in the oestrous cycle or pregnancy but the number increases to a maximum on day 12 at which time they comprise about 45% of the steroidogenic cells. This observation was particularly interesting since the large luteal cells contain essentially all the receptors for $PGF_{2\alpha}$ and PGE_2 (Fitz et al., 1982). The lack of responsiveness of ewes to the luteolytic action of $PGF_{2\alpha}$ early in the oestrous cycle (Hearnshaw et al., 1973) may be due to the lack of large luteal cells, and therefore receptors for $PGF_{2\alpha}$, at this time. During the critical period for luteolysis and maternal recognition of pregnancy (days 12-13 post-oestrus) there are more large cells than at any other time. Therefore, experiments presented here were designed to: 1) characterize the effects of $PGF_{2\alpha}$ and PGE_2 on large luteal cells in vitro; and 2) determine if there is a difference between pregnant and non-pregnant ewes in responsiveness to the luteolytic action of $PGF_{2\alpha}$ in vivo. The results of these experiments will be discussed in light of other recent observations and a working model for the regulation of luteal function during the critical period for luteolysis and maternal recognition of pregnancy will be presented.

MATERIALS AND METHODS

Effects of $PGF_{2\alpha}$ and PGE_2 on large luteal cells *in vitro*

Preparations of large luteal cells were obtained as described previously (Fitz et al., 1982). Briefly, corpora lutea were collected from superovulated ewes on days 10-12 after injection of human chorionic gonadotropin. Tissue was dissociated into a suspension of single cells and separated into large and small steroidogenic cell populations by centrifugal elutriation. Cells were allowed to attach to sterile plastic dishes (~75,000 cells/dish) for at least 12 h before treatments were initiated.

To study the effects of $PGF_{2\alpha}$ and PGE_2, large luteal cells were incubated with the following treatments: 0, $PGF_{2\alpha}$ (250 ng/ml), PGE_2 (250 ng/ml), or $PGF_{2\alpha}$ plus PGE_2 for 6 h. In five replicates of this experiment, incubation medium was changed at hourly intervals over the 6-h incubation period. In four replicates the medium was not changed over this period. The concentration of progesterone in the media was measured by radioimmunoassay (Niswender, 1973) and cell viability was assessed by trypan blue exclusion.

A second experiment was designed to more completely characterize the effects of $PGF_{2\alpha}$ and PGE_2 on large luteal cells. Cells were incubated with 0, 0.1, 1, 10, 100 or 1000 ng of $PGF_{2\alpha}$ or PGE_2/ml of incubation medium for 6 h. In these experiments, incubation medium was not changed over the 6-h incubation period. The concentration of progesterone was measured in the incubation medium by radioimmunoassay.

Responsiveness of pregnant and non-pregnant ewes to $PGF_{2\alpha}$ *in vivo*

Bred and non-pregnant ewes were assigned randomly to receive a single i.m. injection of $PGF_{2\alpha}$: 0, 2, 4, 6, 8 or 10 mg/58 kg body weight on day 13 post-oestrus. Jugular blood samples were collected at 3 h intervals for 24 h beginning at the time of $PGF_{2\alpha}$ injection. Ewes were laparotomized 24 h postinjection, corpora lutea were removed and embryos collected to confirm pregnancy. The concentration of progesterone was quantified in serum and luteal tissue by radioimmunoassy.

RESULTS

Effects of $PGF_{2\alpha}$ and PGE_2 on large luteal cells in vitro

Prostaglandin $F_{2\alpha}$ inhibited secretion of progesterone by large luteal cells ($P < 0.02$) when the incubation medium was not changed over the 6-h incubation period (Table I). There was no effect of $PGF_{2\alpha}$ when the incubation medium was changed at hourly intervals. On the other hand, PGE_2 stimulated secretion of progesterone by large luteal cells when the medium was changed ($P < 0.02$) and tended to do so when the medium was not, although this effect was not significant ($P = 0.18$). In two of the replicates, in which the medium was not changed, viability of the cells was determined at the conclusion of the incubation period. Large luteal cells treated with $PGF_{2\alpha}$ for 6 h were 42% less viable than control cells (Table I). Treatment with PGE_2 alone had no effect on cell viability but inclusion of PGE_2 along with $PGF_{2\alpha}$ prevented the cell death associated with $PGF_{2\alpha}$ treatment.

Table I. Effect of prostaglandins E_2 and $F_{2\alpha}$ on secretion of progesterone by large luteal cells in vitro

	Accumulation of progesterone (% of control)[a]				Viability (%)	
	Medium changed hourly		No medium changes			
Control	100	(5)[b]	100	(4)	82	(2)
$PGF_{2\alpha}$ (250 ng/ml	129 ± 12[c]	(5)	65 ± 9[e]	(3)	40	(2)
PGE_2 (250 ng/ml)	192 ± 47[d]	(4)	152 ± 35	(4)	83	(2)
PGE_2 + $PGF_{2\alpha}$	148 ± 16	(4)	106 ± 14[e]	(3)	79	(2)

[a] Data are presented as % of control value for ease of comparison. Analysis of variance was performed on raw data.

[b] number of replicates.

[c] mean ± SEM (n).

[d] significantly different from control ($P < 0.05$; Dunnett's mean comparison procedure).

[e] Main effect of $PGF_{2\alpha}$ was significant ($P < 0.02$).

In the subsequent experiment, the dose-dependent effects of both prostaglandins were examined more closely. Secretion of progesterone by large luteal cells was progressively increased by doses of PGE2 from 0.1 to 10 ng/ml (Fig. 1). However, higher doses of PGE2 resulted in a progressive decrease in the secretion of progesterone. The negative effect of $PGF_{2\alpha}$ on progesterone secretion was shown to be dose-dependent (Fig. 2). There was no direct inhibitory effect of $PGF_{2\alpha}$ on either secretion of progesterone by, or viability of, small luteal cells incubated under similar conditions.

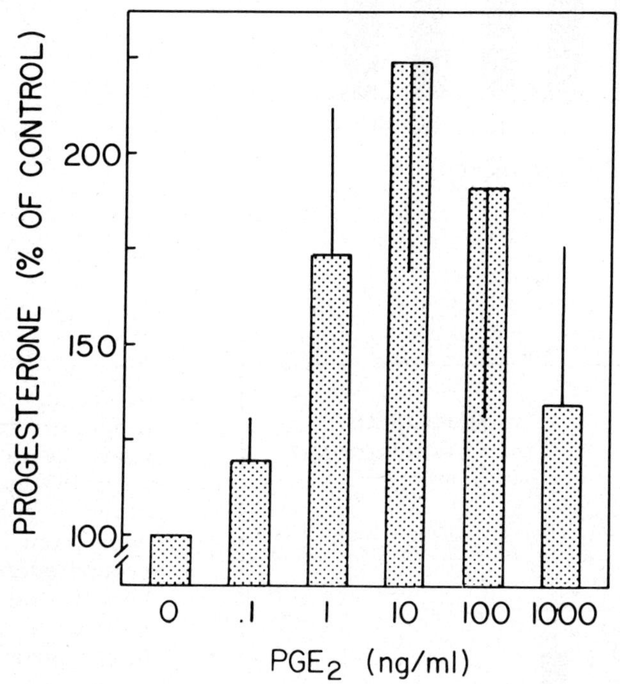

Fig. 1 The dose-dependent effect of PGE_2 on secretion of progesterone by large luteal cells in vitro.

Responsiveness of pregnant and non-pregnant ewes to $PGF_{2\alpha}$ in vivo

The minimum dose of $PGF_{2\alpha}$ required to cause a significant reduction in serum levels of progesterone in non-pregnant ewes was 4 mg/58 kg body weight (Fig. 3) while in pregnant ewes the lowest effective dose was 6 mg/58 kg body weight. If pregnant ewes were separated according to the number of corpora

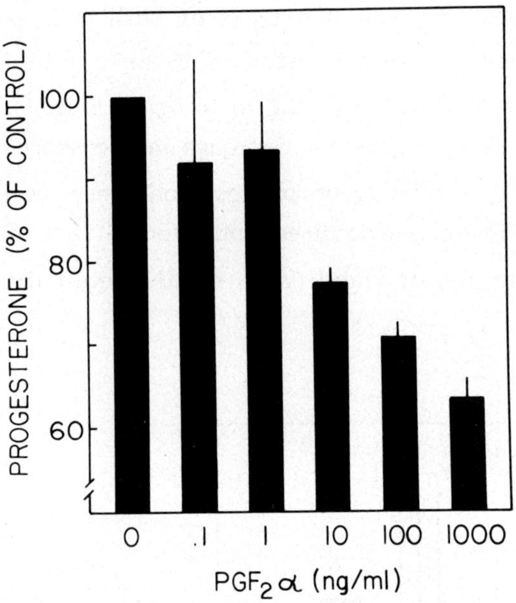

Fig. 2. The dose-dependent effect of $PGF_{2\alpha}$ on secretion of progesterone by large luteal cells in vitro.

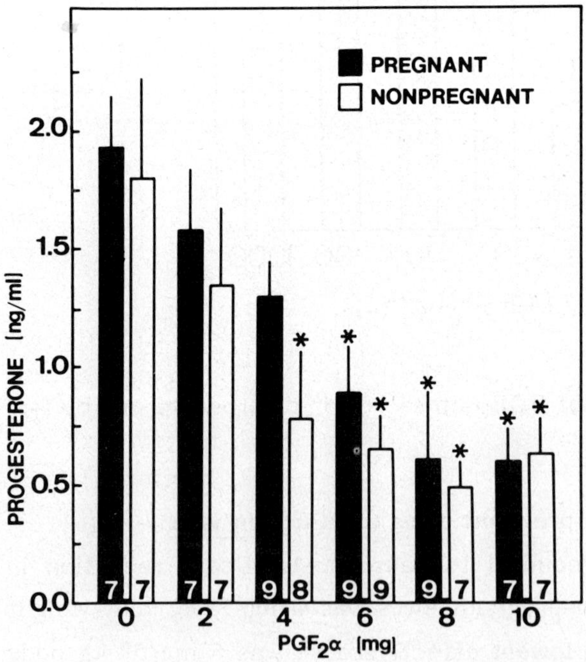

Fig. 3. The concentration of progesterone in serum of pregnant and non-pregnant ewes 24 h after injection with $PGF_{2\alpha}$. Asterisk indicates a significant difference from the corresponding group receiving 0 mg. Numbers at the base of each bar indicate the number of ewes in that group.

lutea present, 6 mg/58 kg body weight was required for significant reduction of serum levels of progesterone in those with one corpus luteum while those with 2 corpora lutea required 10 mg (Fig. 4). There was no effect of number of corpora lutea in non-pregnant ewes; therefore, the enhanced resistence in pregnant ewes with 2 corpora lutea was probably due to the presence of two embryos. The number of embryos recovered was equal to the number of corpora lutea present in 45 of 48 pregnant ewes.

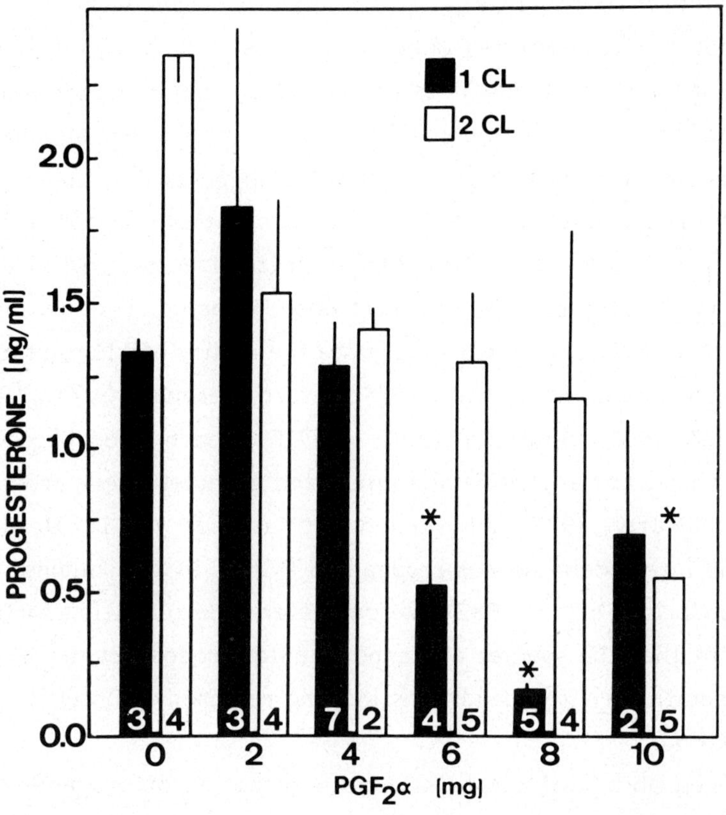

Fig. 4. The concentration of progesterone in serum of pregnant ewes with one or two corpora lutea, 24 h after injecction with $PGF_{2\alpha}$. Asterisk indicates a significant difference from the corresponding group receiving 0 mg. Numbers at the base of each bar indicate the number of ewes in that group.

DISCUSSION

In the ewe, luteolysis is believed to be initiated by $PGF_{2\alpha}$ (for a recent review see Inskeep and Murdoch, 1980) but the mechanism by which it does so remains in question. Treatment of large luteal cells with $PGF_{2\alpha}$ suppressed basal secretion of progesterone in a dose-dependent fashion (Fig. 2). Associated with the reduced secretion of progesterone was an increase in mortality of the cells (Table I). This cytotoxic action of $PGF_{2\alpha}$ may account for the ability of $PGF_{2\alpha}$ to suppress basal secretion of progesterone by lapine and porcine tissue in vitro (O'Grady et al., 1972; Henderson and McNatty, 1977).

The inhibitory effect of $PGF_{2\alpha}$ on secretion of progesterone by, and viability of large luteal cells is observed only after at least 6 h of incubation without changing the incubation media. Prostaglandin $F_{2\alpha}$ had no effect when there was periodic replacement of the incubation medium with fresh medium containing the same concentration of $PGF_{2\alpha}$. This finding suggests that under the influence of $PGF_{2\alpha}$, the large luteal cells secrete some cytotoxic factor which must accumulate to a critical level before it can exert its effects.

Other in vitro effects of $PGF_{2\alpha}$ have been reported. Pretreatment of ovine luteal tissue with $PGF_{2\alpha}$ in vitro suppresses the ability of LH to stimulate secretion of progesterone (Evrard et al., 1978; Evrard-Herouard et al., 1981; Fletcher and Niswender, 1982). Similar effects of $PGF_{2\alpha}$ have been observed with rat luteal cells (Thomas et al., 1978) and luteinized granulosa cells from swine (Henderson and McNatty, 1977) and women (McNatty et al., 1975). However, dispersed bovine luteal cells do not respond to $PGF_{2\alpha}$ in this manner (Weston and Hixon, 1980; Carlson et al., 1982). We can offer no explanation for these discrepancies except that the species and experimental procedures used by each group differ and specific in vitro conditions may be required to observe the inhibitory effects of $PGF_{2\alpha}$.

The reduced ability of LH to stimulate secretion of progesterone by luteal tissue after treatment with $PGF_{2\alpha}$ appears to involve an inhibition in the ability of LH to stimulate adenylate cyclase. Treatment of ovine luteal slices or dissociated rat luteal cells with $PGF_{2\alpha}$ prevented the stimulation of adenylate cyclase by LH (Thomas et al., 1978; Khan and Rosberg, 1979; Fletcher and Niswender, 1982). However, the $PGF_{2\alpha}$ treated cells retained their ability to respond to PGE_2 with enhanced adenylate cyclase activity (Khan et al., 1979; Fletcher and

Niswender, 1982), and secretion of progesterone (McNatty et al., 1975). Therefore, the inhibitory effect of $PGF_{2\alpha}$ was at least partially specific for LH.

As previously noted, receptors for LH and LH-stimulable adenylate cyclase activity and progesterone secretion (Hoyer et al., submitted) are essentially restricted to small luteal cells while the majority of receptors for $PGF_{2\alpha}$ are found on large luteal cells. Therefore, in order for $PGF_{2\alpha}$ to act on small luteal cells, it seems likely that it must stimulate the large cells to produce an inhibitory factor. Although the nature of this factor is not known, we have some clues as to its mechanism of action. One change reported to be associated with both spontaneous and $PGF_{2\alpha}$-induced luteal regression in bovine and rat luteal tissue is a reduction in the fluidity of the plasma membrane (Buhr et al., 1979; Carlson et al., 1982; Goodsaid-Zolduondo et al., 1982). Reduced membrane fluidity may hinder the interaction of the LH-receptor complex with adenylate cyclase. Another possible mechanism by which LH-stimulable adenylate cyclase may be inhibited is through induction of Ca^{++} influx. Dorflinger (1978) showed that treatment of dissociated rat luteal cells with a Ca^{++} ionophore reduced the ability of LH to stimulate secretion of progesterone and accumulation of intracellular cAMP.

At the cellular level, luteolysis is very complex and apparently involves two effects. The first is a direct inhibitory effect of $PGF_{2\alpha}$ on secretion of progesterone and cell viability in large luteal cells. The second effect of $PGF_{2\alpha}$ appears to be an inhibition of the ability of LH to stimuate adenylate cyclase in the small luteal cell. In the sheep, LH is required for normal luteal function as demonstrated in studies involving hypophysectomy (Denamur, 1968; Kaltenbach et al., 1968) and passive immunization to LH (Dermody and Foote, 1969; Fuller and Hansel, 1970; Reimers and Niswender, 1975). Inhibition of the steroidogenic properties of LH may be accompanied by an inhibition of its luteotropic properties as well. The effects of $PGF_{2\alpha}$ on both large and small luteal cells appear to be mediated by inhibitory factors secreted by the large luteal cells.

Finally, since the stage of the oestrous cycle influences the relative numbers of large and small luteal cells present in the corpus luteum and, therefore, its responsiveness to $PGF_{2\alpha}$, and since precise in vitro conditions appear to be required for $PGF_{2\alpha}$ to exert its effects, it is not suprising that the scientific record appears inconsistent with respect to the actions of $PGF_{2\alpha}$ in vitro.

Maternal recognition of pregnancy

The primary mechanism by which the embryo prevents luteolysis in the ewe is not known. In particular, the question of secretion of $PGF_{2\alpha}$ in pregnant vs. non-pregnant ewes remains controversial. If secretion of $PGF_{2\alpha}$ is monitored by measurement of 15-ketodihydro-$PGF_{2\alpha}$ in jugular venous plasma, then it appears that secretion is suppressed in pregnant ewes (Peterson et al., 1976), cows (Kindahl et al., 1976) and goats (Homeida and Cooke, 1982). However, when concentrations of $PGF_{2\alpha}$ in uterine venous plasma are determined directly, most researchers have found no differences in secretion between pregnant and non-pregnant ewes (Pexton et al., 1975; Nett et al., 1976b; Lewis et al., 1977; Ellinwood et al., 1979b). It is not obvious why the two approaches to quantification of $PGF_{2\alpha}$ secretion appear to yield different results. However, extrapolating information from systemic measurements of 15-ketodihydro-$PGF_{2\alpha}$ may have two problems. First, systemic measurements of 15-ketodihydro-$PGF_{2\alpha}$ reflect the status of $PGF_{2\alpha}$ secretion throughout the body, not just the uterus. While jugular venous concentrations of 15-ketodihydro-$PGF_{2\alpha}$ and uterine venous concentrations of $PGF_{2\alpha}$ have been shown to be somewhat correlated ($r = 0.57$, $P < 0.01$; Louis et al., 1977) in one laboratory, few other groups have attempted similar validation procedures. Second, the rate or products of metabolism of $PGF_{2\alpha}$ in pregnant and non-pregnant ewes may differ. Both ovine placental (Keirse et al., 1976) and endometrial tissue from cows during early pregnancy (Lewis et al., 1982) possess the ability to metabolize $PGF_{2\alpha}$. The levels of 15-hydroxy-prostaglandin dehydrogenase in rat placental, lung and kidney tissues fluctuate during pregnancy (Blackwell and Flower, 1976; Lerner and Carminati, 1976). It is possible that a similar phenomenon may occur in the ewe. It is equally possible that the primary metabolic product in pregnant ewes is not 15-ketodihydro-$PGF_{2\alpha}$. Until these problems are resolved, we feel that direct measurements of $PGF_{2\alpha}$ are better estimates of uterine secretory activity.

Assuming that the secretion of $PGF_{2\alpha}$ does not differ between pregnant and non-pregnant ewes, how does the embryo prevent luteolysis? The dose of $PGF_{2\alpha}$ required to inhibit progesterone secretion in pregnant ewes was greater than that required to do the same in non-pregnant ewes (Fig. 3). Within the pregnant ewes, those with 2 corpora lutea (and embryos) were more resistent to $PGF_{2\alpha}$ than those with one (Fig. 4). These effects were observed on day 13, during the critical period for establishment of pregnancy, and are compatable with previous observations (Inskeep et al., 1975; Mapletoft et al., 1976; Pratt et al., 1977; Nancarrow et al., 1982). Nancarrow et al. (1982), using superovulated pregnant ewes on day 20 post-oestrus, demonstrated that the ability of cloprostenol to cause luteal regression was inversely related to the number of embryos present, which also supports our observations that higher levels of $PGF_{2\alpha}$ were required to induce luteolysis in ewes with two corpora lutea and two embryos.

The mechanism by which the embryo confers this resistence to $PGF_{2\alpha}$ by the corpus luteum is not known. Several changes occur in the uterus in response to the embryo at this time. The rate of uterine protein synthesis (Findlay et al., 1981), uterine blood flow (Greiss and Anderson, 1970) and secretion of PGE_2 (Ellinwood et al., 1979b; Marcus, 1981; Silvia et al., 1981; LaCroix and Kann, 1982) all increase during this period. The increased secretion of PGE_2 may have the most significance with respect to maternal recognition of pregnancy. Treatment of ewes with PGE_2 delays luteal regression in non-pregnant ewes (Pratt et al., 1977; Magness et al., 1981). In addition, PGE_2 blocks the luteolytic action of $PGF_{2\alpha}$ when the two compounds are administered simultaneously (Henderson et al., 1977; Mapletoft et al., 1977; Reynolds et al., 1981).

In the experiments presented here, PGE_2 stimulated secretion of progesterone by large luteal cells in a dose-dependent manner (Fig. 1). Secretion of progesterone increased to reach a maximum at 10 ng PGE_2/ml but higher doses resulted in a progressive decrease. A similar dose-response pattern has been described for the ability of PGE_2 to prevent luteolysis induced with $PGF_{2\alpha}$ in vivo (Ginther, 1981).

The dose of PGE_2 used in our study to examine the interaction of the two prostaglandins (250 ng/ml) falls in the range over which the stimulation of progesterone secretion by large luteal cells in response to PGE_2 decreases. This may have contributed to the variability in secretion of progesterone, which was observed in that study. In uterine venous samples, the levels of PGE_2 never exceeded those for $PGF_{2\alpha}$, even during early pregnancy (Ellinwood et al., 1979b; Silvia et al., 1981). Therefore, in subsequent studies of the effects of these hormones in vitro, it may be more appropriate to mimic the endogenous ratio of $PGF_{2\alpha}$ to PGE_2.

It is clear that uterine secretion of PGE_2 increases during the critical period for maternal recognition of pregnancy. We do not know if the PGE_2 secreted at this time is of uterine or embryonic origin. When ovine embryos were collected on day 14 of pregnancy and incubated in vitro they secreted significant quantities of both PGE_2 and $PGF_{2\alpha}$ (Table II). Similar results were obtained by Hyland et al. (1982) and LaCroix and Kann (1982). Bovine blastocysts secrete both $PGF_{2\alpha}$ and PGE_2 when incubated in vitro (Shemesh et al., 1979; Lewis et al., 1982) and the bovine conceptus has a limited ability to metabolize $PGF_{2\alpha}$ (Lewis et al., 1982).

Table II. Embryonic secretion and content of prostaglandins E_2 and $F_{2\alpha}$ in vitro

	PGE_2	$PGF_{2\alpha}$
Secreted (ng/embryo/2 h)	196.7 ± 112.9*	531.6 ± 170.4
Content (ng/embryo)	6.7 ± 1.3	110.4 ± 35.5

* mean ± SEM of 3 incubations

Synthesis and secretion of PGE_2 by the endometrium is enhanced as a result of pregnancy. Endometrial slices and dissociated cell preparations from pregnant ewes secrete more PGE_2 in vitro than similar preparations from non-pregnant ewes (Ellinwood et al., 1979b; Marcus, 1981; LaCroix and Kann, 1982). Homogenates of ovine embryos prolong luteal function if they are infused into the uterine lumen (Rowson and Moor, 1967; Martal et al., 1979; Ellinwood et al., 1979a). These homogenates are not effective if infused into the uterine vein and, unlike PGE_2, they have no direct stimulatory effect on secretion of progesterone by ovine luteal cells in vitro (Ellinwood et al., 1979a). The ability of embryonic homogenates to maintain luteal function is eliminated by treatment with heat or proteolytic enzymes (Rowson and Moor, 1967; Martal et al., 1979). Thus, a case can be made, that the factor secreted by the embryo, which ultimately results in maintenance of the corpus luteum, is a protein, which must act through the uterus to exert its effect. A specific protein which is secreted by the ovine trophoblast during early pregnancy has been isolated and localized immunocytochemically in endometrial tissue of the pregnant ewe (Godkin et al., 1982). These researchers have suggested that the proteinbinds to the endometrium and somehow alters uterine secretory activity. It is possible, that such a protein may be inducing endometrial synthesis and secretion of PGE_2.

The results of the present study and others indicate that the corpus luteum is resistent to the luteolytic action of $PGF_{2\alpha}$ during the period of early pregnancy. Uterine secretion of PGE_2 increases at this time as well. Prostaglandin E_2 can counteract the luteolytic actions of $PGF_{2\alpha}$ both in vivo and in vitro. The mechanism, by which PGE_2 protects the corpus luteum from the luteolytic action of $PGF_{2\alpha}$, is not known. It is possible, that it prevents $PGF_{2\alpha}$ from inducing the release of inhibitory factors from the large luteal cells. It may serve as an alternate luteotropin substituting for LH as a stimulator of adenylate cyclase in small luteal cells. It still remains to be demonstrated, that endogenous secretion of PGE_2 is required for maternal recognition of pregnancy to occur.

Throughout this discussion we have concentrated on direct effects of $PGF_{2\alpha}$ and PGE_2 on the steroidogenic cells of the corpus luteum. However, it is important to remember that these hormones influence the life span and function of the corpus luteum in vivo. Thus, these compounds may also exert effects on the corpus luteum as a result of their vaso-active properties. In the ewe, blood flow to the corpus luteum and secretion of progesterone decrease simultaneously during spontaneous luteal regression (Bruce and Moor, 1976; Niswender et al., 1976; Ford et al., 1979). Induction of luteolysis with $PGF_{2\alpha}$ is also accompanied by a rapid reduction in blood flow to the luteal ovary (Nett et al., 1976a). Thus, there is good evidence, that luteal blood flow is reduced during luteolysis, which is consistent with the vasoconstrictive properties of $PGF_{2\alpha}$ (Bergström et al., 1968). On the other hand, PGE_2 is a vasodilator (Bergström et al., 1968) and may protect the corpus luteum during early pregnancy by maintaining luteal blood flow in the face of $PGF_{2\alpha}$. It seems likely that effects of prostaglandins on blood flow, as well as direct effects on luteal cells are all important in the regulation of the lifespan and function of the corpus luteum.

A hypothetical model, which summarizes the data regarding the effects of $PGF_{2\alpha}$ and PGE_2 at the cellular level, is presented in Fig. 5. In the non-pregnant ewe $PGF_{2\alpha}$ is secreted by the endometrium and induces secretion of a factor(s) from large luteal cells, which has at least two effects: 1) a direct cytotoxic

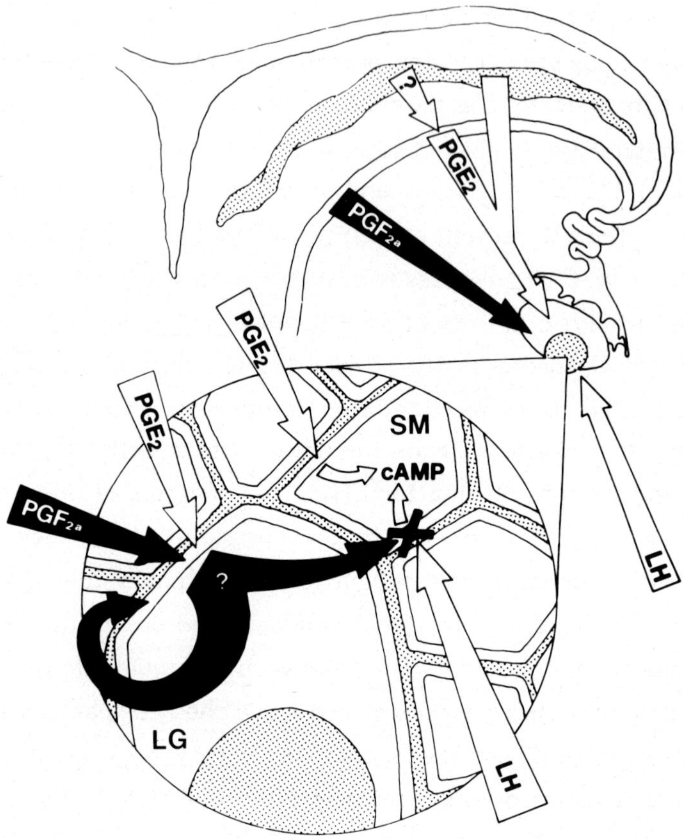

Fig. 5. A hypothetical model depicting the interaction of the embryo, uterus and corpus luteum during luteolysis and early pregnancy. Large luteal cells (LG); Small luteal cells (SM).

effect on the large luteal cells; and 2) a disruption of the ability of LH to stimulate adenylate cyclase in the small luteal cells. Luteolysis may also involve a suppression of blood flow to the corpus luteum. In early pregnancy uterine seretion of $PGF_{2\alpha}$ is not suppressed, however, secretion of PGE_2 is increased. Prostaglandin E_2 may be synthesized by the endometrium under the influence of the embryo or by the embryo itself. It may prevent luteal regression by inhibiting release of the luteolytic factor(s) from large luteal cells, by stimulating adenylate cyclase in the small luteal cells, and/or stimulating blood flow to the ovary. Additional research is required before the appropriateness of the model can be confirmed or refuted.

REFERENCES

Baird, D.T., Swanston, I. and Scaramuzzi, R.J., 1976. Pulsatile release of LH and secretion of ovarian steroids in sheep during the luteal phase of the estrous cycle. Endocrinology, 98: 1490-1496.

Bergström, S., Carlson, L.A. and Weeks, J.R., 1968. The prostaglandins: A family of biologically active lipids. Pharmacol. Rev., 20: 1-48.

Blackwell, G.J. and Flower, R.J., 1976. Effects of steroid hormones on tissue levels of 15-hydroxydehydrogenase in the rat. Brit. J. Pharmacol., 56: 343P.

Bruce, N.W. and Moor, R.M., 1976. Capillary blood flow to ovarian follicles, stroma and corpora lutea of anesthetized sheep. J. Reprod. Fertil., 46: 299-304.

Buhr, M.M., Carlson, J.C. and Thompson, J.E., 1979. A new perspective on the mechanism of corpus luteum regression. Endocrinology, 105: 1330-1335.

Carlson, J.C., Buhr, M.M., Wentworth, R. and Hansel, W., 1982. Evidence of membrane changes during regression in the bovine corpus luteum. Endocrinology, 110: 1472-1476.

Denamur, R., 1968. Formation and maintenance of corpora lutea in domestic animals. J. Anim. Sci., 27 (Suppl. 1): 163-180.

Dermody, W.C. and Foote, W.C., 1969. Effect of rabbit anti-ovine LH on corpus luteum function in the ovine. J. Anim. Sci., 29: 187.

Dorflinger, L.J., 1978. Evidence for the role of Ca^{++} in the acute effect of $PGF_{2\alpha}$ on LH-stimulated adenylate cyclase activity. Biol. Reprod., 18 (Suppl. 1): 62A.

Ellinwood, W.E., Nett, T.M. and Niswender, G.D., 1979a. Maintenance of the corpus luteum of early pregnancy in the ewe. I. Luteotropic properties of embryonic homogenates. Biol. Reprod., 21: 281-288.

Ellinwood, W.E., Nett, T.M. and Niswender, G.D., 1979b. Maintenance of the corpus luteum of early pregnancy in the ewe. II. Prostaglandin secretion by the endometrium in vitro and in vivo. Biol. Reprod., 21: 845-856.

Evrard, M., Leboulleux, P. and Hermier, C., 1978. Role of prostaglandin $F_{2\alpha}$ in modulation of LH stimulated steroidogenesis in vitro in different types of rat and ewe corpora lutea. Prostaglandins, 16: 491-502.

Evrard-Herouard, M., de la Llosa-Hermier, M.P., Martinet, J., Mauleon, P., de la Llosa, P. and Hermier, C., 1981. LH receptors in ovine corpora lutea in relation to various physiological states and effects of $PGF_{2\alpha}$ on LH induced steroidogenesis in vitro. J. Reprod. Fertil., 61: 225-233.

Findlay, J.K., Ackland, N., Burton, R.D., Davis, A.J., Maule Walker, F.M., Walters, D.E. and Heap, R.B., 1981. Protein, prostaglandin and steroid synthesis in caruncular and intercaruncular endometrium of sheep before implantation. J. Reprod. Fertil., 62: 361-377.

Fitz, T.A. and Sawyer, H.R., 1982. Changes in the quantity and size of steroidogenic cells in ovine corpora lutea during the estrous cycle and early pregnancy. Biol. Reprod., 26 (Suppl. 1): 54A.

Fitz, T.A., Mayan, M.H., Sawyer, H.R. and Niswender, G.D., 1982. Characterization of two steroidogenic cell types in the ovine corpus luteum. Biol. Reprod., 22: 703-711.

Fletcher, P.W. and Niswender, G.D., 1982. Effect of $PGF_{2\alpha}$ on progesterone secretion and adenylate cyclase activity in ovine luteal tissue. Prostaglandins, 23: 803-818.

Ford, S.P., Christenson R.K. and Chenault, J.R., 1979. Patterns of blood flow to the uterus and ovaries of ewes during the period of luteal regression. J. Anim. Sci., 49: 1510-1516.

Fuller, G.B. and Hansel, W., 1970. Regression of sheep corpora lutea after treatment with anti-bovine luteinizing hormone. J. Anim. Sci., 31: 99-103.

Ginther, O.J., 1981. Local versus systemic uteroovarian relationships in farm animals. Acta vet. scand., Suppl. 77: 103-115.

Godkin, J.A., Bazer, F.W., Moffatt, J., Sessions, F. and Roberts, R.M., 1982. Purification and properties of a major, low molecular weight protein released by the trophoblast of sheep blastocysts at day 13-21. J. Reprod. Fertil., 65: 141-150.

Goodsaid-Zalduondo, F., Rintoul, D.A., Carlson, J.C. and Hansel, W., 1982. Luteolysis-induced changes in phase composition and fluidity of bovine luteal cell membranes. Proc. Nat. Acad. Sci., (USA) 79: 4332-4336.

Griess, F.D. and Anderson, S.G., 1970. Uterine blood flow during early ovine pregnancy. Am. J. Obstet. Gynecol., 106: 30-38.

Hearnshaw, H., Restall, B.J. and Gleeson, A.R., 1973. Observations on the luteolytic effects of prostaglandin $F_{2\alpha}$ during the oestrous cycle and early pregnancy in the ewe. J. Reprod. Fertil., 32: 322-323.

Henderson, K.M., and McNatty, K.P., 1977. A possible interrelationship between gonadotrophin stimulation and prostaglandin $F_{2\alpha}$ inhibition of steroidogenesis by granulosa-luteal cells in vitro. J. Endocrinol., 73: 71-78.

Henderson, K.M., Scaramuzzi, R.J. and Baird, D.T., 1977. Simultaneous infusion of prostaglandin E_2 antagonizes the luteolytic action of prostaglandin $F_{2\alpha}$ in vivo. J. Endocrinol., 72: 379-383.

Homeida, A.M. and Cooke, R.G., 1982. Peripheral plasma concentrations of 13,14-dihydro-15-keto-prostaglandin $F_{2\alpha}$ and progesterone around luteolysis and during early pregnancy in the goat. Prostaglandins, 24: 313-317.

Hoyer, P.B., Fitz, T.A. and Niswender, G.D., 1983. Hormone-independent activation of adenylate cyclase in large steroidogenic ovine luteal cells does not result in increased progesterone secretion. Endocrinology (submitted).

Hyland, J.H., Manns, J. G. and Humphrey, W.D., 1982. Prostaglandin production by ovine embryo and endometrium in vitro. J. Reprod. Fertil., 65: 299-304.

Inskeep, E.K., Smutny, W.J., Butcher, R.L. and Pexton, J.E. 1975. Effects of intrafollicular injections of prostaglandins in non-pregnant and pregnant ewes. J. Anim. Sci., 41: 1098-1104.

Inskeep, E.K. and Murdoch, W.J., 1980. Relationship of ovarian functions to uterine and ovarian secretion of prostaglandins during the estrous cycle and early pregnancy in the ewe and cow. In: R.O. Greep (Editor), International Review of Physiology, Vol. 22. University Park Press, Baltimore, MD, pp. 325-356.

Kaltenbach, C.C., Graber, J.W., Niswender, G.D. and Nalbandov, A.V., 1968. Luteotropic properties of some pituitary hormones in nonpregnant and pregnant hypophysectomized ewes. Endocrinology, 82: 818-824.

Keirse, M.J., Hicks, B.R. and Turnbull, A.C., 1976. Metabolism of prostaglandin $F_{2\alpha}$ in fetal and maternal cotyledons of sheep. J. Reprod. Fertil., 46: 417-420.

Kindahl, H., Edqvist, L.-E., Bane, A. and Granström, E., 1976. Blood levels of progesterone and 15-keto-13,14-dihydro-prostaglandin $F_{2\alpha}$ during the normal oestrous cycle and early pregnancy in heifers. Acta endocrinol., 82: 134-149.

Khan, M.I. and Rosberg, S., 1979. Acute suppression by $PGF_{2\alpha}$ on luteinizing hormone, epinephrine, and fluoride stimulation of adenylate cyclase in rat luteal tissue. J. Cycl. Nucl. Res., 5: 55-63.

Khan, M.I., Rosberg, S., Lahav, M., Lamprecht, S.A., Selstam, G., Herlitz, H. and Ahren, K., 1979. Studies on the mechanism of action of the inhibitory effect of prostaglandin $F_{2\alpha}$ on cyclic AMP accumulation in rat corpora lutea of various ages. Biol. Reprod., 21: 1175-1183.

LaCroix, M.C. and Kann, G., 1982. Comparative studies of prostaglandins $F_{2\alpha}$ and E_2 in late cyclic and early pregnant sheep; In vitro synthesis by endometrium and conceptus. Effects of in vivo indomethacin treatment on establishment of pregnancy. Prostaglandins, 23: 507-526.

Lerner, L.J. and Carminati, P., 1976. Effect of L-10503 (a novel antifertility compound) on the synthesis and metabolism of prostaglandins in vivo and in vitro in the pregnant rat placenta, ovary, kidney and lung and in rat deciduoma. In: Advances in Prostaglandin and Thromboxane Research, Vol. 2, (Eds. B. Samuelsson and R. Paoletti), Raven Press, New York, pp. 645-653.

Lewis, G.S., Wilson, Jr., L., Wilks, J.W., Pexton, J.E., Fogwell, R.P., Ford, S.P., Butcher, R.L., Thayne, W.V. and Inskeep, E.K., 1977. $PGF_{2\alpha}$ and its metabolites in uterine and jugular venous plasma and endometrium of ewes during early pregnancy. J. Anim. Sci., 45: 320-327.

Lewis, G.S., Thatcher, W.W., Bazer, F.W. and Curl, J.S., 1982. Metabolism of arachidonic acid in vitro by bovine blastocysts and endometrium. Biol. Reprod., 26: 431-439.

Louis, T.M., Parry, D.M., Robinson, J.S., Thorburn, G.D. and Challis, J.R.G., 1977. Effects of exogenous progesterone and oestradiol on prostaglandin F and 13,14-dihydro-15-oxo prostaglandin $F_{2\alpha}$ concentrations in uteri and plasma of ovariectomized ewes. J. Endocrinol., 73: 427-439.

Magness, R.R., Huie, J.M. and Weems, C.W., 1981. Effect of chronic ipsilateral or contralateral intrauterine infusion of prostaglandin E_2 (PGE_2) on luteal function of unilaterally ovariectomized ewes. Prostaglandins and Medicine, 6: 389-401.

Mapletoft, R.J., Lapin, D.R. and Ginther, O.J., 1976. The ovarian artery as the final component of the local pathway between a gravid uterine horn and ovary in ewes. Biol. Reprod., 15: 414-421.

Mapletoft, R.J., Miller, K.F. and Ginther, O.J., 1977. Effects of $PGF_{2\alpha}$ and PGE_2 on corpora lutea in ewes. J. Anim. Sci., 45 (Suppl. 1): 185.

Marcus, G.J., 1981. Prostaglandin formation by the sheep embryo and endometrium as an indication of maternal recognition of pregnancy. Biol. Reprod., 25: 56-64.

Martal, J., Lacroix, M.C., Loudes, C., Saunier, M. and Wintenberger-Torres, C., 1979. Trophoblastin, an antiluteolytic protein present in early pregnancy in sheep. J. Reprod. Fertil., 56: 63-73.

McNatty, K.P., Henderson, K.M. and Sawers, R.S., 1975. Effects of prostaglandin $F_{2\alpha}$ and E_2 on the production of progesterone by human granulosa cells in tissue culture. J. Endocrinol., 67: 231-240.

Nancarrow, C.D., Evinson, B.M. and Connell, P.J., 1982. Effect of embryo on luteolysis and termination of early pregnancy in sheep with cloprostenol. Biol. Reprod., 26: 263-269.

Nett, T.M., McClellan, M.C. and Niswender, G.D., 1976a. Effects of prostaglandins on the ovine corpus luteum: Blood flow, secretion of progesterone and morphology. Biol. Reprod., 15: 66-78.

Nett, T.M., Staigmiller, R.B., Akbar, A.M., Diekman, M.A., Ellinwood, W.E. and Niswender, G.D., 1976b. Secretion of prostaglandin $F_{2\alpha}$ in cycling and pregnant ewes. J. Anim. Sci., 42: 876-880.

Niswender, G.D., 1973. Influence of the site of conjugation on the specificity of antibodies to progesterone. Steroids, 22: 413-424.

Niswender, G.D., Reichert, L.E., Midgley, Jr., A.R. and Nalbandov, A.V., 1969. Radioimmunoassay for bovine and ovine luteinizing hormone. Endocrinology, 84: 1166-1173.

Niswender, G.D, Reimers, T.J., Diekman, M.A. and Nett, T.M., 1976. Blood flow: A mediator of ovarian function. Biol. Reprod., 14: 64-81.

Niswender, G.D., Sawyer, H.R., Chen, T.T. and Enders, D.B., 1980. Action of luteinizing hormone at the luteal cell level. In: Advances in Sex Hormone Research, vol. 4 (Eds. J.A. Thomas and R.L. Singhal), Urban and Schwarzenberg, Baltimore, MD, pp. 153-185.

O'Grady, J.P., Kohorn, E.I., Glass, R.H., Caldwell, B.J., Brock, W.A. and Speroff, L., 1972. Inhibition of progesterone synthesis in vitro by $PGF_{2\alpha}$. J. Reprod. Fertil., 30: 153-156.

Peterson, A.J., Tervit, H.R., Fairclough, R.J., Havik, P.G., Smith, J.F., 1976. Jugular levels of 13,14-dihydro-15-keto-prostaglandin F and progesterone around luteolysis and early pregnancy in the ewe. Prostaglandins, 12: 551-558.

Pexton, J.E., Weems, C.M. and Inskeep, E.K., 1975. Prostaglandins F in uterine and ovarian venous plasma from non-pregnant and pregnant ewes collected by cannulation. Prostaglandins, 9: 501-509.

Pratt, B.R., Butcher, R.L. and Inskeep, E.K., 1977. Antiluteolytic effect of the conceptus and of PGE_2 in ewes. J. Anim. Sci., 46: 784-791.

Reimers, T.J. and Niswender, G.D., 1975. Effects of specific neutralization of luteinizing hormone and prolactin on ovarian blood flow and progesterone secretion. In: Immunization with Hormones in Reproduction Research (Ed. E. Nieshlag), North Holland Publ. Co., Amsterdam, pp. 95-105.

Reynolds, L.P., Stigler, J., Hoyer, G.L., Magness, R.R., Huie, J.M., Huecksteadt, T.P., Whysong, G.L., Behrman, H.R. and Weems, C.W., 1981. Effect of PGE_1 or PGE_2 on $PGF_{2\alpha}$-induced luteolysis in nonbred ewes. Prostaglandins, 21: 957-972.

Rowson, L.E.A. and Moor, R.M., 1967. The influence of embryonic tissue homogenates infused into the uterus, on the life-span of the corpus luteum in the sheep. J. Reprod. Fertil., 13: 511-516.

Shemesh, M., Milaguir, F., Ayalon, N. and Hansel, W., 1979. Steroidogenesis and prostaglandin synthesis by cultured bovine blastocysts. J. Reprod. Fertil., 56: 181-185.

Silvia, W.J., Ottobre, J.S. and Inskeep, E.K., 1981. Concentrations of prostaglandins E_2, $F_{2\alpha}$ and 6-keto-prostaglandin $F_{1\alpha}$ in the utero-ovarian venous plasma of early pregnant and cycling ewes. Biol Reprod., 24 (Suppl. 1): 99A.

Thomas, J.P., Dorflinger, L.J. and Behrman, H.R., 1978. Mechanism of the rapid anti-gonadotropic action of prostaglandins in cultured luteal cells. Proc. Nat. Acad. Sci. (USA), 75: 1344-1348.

Weston, P.G. and Hixon, J.E., 1980. Effects of in vivo prostaglandin $F_{2\alpha}$ administration on in vitro progesterone synthesis by bovine corpora lutea. Biol. Reprod., 22: 259-268.

Yuthasastrakosol, P., Palmer, W.M. and Howland, B.E., 1977. Release of LH in anoestrous and cyclic ewes. J. Reprod. Fertil., 50: 319-321.

Animal Reproduction Science, 7 (1984) 75—100
Elsevier Science Publishers B.V., Amsterdam — Printed in The Netherlands

ASPECTS OF REGULATION OF UTERINE SECRETION OF PROSTAGLANDINS DURING THE OESTROUS CYCLE AND EARLY PREGNANCY

J.S. Ottobre, D.L. Vincent, W.J. Silvia and E.K. Inskeep

Division of Animal and Veterinary Sciences, West Virginia University,

Morgantown, West Virginia 26506, (USA).

ABSTRACT

Ottobre, J.S., Vincent, D.L., Silvia, W.J. and Inskeep, E.K., 1984. Aspects of regulation of uterine secretion of prostaglandins during the oestrous cycle and early pregnancy. Anim. Reprod. Sci., 7: 75-100.

Evidence that the pattern of secretion of prostaglandin E_2 (PGE_2) by the uterus: conceptus unit is consistent with that compound playing a role in the local antiluteolytic effect of the embryo is reviewed briefly. Utero-ovarian venous concentrations of PGE_2 increased on day 13 in pregnant compared to non-pregnant ewes, both absolutely and in relation to prostaglandin $F_{2\alpha}$ ($PGF_{2\alpha}$).

The roles of oestrogen and progesterone in regulation of utero-ovarian venous concentrations of $PGF_{2\alpha}$ were examined during two intervals of the oestrous cycle and on day 13 of the cycle or pregnancy. The first interval (Experiment 1, days 10 to 14) was chosen to study the initial rises of $PGF_{2\alpha}$ that precede luteolysis. At laparotomy on day 9, a catheter was inserted into a utero-ovarian vein. Eight ewes received no further treatment. At surgery, 12 ewes were given 10 mg of progesterone (i.m.) along with an intravaginal pessary containing 30 mg flurogestone acetate and were either ovariectomized (n=5) or lutectomized (n=7). Five ewes were ovariectomized only. The second interval (Experiment 2, days 14 to 17) was chosen to study the maximal rises of $PGF_{2\alpha}$ associated with the completion of luteal regression and the preovulatory increase of oestrogen. Ewes were laparotomized and catheterized on day 14 and either received no further treatment (n=5), or were lutectomized (n=6), ovariectomized (n=5) or ovariectomized and implanted with oestradiol-17β (n=5). Five utero- ovarian venous samples were collected, every 30 min for 2 hours beginning at 0600 and 1800 on each day in each experiment, and assayed for $PGF_{2\alpha}$. In a third experiment, 15 bred and 12 nonbred ewes were laparotomized and catheterized on day 12. One third of each type remained intact, while two thirds were ovariectomized and half of those received 10 mg of progesterone (s.c.) and an intravaginal sponge containing 30 mg of progesterone on day 12. Utero-ovarian samples were collected at 15-minute intervals for 3 hours on day 13.

Neither the average day of occurrence of the first peak of $PGF_{2\alpha}$ (control, 12.1; lutectomy plus progesterone, 11.6; ovariectomy plus progesterone, 12.0;

ovariectomy, 12.2) nor the mean number of peaks detected in each ewe during the sampling periods on days 10 to 14 (2.8, 3.7, 3.2 and 3.4, respectively) differed among groups in Experiment 1. The mean number of peaks of $PGF_{2\alpha}$ detected in each ewe during days 14 to 17 did not differ among groups in Experiment 2. Neither of the patterns of concentrations of $PGF_{2\alpha}$ studied (mean within sampling period over time and natural logarithm of the variance within sampling period over time) differed among groups in either experiment. Concentrations of $PGF_{2\alpha}$ increased after ovariectomy in Experiment 3 and were restored to values similar to intact ewes by replacement therapy with progesterone. These data do not support the idea that increases in oestrogen are involved in regulating patterns of uterine secretion of $PGF_{2\alpha}$ during the ovine oestrous cycle or the initiation of maternal recognition of pregnancy. In addition, progesterone may not be necessary after day 9 for initial rises of $PGF_{2\alpha}$ to occur on days 12 and 13, but does appear to be necessary to sustain normal patterns of secretion of $PGF_{2\alpha}$ beyond day 14.

INTRODUCTION

It is well accepted that $PGF_{2\alpha}$ is a naturally occurring luteolysin in the ewe that is produced by the uterus (Horton and Poyser, 1976; Inskeep and Murdoch, 1980). The increased secretion of $PGF_{2\alpha}$ by the uterus beginning on about day 12 post-oestrus results in luteolysis in the non-pregnant ewe at the appropriate time for normal oestrous cyclicity to be maintained. In the pregnant ewe, on the other hand, the corpus luteum (CL) is maintained in spite of a very similar pattern of secretion of $PGF_{2\alpha}$ to that usually observed in the non-pregnant animal, as reviewed by Inskeep and Murdoch (1980) and supported by the data of Lacroix and Kann (1982) and Silvia et al. (1981). Furthermore, exogenously administered $PGF_{2\alpha}$ is less effective as a luteolysin in pregnant than in non-pregnant and hysterectomized ewes (Inskeep et al. 1975; Mapletoft et al., 1976; Pratt et al., 1977; and Nancarrow et al., 1982). Thus in this species, the embryo appears to inhibit the luteolytic effect of $PGF_{2\alpha}$, rather than to suppress its secretion.

In the present paper, two areas of work will be considered. First, the recent evidence that PGE_2 might play an antiluteolytic role as a part of the mechanism of maternal recognition of pregnancy will be reviewed briefly. Second, previously unpublished studies concerning the roles of oestradiol-17β and progesterone in regulating the secretion of $PGF_{2\alpha}$ will be discussed in detail.

Prostaglandin E_2 as an antiluteolysin

The antiluteolytic effect of the embryo is expressed locally. Only the CL ipsilateral to an isolated gravid uterine horn is maintained (Moor and Rowson,

1966). Using various anastomoses from side to side of the uterine venous and ovarian arterial vasculature Mapletoft et al. (1975, 1976) have shown that the antiluteolytic effect is mediated by some blood-borne factor(s) which passes from the uterine vein to the ovarian artery. Given such a local transfer, one might expect this factor(s) to be lipid soluble, of low molecular weight and rapidly metabolizable, properties similar to those of the luteolysin, $PGF2\alpha$.

Prostaglandins E_2 and E_1 fit the criteria cited above and, in many biological systems, have effects opposite to those of $PGF2\alpha$ (Bergström et al., 1968). Prostaglandin E_2 is found in (Wilson et al., 1972) and secreted during in vitro incubations of (Ellinwood et al., 1979; Marcus, 1981; Hyland et al., 1982; Lacroix and Kann, 1982) endometrial tissue. Evidence has been presented for production of PGE_2 by blastocysts of sheep (Hyland et al., 1982), cattle (Shemesh et al., 1979; Lewis et al., 1982), pigs (Geisert et al., 1982) and rabbits (Harper et al., 1983). Secretion of progesterone by bovine luteal tissue incubated in vitro is stimulated by PGE_2 (Marsh, 1970). Spontaneous luteal regression and that induced by oestradiol or intrauterine devices in ewes have been delayed by uterine luminal infusion of PGE_2 or PGE_1 (Pratt et al., 1977; Colcord et al., 1978; Hoyer et al., 1978; Huecksteadt and Weems, 1978). In order to exert this antiluteolytic effect, PGE_2 or PGE_1 must be infused into the uterine horn adjacent to the ovary bearing the corpus luteum (Huie et al., 1981; Magness et al., 1981). Arterial infusions or perivascular injections of $PGF_{2\alpha}$ were less effective in causing luteolysis when PGE_2 or PGE_1 was administered concomitantly (Henderson et al., 1977; Mapletoft et al., 1977; Reynolds et al., 1981).

Lewis et al. (1978) did not detect differences in concentrations of PGE_2 in uterine venous plasma collected via indwelling catheters from pregnant and non-pregnant ewes during five short periods on days 15 and 16 post-oestrus or in endometrial tissue collected on day 16. In contrast, Ellinwood et al. (1979) reported greater concentrations of PGE_2 in single samples of uterine venous serum, uterine flushings and endometrial tissue collected from pregnant ewes on days 15 and 17. On day 13, differences between pregnant and non-pregnant ewes were not significant, except for greater PGE_2 in uterine flushings of pregnant ewes. Slices or dispersed cell preparations of endometrial tissue from pregnant ewes as early as day 12 post-oestrus released more PGE_2 in vitro than similar preparations from non-pregnant ewes (Ellinwood et al., 1979, Marcus, 1981).

Recently in our laboratory, Silvia et al. (1981) utilized indwelling catheters in the utero-ovarian vein to investigate patterns of secretion of PGE_2 and the ratio of PGE_2 to $PGF_{2\alpha}$ on days 10 through 14 post-oestrus in 6 pregnant and 8 non-pregnant ewes. Samples were collected at 30 minutes intervals for 2 hours, twice daily on these days. Mean PGE_2 during the sampling period rose dramatically (3 fold) on the morning of day 13 in the pregnant ewes and remained high through day 14, whereas no change was seen in the non-pregnant ewes. The ratio of PGE_2 to $PGF_{2\alpha}$ also increased in pregnant ewes compared to non-pregnant ewes during this period ($P < 0.05$; Fig. 1). These results agree reasonably well with the findings in single samples and endometrium cited above. They do not separate the embryo and endometrium as sources of PGE_2. Hyland et al. (1982) found that addition of endometrium to embryos in culture did not markedly increase release per unit tissue of either PGE_2 or $PGF_{2\alpha}$ into the culture medium, but pointed out that the endometrium is much larger than the embryo, so that its contribution in vivo might well be substantial.

Considering all reports to date, it is not possible to ascribe a specific, obligatory role in the maintenance of pregnancy, or of the corpora lutea of pregnancy, to PGE_2. Because of the limited effect in maintaining corpora lutea in some studies (Pratt et al., 1977) and the observation by Lacroix and Kann (1982) that indomethacin from day 7 to 22 did not interfere with pregnancy in the ewe (despite lowering both PGE_2 and $PGF_{2\alpha}$ in conceptus and endometrium), it is tempting to propose that the role of PGE_2 is specifically to inhibit the luteolytic action of $PGF_{2\alpha}$ for a short period. This would be consistent with the longer action of an antiluteolytic protein from trophoblasts on day 14 to 16 reported by Martal et al. (1979), although they proposed that one route of its action might be to stimulate endometrial secretion of PGE_2.

Effects of progesterone and oestradiol on uterine secretion of $PGF_{2\alpha}$

Despite a considerable effort to date, the roles of ovarian steroids in control of the patterns of release of uterine $PGF_{2\alpha}$ during the cycle and early pregnancy are not clear. Active immunization of intact ewes against progesterone did not provide useful evidence as to its role in luteal regulation (French and Spennetta, 1981). Some ewes showed normal cycles; cycles in other ewes were prolonged and others were shortened. Progesterone in serum was increased tremendously, but

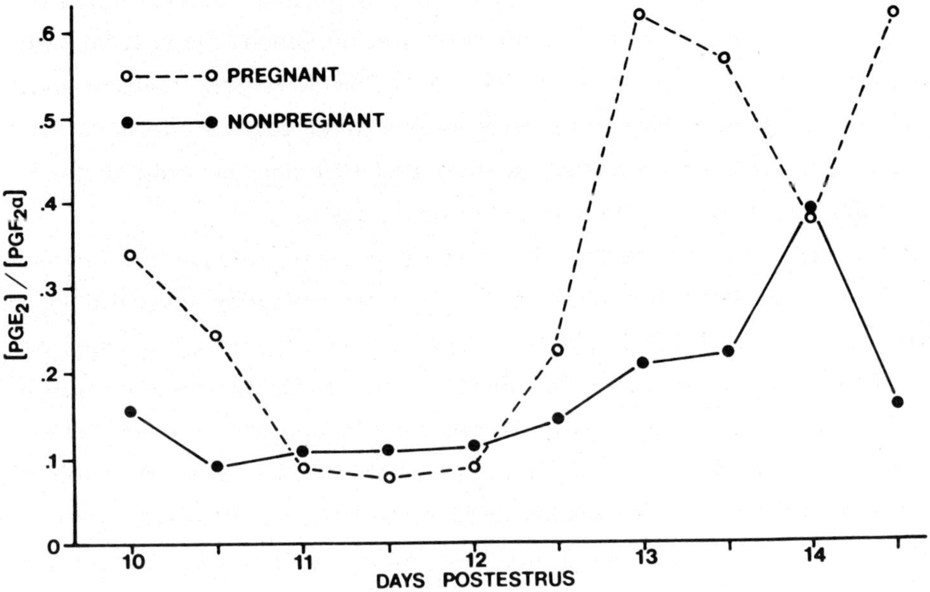

Fig. 1. Patterns of the ratio of mean concentrations of PGE_2 to $PGF_{2\alpha}$ in pregnant and non-pregnant ewes during two daily sampling periods of 2 hours each on days 10 through 14 post-oestrus. Each point represents the least square mean for n observations (for pregnant ewes, n = 2, 4, 6, 6, 5, 6, 6, 5, 5, 5 and for non-pregnant ewes, n = 4, 5, 8, 8, 8, 8, 8, 5, 4, 2) on days 10, 10.5 . . . . 14.5, respectively. Patterns differed ($P < 0.05$).

whether it was biologically available could have varied greatly among animals and over time within animal. In ovariectomized ewes, exogenous progesterone for several days increased the uterine release (Scaramuzzi et al., 1977) or caruncular concentrations (Louis et al., 1977) of $PGF_{2\alpha}$ in treated ewes over values for control ewes. Similar results were found for intercaruncular production of PGF in the presence of arachidonic acid in vitro (Findlay et al., 1981).

Baird et al. (1976) and Ottobre et al. (1980) have proposed that a period of progestational influence regulates the timing of the initial rises in release of $PGF_{2\alpha}$. The latter authors showed that early initiation of this progestational influence, by administration of progesterone on days 0 and 1, advanced the time of occurrence of the initial rises of $PGF_{2\alpha}$ to day 8, compared to the usual time of day 12 in control ewes. Similar results were observed by McCracken (1980) in two ewes infused with progesterone for 10 days beginning immediately after oestrus. He reported peaks of $PGF_{2\alpha}$ of low magnitude "around day 9 or 10 of continuous P infusion" in hourly samples of uterine venous blood from ewes with utero-ovarian transplants to a skinfold in the neck region. Lawson and Cahill (1983) have now shown that progesterone early after oestrus advances the entire uterine preparation for pregnancy so that day 10 embryos could be transferred successfully into progesterone-treated ewes on day 6.

Administration of oestradiol-17β increased uterine release and (or) caruncular or endometrial concentrations of PGF if given following several days of exogenous (Ford et al., 1975; Louis et al., 1977; Scaramuzzi et al., 1977) or endogenous (Barcikowski et al., 1974; Ford et al., 1975) progesterone. However, oestradiol alone was only slightly stimulatory in anoestrous ewes (Sharma and Fitzpatrick, 1974) and its initial stimulatory effect (24 h after first injection of 12.5 µg at 12-hour intervals) was lost before the fifth day in ovariectomized ewes (Ford et al., 1975). In ewes which had been ovariectomized for 3 weeks, Findlay et al. (1981) gave oestradiol for 15 days and observed reductions in basal release of PGF by intercaruncular endometrium and of net production of PGF by caruncular endometrium. However, synthesis of PGF by intercaruncular endometrium in the presence of arachidonic acid was increased. Higher concentrations of oestradiol in plasma have been associated temporally with the release of $PGF_{2\alpha}$ in several species (Shaikh et al., 1973; Cox et al., 1974; Horton and Poyser, 1976; Ottobre et al., 1980). This has been particularly true for the maximal utero-ovarian concentrations of $PGF_{2\alpha}$ during luteal regression in the ewe, either during the cycle (Scaramuzzi et al., 1977; Baird et al., 1976; Ottobre et al., 1980) or at induced parturition (Thorburn et al., 1972). Results such as these have led some authors to imply or suggest that endogenous oestrogen stimulates the production of $PGF_{2\alpha}$ by the uterus, after a period of prior progestational influence in the non-pregnant ewe (Horton and Poyser, 1976). On the other hand, Baird et al. (1976) and Ottobre et al. (1980) found no relationship of the initial rises in $PGF_{2\alpha}$ to increases in oestradiol-17β during the mid-luteal phase.

Removal of the main source of oestrogen by destruction of follicles (Karsch et al., 1970; Ginther, 1971) only delayed, but did not prevent luteal regression in the ewe. Antisera to oestradiol, which prevented behavioral oestrus, did not delay luteal regression (Fairclough et al., 1976 a,b). Thus, a rise in oestrogen may not be necessary for the secretion of luteolytic concentrations of $PGF_{2\alpha}$ to occur. The effects of removal of oestrogen, at strategic times during the oestrous cycle, on secretion of PGF2α into the utero-ovarian vein have not been tested.

Three experiments were done to examine the roles of progesterone and oestrogen in the regulation of concentrations of PGF2α in the utero-ovarian vein of the ewe. Specifically, Experiment 1 was designed to test the hypothesis that removal of ovarian oestrogen, removal of progesterone plus oestrogen (ovariectomy), or removal of the luteal tissue (while maintaining the steroid status by replacement therapy with progestagen) on day 9 would not change the time of occurrence of the initial peaks in concentrations of PGF2α in utero-ovarian venous plasma, normally seen on day 12, or the overall pattern of PGF2α in utero-ovarian venous plasma from days 10 through 14. Experiment 2 was designed to determine whether the maximal increases in concentrations of PGF2α, which have been observed during and immediately following decline in concentrations of progesterone in peripheral plasma at the end of the oestrous cycle, would occur in the absence of the ovaries and if not, whether they could be reinstituted by replacement therapy with a constant amount of oestradiol-17β. Experiment 3 was designed to test whether removal of ovarian tissue on day 12 with or without replacement therapy with progesterone begun on day 12 would change the patterns of utero-ovarian concentration of PGF2α on day 13 in pregnant and non-pregnant ewes.

GENERAL METHODS

Mature ewes were kept with vasectomized, brisket-painted rams for detection of oestrus (oestrus = day 0 of the oestrous cycle). Marked ewes were removed twice daily and rechecked for standing oestrus with other vasectomized rams. Midventral laparotomies were performed under sodium pentobarbital anesthesia and the reproductive organs were exposed. Samples of utero-ovarian venous plasma were obtained from each ewe through a polyvinyl catheter positioned in the utero-ovarian vein and exteriorized through a small incision in the flank (Ottobre et al., 1980). Samples were collected from these catheters during two sampling periods each day within the specified interval in Experiments 1 and 2. During a sampling period, which began at 0600 or 1800 each day, samples were collected every 30 minutes for 2 hours (5 samples/period). In Experiment 3, samples were collected during one period each day, every 15 minutes for 3 hours, between 0600 and 1100. After each sample was drawn, the catheter was flushed with sterile physiological saline containing 80 IU of sodium heparin and 20,000 IU of potassium penicillin G/ml. All samples were processed and assayed for concentrations of $PGF_{2\alpha}$ as described by Lewis et al. (1978). Radioimmunoassays were used for quantification of progesterone (Berardinelli et al., 1980) and oestradiol-17β (Fogwell et al., 1977) in certain samples.

Sensitivities (lowest detectable masses) for each assay have been estimated as the mass equivalent to a hypothetical sample with counts per minute 2 standard deviations less than the mean of the tubes containing zero mass. The assay for progesterone had a sensitivity of 10 pg and coefficients of variation within and between assays of 11.2 and 17.7%, respectively. The volume of plasma assayed was 100 µl. The assay for oestradiol-17β had a sensitivity of 2.5 pg, and coefficients of variation within and between assays were 13.9 and 18.7%, respectively. The volumes of plasma assayed were 1 to 3 ml. The assay for $PGF_{2\alpha}$ had a sensitivity of 30 pg, and coefficients of variation within and between assays of 14.2 and 17.4%, respectively. The volumes of plasma assayed ranged from 50 to 200 µl.

Experiment 1

Ewes received utero-ovarian venous catheters at laparotomy on day 9, at which time each ewe was assigned at random to one of four treatment groups. Ewes in an intact control group received no further treatment (n=8). Utero-ovarian venous concentrations of $PGF_{2\alpha}$ were compared among these ewes and those in three experimental groups: 1) the removal of both progesterone and oestrogen (bilateral ovariectomy, n=5), 2) the removal of oestrogen only (bilateral ovariectomy plus immediate replacement therapy with progestagen, consisting of an i.m. injection of 10 mg progesterone plus an intravaginal sponge containing 30 mg flurogestone acetate (n=5) and 3) the absence of luteal tissue, but the presence of oestrogen and progestagen (lutectomy plus immediate replacement therapy with progestagen as above, n=7). The latter group had concentrations of oestradiol essentially duplicating those in control ewes (Table I) and allowed detection of effects of endogenous oestrogen on concentrations of $PGF_{2\alpha}$, by comparison with ovariectomy plus progestagen.

Table I. Effect of ovariectomy and lutectomy plus progestagen replacement and ovariectomy alone on utero-ovarian venous concentrations of oestradiol-17β, day of first peak of $PGF_{2\alpha}$ and number of peaks of $PGF_{2\alpha}$ (Experiment 1)

Treatment[a]	Number of ewes	Mean daily utero-ovarian venous oestradiol-17β (pg/ml)[b]	Mean day of first peak of $PGF_{2\alpha}$[b]	Mean number of peak values of $PGF_{2\alpha}$ within ewe[b]
Lutectomy plus progestagen	7	91.6 ± 26.1	11.6 ± 0.5	3.7 ± 0.3
Ovariectomy plus progestagen	5	16.7 ± 7.7	12.0 ± 0.4	3.2 ± 0.4
Ovariectomy	5	8.4 ± 1.5	12.2 ± 0.5	3.4 ± 0.4
Control	8	114.4 ± 31.6	12.1 ± 0.4	2.6 ± 0.8

[a] Surgeries were done and hormone replacement therapy was begun on day 9. The progestagen-treated groups received 10 mg of progesterone i.m. plus an intravaginal sponge containing 30 mg Cronolone.

[b] $\bar{x}$ ± SEM

Each ewe was injected with 10,000 IU of sodium heparin every 6 h to maintain patency of catheters. Beginning on day 10, which started at 0600 or 1800 hr depending on the time of day at which the onset of oestrus had occurred in each ewe, samples were collected from catheters, as described above, until day 14 or until interruption of patency of the catheter. All samples were assayed for $PGF_{2\alpha}$ and one sample collected each day was assayed for oestradiol-17β.

Experiment 2

Ewes were laparotomized and received utero-ovarian and jugular venous catheters on day 14. At this time, each ewe was assigned at random to one of four treatment groups: 1) removal of oestrogen and progesterone (bilateral ovariectomy, n=5), 2) removal of progesterone, but exogenous oestrogen present (ovariectomy plus insertion s.c. of a 1.5 cm silastic implant, .132 mm id x .183 mm od filled with oestradiol-17β, (n=5), 3) removal of progesterone and luteal tissue, but endogenous oestrogen present (lutectomy, n=6) and 4) intact control ewes (n=5). In the latter group, bilaterally ovulating ewes had corpora lutea removed from the ovary on the side adjacent to the catheter to prevent possible maintenance of these corpora lutea due to any local interference of the catheter with the proposed utero-ovarian luteolytic pathway (Ginther, 1974). This modification was considered acceptable since utero-ovarian venous concentrations of $PGF_{2\alpha}$ ipsilateral and contralateral to the corpus luteum have not differed in previous studies (Lewis et al., 1977). Following surgery, all ewes received a constant infusion of 200 ml/day of sterile saline containing 250 IU/ml sodium heparin, into jugular catheters via Buchler infusion pumps.

Control and lutectomized ewes were checked for oestrus twice daily with vasectomized rams. Beginning at the first sampling period after surgery (day 14.5), samples were collected from catheters as described in the protocol for Experiment 1 until day 17 or, in the control and lutectomized ewes, until the onset of oestrus, whichever came last. This procedure assured that samples were collected during the preovulatory rise in oestrogen because its occurrence precedes oestrus. At each sampling period, a single blood sample was obtained from the jugular vein. Aliquots of plasma from the samples were assayed for oestradiol-17β and progesterone.

Experiment 3

Ewes were assigned randomly to either bred (n=15) or nonbred (n=12) groups. Ewes assigned to the bred group were penned with a brisket-painted fertile ram on the day of oestrus. Ewes assigned to the nonbred group remained with the main flock. Ewes were laparotomized and received utero-ovarian venous catheters on day 12. At this time bred and nonbred ewes were assigned to one of three groups 1) bilateral ovariectomy (n=9), 2) bilateral ovariectomy with immediate progesterone replacement therapy (consisting of a s.c. injection of 10 mg of progesterone and an intravaginal sponge containing 30 mg progesterone, n=9) or 3) intact (n=9). In the intact control ewes, bilaterally ovulating ewes had corpora lutea removed from the ovary ipsilateral to the catheter. Following surgery all ewes received a constant infusion of 200 ml/day of 100 IU sodium heparin/ml sterile saline into the utero-ovarian catheters via Buchler infusion pumps. The day after surgery (day 13), utero-ovarian blood samples were collected from the catheters at 15-minute intervals for 3 hours beginning between 0600 and 1100. Bred ewes were killed and catheter placement confirmed on day 17 or 18. Uteri were removed and examined for the presence of embryos to confirm pregnancy. All utero-ovarian venous plasma samples were assayed for concentrations of $PGF_{2\alpha}$. Samples taken at 0, 90, and 180 minutes were assayed for utero-ovarian venous plasma concentrations of progesterone.

Statistical analyses

Patterns over time of mean concentrations of $PGF_{2\alpha}$ within sampling period were compared among groups in experiments 1 and 2 using analysis of variance for a completely randomized split-plot design with treatment as the main plot and period as the subplot. Because differences in patterns of secretion of $PGF_{2\alpha}$ may be missed by examination of only means within period and because differences in amplitudes of peaks might occur as a result of treatment, patterns of fluctuations in concentrations of $PGF_{2\alpha}$ among groups were compared. This was done by determining the variances for the five samples collected from each ewe within each period. To reduce the large range of these variances, they were transformed to their natural logarithms. D.H. Keisler (unpublished observations, West Virginia University) has shown that natural logarithms of variances of frequent samples provide a close reflection of changes in amplitude of episodic secretion. Patterns over periods of natural logarithms of variances (ln variance) of concentrations of $PGF_{2\alpha}$ within period were compared among groups using analysis of variance for a split-plot design, exactly as was done for means.

In Experiment 2, it was assumed that basal concentration of $PGF_{2\alpha}$ would be distributed normally and that peaks would be deviations from this distribution (Brinkley et al., 1973). A normal distribution of concentrations of $PGF_{2\alpha}$ within each ewe was established by sequentially removing all extreme values until there was neither skewness nor kurtosis in the data set. Peaks were defined as single values at least 2 standard deviations greater than the mean of the normally distributed values for each ewe. In Experiment 1, many samples contained concentrations of $PGF_{2\alpha}$ that were undetectable and these were given the value corresponding to the sensitivity of the assay for the highest volume of plasma used. Thus, the distributions of values within many ewes were skewed toward the lower end of the range. Removal of these values to achieve normal distribution would have increased the apparent baseline and rendered the detection of peaks extremely insensitive. Therefore, in Experiment 1, peaks were defined as single values at least 2 SD above the mean concentration of $PGF_{2\alpha}$ for each ewe. Comparisons of time of occurrence of first peaks of $PGF_{2\alpha}$ (Experiment 1) and number of peaks of $PGF_{2\alpha}$ within ewe from days 10 to 14 (Experiment 1) and from days 14 to 17 (Experiment 2) were made among groups using analysis of variance.

In Experiment 3, patterns of concentrations of $PGF_{2\alpha}$ and progesterone were compared among groups using analysis of variance for a completely randomized split-plot design with treatment as the main plot and time as the subplot. Treatments were arranged in a 2 x 3 factorial design with breeding or nonbreeding and ovarian status as main effects. Single degree of freedom orthogonal comparisons were used to compared main effects of ovarian status.

RESULTS

Experiment 1

Mean daily concentrations of oestradiol-17β in utero-ovarian venous plasma decreased dramatically following ovariectomy, thus fulfilling the prescribed conditions of treatment (Table I). The fluctuations in utero-ovarian venous concentrations of $PGF_{2\alpha}$ that occurred within sampling period are shown for two ewes in each group in Fig. 2, illustrating the peaks that occurred at different times in individual ewes in all groups. Neither patterns of mean concentrations of PGF2α nor patterns of the ln variance of concentrations of PGF2α differed among groups (Fig. 3). The mean day of first peak of PGF2α and the mean number of peaks within ewe did not differ among groups (Table I).

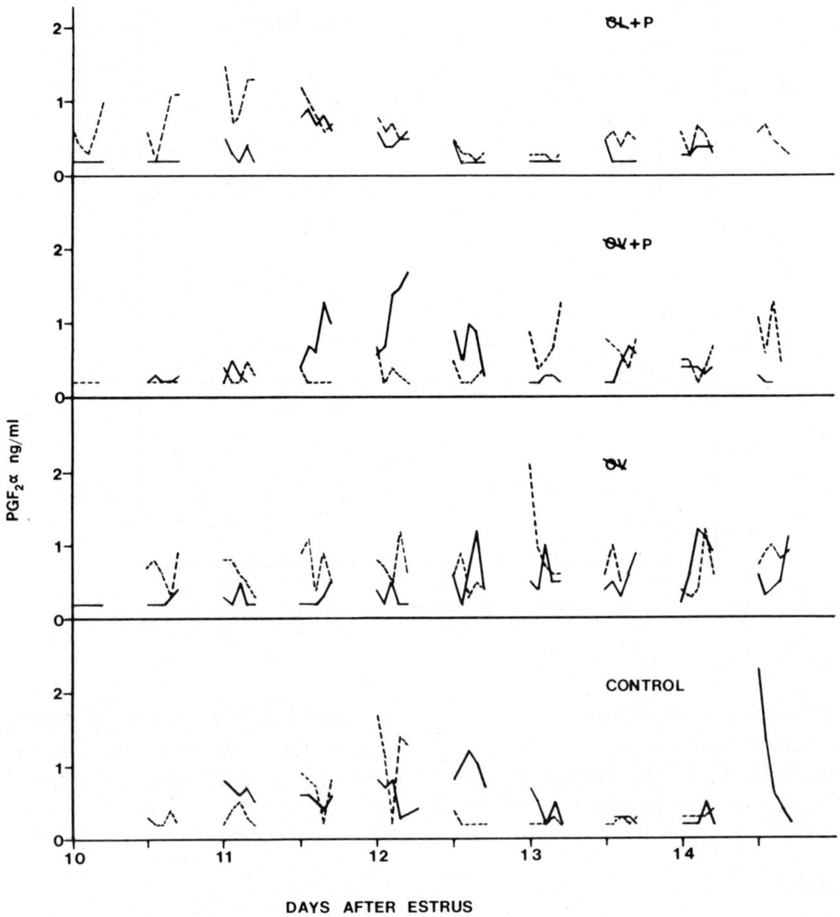

Fig. 2. Patterns of concentrations of PGF2α in individual samples collected each 30 minutes during two periods of two hours per day from each of two ewes per group in Experiment 1 (solid lines represent one ewe and dashed lines another). Groups are coded as OV (ovariectomy), OV + P (ovariectomy plus progestagen replacement therapy), CL + P (lutectomy plus progestagen) and control. The high degree of individual variation in times of release of PGF2α is readily apparent, but it is clear that episodic release is occurring throughout the experimental period in all groups.

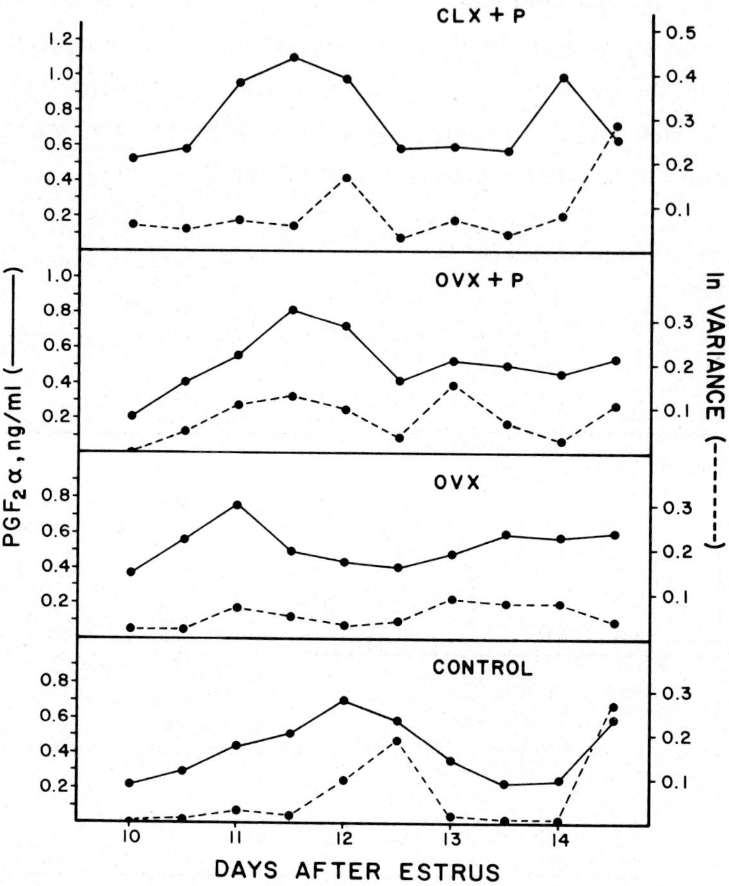

Fig. 3. Mean concentrations (——) and ln variances (----) of $PGF_{2\alpha}$ within sampling period for each group in Experiment 1. Groups are coded as OVX (ovariectomy), OVX + P (ovariectomy plus progestagen replacement therapy), CLX + P (lutectomy plus progestagen) and control.

Experiment 2

Mean peripheral concentrations of oestradiol-17β within period decreased following ovariectomy and replacement therapy with oestradiol-17β restored these concentrations to the level of controls, fulfilling the prescribed conditions of treatment (Table II). The mean number of peaks of $PGF_{2\alpha}$ within ewe did not differ among groups (Table II). As in Experiment 1, patterns of mean concentrations of $PGF_{2\alpha}$ within period (Fig. 4) and patterns of the ln variance of concentrations of $PGF_{2\alpha}$ within period (Fig. 4) did not differ among groups. There was a

Table II. Effects of lutectomy alone and ovariectomy with or without oestradiol-17β replacement on peripheral concentrations of oestradiol-17β and numbers of peak values of $PGF_{2\alpha}$ (Experiment 2)

Treatment[a]	Number of ewes	Mean peripheral concentrations of oestradiol-17β (pg/ml)[b]	Mean number of peak values of $PGF_{2\alpha}$ within ewe[b]
Lutectomy	6	5.2 ± 0.4	8.0 ± 0.8
Ovariectomy	5	2.9 ± 0.5	6.8 ± 2.1
Ovariectomy plus oestradiol-17β	5	8.2 ± 1.2	6.4 ± 1.6
Control	5	6.3 ± 0.3	7.8 ± 1.6

[a] Surgery was done and hormonal replacement therapy was begun on day 14. Oestradiol-17β was administered s.c. to ewes in one group via a 1.5-cm silastic implant (.132 id x .183 od).

[b] $\bar{x}$ ± SEM

main effect of treatment ($P < 0.05$) on the ln variance over all sampling periods, the mean ln variance being greater in the control ewes than in the other three groups. This appeared to be due to variation in the timing of the decline in concentrations of progesterone in the control ewes since $PGF_{2\alpha}$ declined soon after the fall in progesterone in each ewe in all groups.

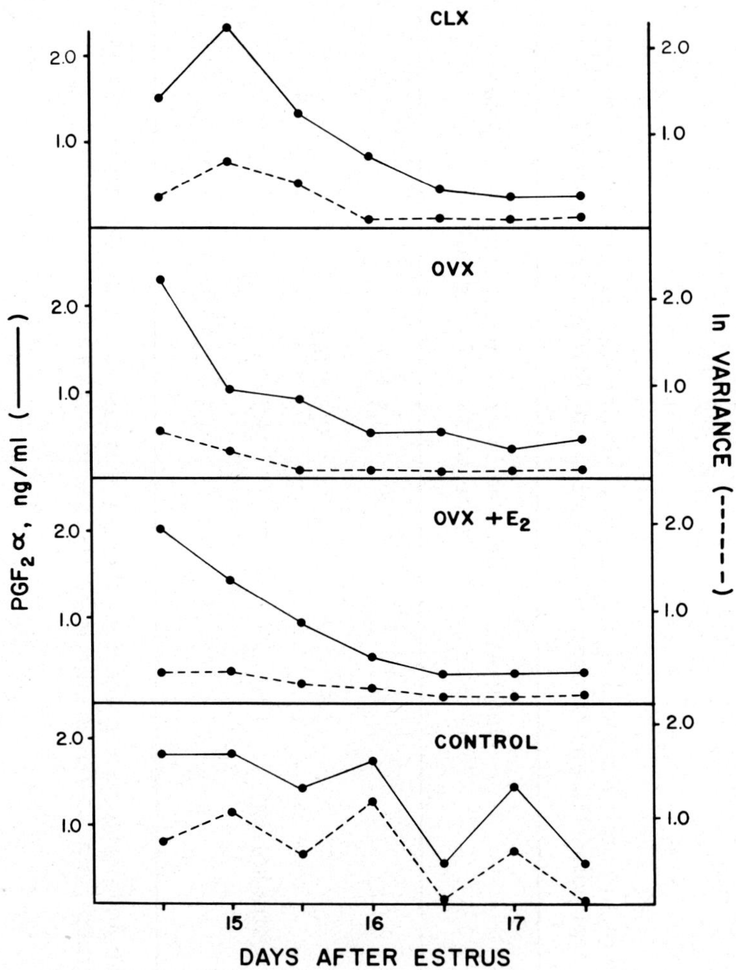

Fig. 4. Mean concentrations (——) and ln variances (----) of $PGF_{2\alpha}$ within sampling period for each group in Experiment 2. Groups are coded as CLX (lutectomy), OVX (ovariectomy), OVX + E_2 (ovariectomy plus oestradiol-17β implant) and control.

Experiment 3

Pregnancy was confirmed in 5 bred, intact ewes and 5 bred, bilaterally ovariectomized ewes with immediate progesterone replacement. No embryos or embryonic fragments were observed in 5 bred ewes ovariectomized on day 12.

Concentrations of $PGF_{2\alpha}$ in utero-ovarian venous plasma on day 13 were greater in bred ewes than in nonbred ewes ($P < 0.05$, Fig. 5A). Main effects of ovarian status on the patterns of concentrations of $PGF_{2\alpha}$ were also significant ($P < 0.05$, Fig. 5B). Ovariectomized ewes had greater concentrations of $PGF_{2\alpha}$ than either intact ewes ($P < 0.025$) or ovariectomized ewes with progesterone replacement therapy ($P < 0.05$).

Concentrations of progesterone in utero-ovarian venous plasma on day 13 were greatest in intact ewes, intermediate in ovariectomized ewes with progesterone replacement therapy and least in ovariectomized ewes.

DISCUSSION

Removal of the ovaries or corpora lutea on day 9 with or without replacement therapy with progestagen did not alter the patterns of mean utero-ovarian concentrations of $PGF_{2\alpha}$, the patterns of ln variances in concentrations of $PGF_{2\alpha}$ or the time of occurrence of the first peaks of $PGF_{2\alpha}$ as compared to intact ewes or lutectomized ewes receiving progestagen. This implies that ovarian steroids secreted during days 9 to 14 are not involved in regulating utero-ovarian venous concentrations of $PGF_{2\alpha}$ during that same period.

Progesterone appears to be involved in the timing of the initial peaks in release of $PGF_{2\alpha}$ because early initiation of progestational influence has been shown to advance the release of $PGF_{2\alpha}$ (McCracken, 1980; Ottobre et al., 1980). Treatment of ovariectomized ewes for several days with progesterone caused an increase in the release (Scaramuzzi et al., 1977) and caruncular concentrations (Louis et al., 1977) of $PGF_{2\alpha}$ or made the transplanted uterus more sensitive to release of $PGF_{2\alpha}$ by a combination of oestradiol and oxytocin (McCracken, 1980). Finally, exogenous oestrogen was luteolytic and caused an increase in uterine venous $PGF_{2\alpha}$ only when preceded by endogenous or exogenous progesterone (Ford et al., 1975).

Curiously, the secretion of $PGF_{2\alpha}$ did not appear to depend on continued presence of the ovaries from day 9 through 14. In fact, utero-ovarian concentra-

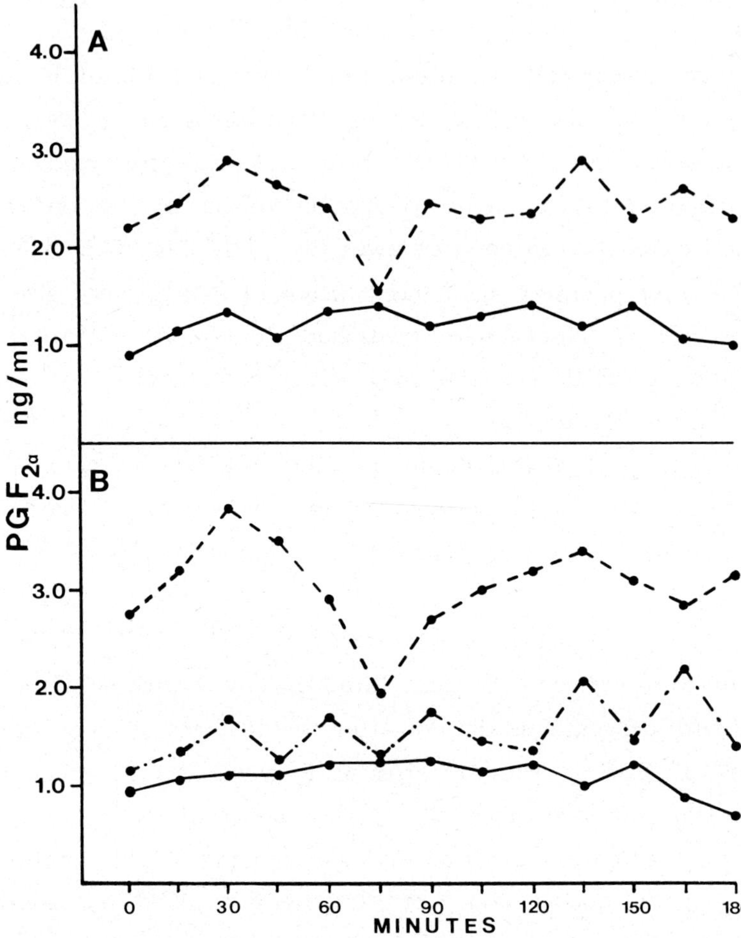

Fig. 5. In the top panel are shown the patterns of mean concentrations of $PGF_{2\alpha}$ in bred (----) and non-bred (——) ewes on day 13 (in Experiment 3). Main effects of breeding vs. non-breeding differ ($P \leq 0.05$). In the bottom panel are shown the patterns of mean concentrations of $PG\overline{F}_{2\alpha}$ of ovariectomized ewes (----), ovariectomized ewes with progesterone replacement (-·-·-·-), and intact ewes (——). Patterns differ ($P < 0.05$) with ovariectomized ewes greater than intact ($P < 0.01$) or ovariectomized ewes with progesterone replacement ($P < 0.05$).

tions of $PGF_{2\alpha}$ were increased on day 13 by ovariectomy on day in both pregnant and non-pregnant ewes (Experiment 3). Perhaps the timing of initiation of peaks of $PGF_{2\alpha}$ on day 11 or 12 was already set by exposure to progesterone until day 9. It is not known whether progestational influence can be removed or oestrogenic influence reduced earlier in the cycle than day 9 without altering the time of initiation of the increased release of $PGF_{2\alpha}$.

Removal of the ovaries on day 14 (Experiment 2) did not alter the patterns of mean concentrations of $PGF_{2\alpha}$ within period, the patterns of ln variance of $PGF_{2\alpha}$ within period or the mean number of peaks of $PGF_{2\alpha}$/ewe during days 14 to 17 from those observed in untreated and lutectomized ewes. However, there appeared to be more fluctuations in the secretion of $PGF_{2\alpha}$, as reflected by a greater mean ln variance of $PGF_{2\alpha}$ within period over all times, in the control ewes than in ewes in the other groups. It was reasoned that this may have occurred because the secretion of $PGF_{2\alpha}$ showed final peaks and then declined shortly after removal of progesterone, and the removal of progesterone in some control ewes (luteal regression) occurred at later times than in ewes in other groups. The greater concentrations of $PGF_{2\alpha}$ on day 13 after ovariectomy on day 12 (Experiment 3) were suggestive of the final peaks seen after ovariectomy on day 14 or normal luteal regression (Experiment 2). Concentrations of $PGF_{2\alpha}$ on day 13 were restored to levels seen in intact ewes by replacement therapy with progesterone. While secretion of $PGF_{2\alpha}$ continued through day 14 after ovariectomy on day 9 (Experiment 1), the apparent increase in ln variance during the final sampling period on day 14 in the ewes with intact ovaries was not seen in the ovariectomized ewes (Fig. 3). This observation, the rise in $PGF_{2\alpha}$ on day 13 after ovariectomy on day 12 (Experiment 3), and the fall in $PGF_{2\alpha}$ soon after lutectomy or ovariectomy on day 14 (Experiment 2), lead one to conclude that continued secretion of progesterone is necessary for the secretion of $PGF_{2\alpha}$ to continue beyond day 14 of the oestrous cycle. This conclusion is supported by work of Kindahl et al. (1980) in heifers. They observed that lutectomy terminated and progesterone implants extended pulsatile release of $PGF_{2\alpha}$ as reflected by peripheral concentrations of 15-keto-13,14-dihydro-$PGF_{2\alpha}$.

Pregnancy increased concentrations of $PGF_{2\alpha}$ in utero-ovarian plasma on day 13 (Experiment 3). This result is in good agreement with the conclusion discussed earlier, that early pregnancy does not suppress uterine secretion of $PGF_{2\alpha}$. There was no evidence that the role of progesterone in regulating $PGF_{2\alpha}$ differed in pregnant and non-pregnant ewes.

In the present study, exogenous oestradiol did not increase plasma concentrations of $PGF_{2\alpha}$ as had been seen in previous studies (Barcikowski et al., 1974;

Ford et al., 1975, Louis et al., 1977; Scaramuzzi et al., 1977). The dose of oestradiol delivered in Experiment 2 seems to be within the range used in at least two of the earlier studies. However, those studies did not include removal of the ovaries at the time of treatment with oestrogen. In addition, concentrations of $PGF_{2\alpha}$ were already elevated (compare Fig. 4 to Fig. 3) and frequency of peaks was already high (compare Table II to Table I) when the oestradiol was given in Experiment 2 of the present study.

A similar study of effects of ovariectomy and steroid replacement therapy has been done in pregnant rats near term (Wilson and Stanisic, 1981), an animal in which increases in uterine concentrations of several prostanoids had been seen to occur concomitantly with increases in oestradiol and oestrone (Wilson et al., 1982). In that study, ovariectomy on day 19 increased uterine concentrations of PGF on day 20 over intact controls. Uterine concentrations of PGF in rats ovariectomized and treated with implants containing either oestrogen or oestrogen plus progesterone were not significantly different from those in the ovariectomized controls. However, rats ovariectomized and given implants containing progesterone had uterine concentrations of PGF below ovariectomized controls and similar to intact controls. Thus, an increase in PGF occurred in response to withdrawal of progesterone by ovariectomy. Interestingly, progesterone has been shown to inhibit uterine concentrations of $PGF_{2\alpha}$ (Wilson et al., 1972) and its release in response to an intrauterine device (Murdoch et al., 1978) during early to midcycle in the ewe.

Although the concomitant presence of either steady or increasing concentrations of oestrogen does not appear to be necessary for normal secretion of $PGF_{2\alpha}$ on the days tested in this study, we have not addressed the possibility that it may act earlier to prepare the uterus to release $PGF_{2\alpha}$ in response to oxytocin as reported by McCracken (1980). It must be recognized that oestrogen may have a role in luteolysis that does not involve uterine secretion of $PGF_{2\alpha}$. Evidence for such an action lies in the observation that the combination of $PGF_{2\alpha}$ and oestrogen was more effective in causing luteolysis in hysterectomized ewes than treatments with either $PGF_{2\alpha}$ or oestrogen alone (Gengenbach et al., 1977). Such an action may account for the delay in the completion of luteolysis in ewes (Karsch et al., 1970) and (or) the prevention of luteolysis in cows (Cowley et al., 1979; Villa-Godoy et al., 1981) following destruction of ovarian follicles.

Alternatively, as suggested by C.E. Rexroad, Jr. (personal communication), the major role of oestrogen in uterine synthesis of PGF$_{2\alpha}$ may occur at oestrus before the luteal phase. In both endometrium (Koligian and Stormshak, 1977a) and myometrium (Rexroad, 1981a), cytoplasmic receptors for progesterone were highest at oestrus and immediately thereafter, declining in midluteal phase as progesterone rose. Accordingly, oestradiol increased receptors for progesterone in whole uterus (Stone et al., 1978), endometrium (C.E. Rexroad, Jr., unpublished data, 1982) and myometrium (Rexroad, 1981b) of ovariectomized ewes and progesterone inhibited this effect. One could propose that, after synthesis of receptors for progesterone is induced by the high concentrations of oestrogen associated with oestrus, a period of increased progesterone and of receptor-mediated transfer of progesterone to the nucleus of uterine cells is required in order to establish the appropriate potential for increased secretion of PGF$_{2\alpha}$. In this model, the luteolytic effect of large doses of oestradiol in midluteal phase (Hawk and Bolt, 1970) would be considered a pharmacological effect, in view of the present results and the lowered concentrations of receptors for oestradiol in endometrium at that stage of the cycle (Koligian and Stormshak, 1977b; Rexroad, 1981a; Findlay et al., 1982). Nevertheless, that luteolytic effect may be mediated by synthesis of progesterone receptors. Zelinski et al. (1980) have shown that luteolytic doses of oestradiol increase receptors for both oestrogen and progesterone in the uterus of the midluteal phase ewe.

In summary, the present results cannot be interpreted as support for the hypothesis that rises in oestrogen are involved in the regulation of uterine secretion of PGF$_{2\alpha}$ after day 9. In addition, it does not appear that there is a role for progesterone after day 9 in the initiation of the increased release of PGF$_{2\alpha}$, but it may be required for a high secretion rate to continue beyond day 14.

ACKNOWLEDGEMENTS

This manuscript is published with the approval of the Director of the West Virginia Agricultural and Forestry Experiment Station as Scientific Article No. 1734 and was supported by Hatch Project 224 (NE-72). The authors are indebted to Dr. John Pike, The Upjohn Co., for generous supplies of PGF$_{2\alpha}$ for assay standards, to Dr. Lawrence Levine, Brandeis University, for highly specific antiserum to PGF$_{2\alpha}$ and to Dr. Norman Mason of Eli Lilly and Company for highly specific antiserum to PGE_2. We are grateful to Phyllis Jenkins and Lionel Jones for technical aid, Dr. W.V. Thayne for assistance with statistical analyses and Drs. R.A. Dailey and C.E. Rexroad, Jr. for critical reading of the manuscript.

REFERENCES

Baird, D.T., Land, R.B., Scaramuzzi, R.J. and Wheeler, A.G., 1976. Endocrine changes associated with luteal regression in the ewe; the secretion of ovarian oestradiol, progesterone and androstenedione and uterine prostaglandin $F_{2\alpha}$ throughout the oestrous cycle. J. Endocrinol., 69: 275-286.

Barcikowski, B., Carlson, J.C., Wilson, L. Jr. and McCracken, J.A., 1974. The effect of endogenous and exogenous estradiol-17β on the release of prostaglandin $F_{2\alpha}$ from the ovine uterus. Endocrinology, 95: 1340-1349.

Berardinelli, J.G., Dailey, R.A., Butcher, R.L. and Inskeep, E.K., 1980. Source of circulating progesterone in prepubertal ewes. Biol. Reprod., 22: 233-236.

Bergström, S., Carlson, L.A. and Weeks, J.R., 1968. The prostaglandins: A family of biologically active lipids. Pharmacol. Rev., 20: 1-48.

Brinkley, H.J., Wilfinger, W.W. and Young, E.P., 1973. Statistical analysis of plasma gonadotropin data. J. Anim. Sci., 37: 302A.

Colcord, M.L., Hoyer, G.L. and Weems, C.W., 1978. Effect of prostaglandin E_2 (PGE_2) as an anti-luteolysin in estrogen-induced luteolysis in ewes. J. Anim. Sci., 47 (Suppl. 1): 352.

Cowley, J.L., Ireland, J.J., Wortman, J.A. and Fogwell, R.L., 1979. Luteal function in cows following destruction of ovarian follicles at midcycle. J. Anim. Sci., (Suppl. 1): 289A.

Cox, R.I., Thorburn, G.D., Currie, W.B. and Restall, B.J., 1974. Interrelations between ooestrogens and prostaglandin F concentrations in the uteroovarian plasma of the ewe. J. Reprod. Fertil., 36: 448A.

Ellinwood, W.E., Nett, T.M. and Niswender, G.D., 1979. Maintenance of the corpus luteum of early pregnancy in the ewe. II Prostaglandin secretion by the endometrium in vitro and in vivo.. Biol. Reprod., 21: 845-856.

Fairclough, R.J., Smith, J.F. and Peterson, A.J., 1976a. Passive immunization against oestradiol-17β and its effect on luteolysis, oestrus and ovulation in the ewe. J. Reprod. Fertil., 48: 169-171.

Fairclough, R.J., Smith, J.F., Peterson, A.J. and McGowan, L., 1976b. Effect of oestradiol-17β, progesterone, and prostaglandin $F_{2\alpha}$ antiplasma on luteal function in the ewe. J. Reprod. Fertil., 46: 523A.

Findlay, J.K., Ackland, N., Burton, R.D., Davis, A.J., Maule Walker, F.M., Walters, D.E. and Heap, R.B., 1981. Protein, prostaglandin and steroid synthesis in caruncular and intercaruncular endometrium of sheep before implantation. J. Reprod. Fertil., 62: 361-377.

Findlay, J.K., Clarke, I.J., Swaney, J., Colvin, N. and Doughton, B., 1982. Oestrogen receptors and protein synthesis in caruncular and intercaruncular endometrium of sheep before implantation. J. Reprod. Fertil., 64: 329-339.

Fogwell, R.L., Lewis, G.S., Butcher, R.L. and Inskeep, E.K., 1977. Effects of ovarian bisection on response to intrafollicular injection of $PGF_{2\alpha}$ and on follicular development in ewes. J. Anim. Sci., 45: 328-335.

Ford, S.P., Weems, C.W., Pitts, R.E., Pexton, J.E., Butcher, R.L. and Inskeep, E.K., 1975. Effects of estradiol-17β and progesterone on prostaglandins F in sheep uteri and uterine venous plasma. J. Anim. Sci., 41: 1407-1413.

French, L.R. and Spennetta, B., 1981. Effects of antibodies to progesterone on reproduction in ewes. Theriogenology, 16: 407-418.

Geisert, R.D., Renegar, R.H., Thatcher, W.W., Roberts, R.M. and Bazer, F.W., 1982. Establishment of pregnancy in the pig: 1. Interrelationships between preimplantation development of the pig blastocyst and uterine endometrial secretions. Biol Reprod., 27: 925-939.

Gengenbach, D.R., Hixon, J.E. and Hansel, W., 1977. A luteolytic interaction between estradiol and prostaglandin $F2\alpha$ in hysterectomized ewes. Biol. Reprod., 16: 571-579.

Ginther, O.J., 1971. Response of corpora lutea to cauterization of follicles in sheep. Amer. J. Vet. Res., 32: 59-62.

Ginther, O.J., 1974. Internal regulation of physiological processes through local venoarterial pathways: A review. J. Anim. Sci., 39: 550-564.

Harper, M.J.K., Norris, C.J. and Rajkumar, K., 1983. Prostaglandin release by zygotes and endometria of pregnant rabbits. Biol. Reprod., 28: 350-362.

Hawk, H.W. and Bolt, D.J., 1970. Luteolytic effect of estradiol-17β when administered after midcycle in the ewe. Biol. Reprod., 2: 275-278.

Henderson, K.M., Scaramuzzi, R.J. and Baird, D.T., 1977. Simultaneous infusion of prostaglandin E_2 antagonizes the luteolytic action of prostaglandin $F_{2\alpha}$ in vivo. J. Endocrinol., 72: 379-383.

Horton, E.W. and Poyser, N.L., 1976. Uterine luteolytic hormone: A physiological role for prostaglandin $F_{2\alpha}$. Physiol. Rev., 56: 595-651.

Hoyer, G.L., Colcord, M.L. and Weems, C.W., 1978. Effects of prostaglandin E_1 on estrogen induced luteolysis in ewes. J. Anim. Sci., 47 (Suppl. 1): 367.

Huecksteadt, T.P. and Weems, C.W., 1978. Effect of PGE_1 on normal luteolysis and PGE_1 or PGE_2 in IUD-induced luteolysis in ewes. The Pharmacol., 20: 233-234.

Huie, J.M., Magness, R.R., Reynolds, L.P., Hoyer, G.L., Huecksteadt, T.P., Colcord, M.L., Stalcup, B., Whysong, G.L. and Weems, C.W., 1981. Effect of chronic ipsilateral and contralateral intrauterine infusion of prostaglandin E_1 (PGE_1) on luteal function of unilaterally ovariectomized ewes. Prostaglandins, 21: 945-955.

Hyland, J.H., Manns, J.G. and Humphrey, W.D., 1982. Prostaglandin production by ovine embryos and endometrium in vitro. J. Reprod. Fertil., 65: 299-304.

Inskeep, E.K., Smutny, W.J., Butcher, R.L. and Pexton, J.E., 1975. Effects of intrafollicular injections of prostaglandins in non-pregnant and pregnant ewes. J. Anim. Sci., 41: 1098-1104.

Inskeep, E.K. and Murdoch, W.J., 1980. Relation of ovarian functions to uterine and ovarian secretion of prostaglandins during the estrous cycle and early pregnancy in the ewe and cow. Int. Rev. Physiol., 22: 325-356.

Karsch, F.J., Noveroske, J.W., Roche, J.F., Norton, H.W. and Nalbandov, A.V., 1970. Maintenance of ovine corpora lutea in the absence of ovarian follicles. Endocrinology, 87: 1228-1236.

Kindahl, H., Edqvist, L.-E. and Lindell, J.O., 1980. On the control of prostaglandin release during the bovine estrous cycle. Adv. Prostaglandin and Thromboxane Res., 8: 1351-1355.

Koligian, K.B. and Stormshak, F., 1977a. Nuclear and cytoplasmic oestrogen receptors in ovine endometrium during the estrous cycle. Endocrinology, 101: 524-533.

Koligian, K.B. and Stormshak, F., 1977b. Progesterone inhibition of estrogen receptor replenishment in ovine endometrium. Biol. Reprod., 17; 412-416.

Lawson, R.A.S. and Cahill, L.P., 1983. Modification of the embryomaternal relationship in ewes by progesterone treatment early in the cycle. J. Reprod. Fertil., 67: 473-475.

Lacroix, M.C. and Kann, G., 1982. Comparative studies of prostaglandin $F_{2\alpha}$ and E_2 in late cyclic and early pregnant sheep: In vitro synthesis by endometrium and conceptus; Effects of in vivo indomethacin treatment on establishment of pregnancy. Prostaglandins, 23: 507-526.

Lewis, G.S., Wilson, L., Jr., Wilks, J.W., Pexton, J.E., Fogwell, R.L., Ford, S.P., Butcher, R.L., Thayne, W.V. and Inskeep, E.K., 1977. $PGF_{2\alpha}$ and its metabolites in uterine and jugular venous plasma and endometrium of ewes during early pregnancy. J. Anim. Sci., 45: 320-327.

Lewis, G.S., Jenkins, P.E., Fogwell, R.L. and Inskeep, E.K., 1978. Concentrations of prostaglandins E_2 and $F_{2\alpha}$ and their relationship to luteal function in early pregnant ewes. J. Anim. Sci., 47: 1314-1323.

Lewis, G.S., Thatcher, W.W., Bazer, F.W. and Curl, J.S., 1982. Metabolism of arachidonic acid in vitro by bovine blastocysts and endometrium. Biol. Reprod., 27: 431-439.

Louis, T.M., Parry, D.M., Robinson, J.S., Thorburn, G.D. and Challis, J.R.G., 1977. Effects of exogenous progesterone and oestradiol on prostaglandin F and 13,14-dihydro-15-OXO prostaglandin $F_{2\alpha}$ concentrations in uteri and plasma of ovariectomized ewes. J. Endocrinol., 73: 427-439.

Magness, R.R., Huie, J.M. and Weems, C.W., 1981. Effect of chronic ipsilateral or intrauterine infusion of prostaglandin E_2 (PGE_2) on luteal function of unilaterally ovariectomized ewes. Prostaglandins and Medicine, 6: 389-401.

Mapletoft, R.J., Del Campo, M.R. and Ginther, O.J., 1975. Unilateral luteotropic effect of uterine venous effluent of a gravid uterine horn in sheep. Proc. Soc. Exp. Biol. Med., 150: 129-133.

Mapletoft, R.J., Lapin, D.R. and Ginther, O.J., 1976. The ovarian artery as the final component of the local pathway betweeen a gravid uterine horn and ovary in ewes. Biol. Reprod., 15: 414-421.

Mapletoft, R.J., Miller, K.F. and Ginther, O.J., 1977. Effects of $PGF_{2\alpha}$ and PGE_2 on corpora lutea in ewes. J. Anim. Sci., 45 (Suppl. 1): 185.

Marcus, G.J., 1981. Prostaglandin formation by the sheep embryo and endometrium as an indication of maternal recognition of pregnancy. Biol. Reprod., 25: 56-64.

Marsh, J.M., 1970. The stimulatory effect of PGE_2 on adenyl cyclase in the bovine corpus luteum. FEBS Lett., 7: 283-286.

Martal, J., Lacroix, M.-C., Loudes, G., Saunier, M. and Wintenberger-Torres, S., 1979. Trophoblastin, an antiluteolytic protein present in early pregnancy in sheep. J. Reprod. Fertil., 56: 63-73.

McCracken, J.A., 1980. Hormonal receptor control of prostaglandin $F_{2\alpha}$ secretion by the ovine uterus. Adv. Prostaglandin and Thromboxane Res., 8: 1329-1344.

Moor, R.M. and Rowson, L.E.A., 1966. Local maintenance of the corpus luteum in sheep with embryos transferred to various isolated portions of the uterus. J. Reprod. Fertil., 12: 539-550.

Murdoch, W.J., Lewis, G.S., Inskeep, E.K. and Tillson, S.A., 1978. The effect of a progesterone-releasing intrauterine contraceptive device on uterine secretion of prostaglandin $F_{2\alpha}$ and lifespan of corpora lutea in the ewe. Amer. J. Obstet. Gynecol., 132: 81-86.

Nancarrow, C.D., Evinson, B.M. and Connell, P.J., 1982. Effect of embryo on luteolysis and termination of early pregnancy in sheep with cloprostenol. Biol. Reprod., 26: 263-269.

Ottobre, J.S., Lewis, G.S., Thayne, W.V. and Inskeep, E.K., 1980. Mechanism by which progesterone shortens the estrous cycle of the ewe. Biol. Reprod., 23: 1046-1053.

Pratt, B.R., Butcher, R.L. and Inskeep, E.K., 1977. Antiluteolytic effect of the conceptus and of PGE_2 in ewes. J. Anim. Sci., 46: 784-791.

Rexroad, C.E., Jr., 1981a. Estrogen and progestogen binding in the myometrium of the ewe. I. During the estrous cycle. J. Anim. Sci., 53: 1057-1069.

Rexroad, C.E., Jr., 1981b. Estrogen and progestogen binding in the myometrium of the ewe. II. Regulation by estradiol and progesterone. J. Anim. Sci., 53: 1070-1076.

Reynolds, L.P., Stigler, J., Hoyer, G.L., Magness, R.R., Huie, J.M., Huecksteadt, T.P., Whysong, G.L., Behrman, H.R. and Weems, C.W., 1981. Effect of PGE_1 or PGE_2 on $PGF_{2\alpha}$-induced luteolysis in nonbred ewes. Prostaglandins, 21: 957-972.

Scaramuzzi, R.J., Baird, D.T., Boyle, H.P., Land, R.B. and Wheeler, A.G., 1977. The secretion of prostaglandin F from the autotransplanted uterus of the ewe. J. Reprod. Fertil., 49: 157-160.

Shaikh, A.A., Birchall, K. and Saksena, S.K., 1973. Steroids in the ovarian venous plasma and F prostaglandins in the peripheral plasma during pseudopregnancy and days 1-9 of pregnancy in the golden hamster. Prostaglandins, 4: 17-30.

Sharma, S.C. and Fitzpatrick, R.J., 1974. Effect of oestradiol-17β and oxytocin treatment on prostaglandin F alpha release in the anoestrous ewe. Prostaglandins, 6: 97-105.

Shemesh, M., Milaguir, F., Ayalon, N. and Hansel, W., 1979. Steroidogenesis and prostaglandin synthesis by cultured bovine blastocysts. J. Reprod. Fertil., 56: 181-185.

Silvia, W.J., Ottobre, J.S. and Inskeep, E.K., 1981. Concentrations of prostaglandins E_2, $F_{2\alpha}$ and 6-keto prostaglandin $F_{1\alpha}$ in the uteroovarian venous plasma of early pregnant and cycling ewes. Biol. Reprod., 24 (Suppl. 1): 99A.

Stone, G.M., Murphy, L. and Miller, B.G., 1978. Hormone receptor levels and metabolic activity in the uterus of the ewe: Regulation by oestradiol and progesterone. Australian J. Biol. Sci., 31: 395-403.

Thorburn, G.D., Nicol, D.H., Bassett, J.M., Shutt, D.A. and Cox, R.I., 1972. Parturition in the goat and sheep: Changes in corticosteroids, progesterone, oestrogens, and prostaglandin F. J. Reprod. Fertil., Suppl. 16: 61-84.

Villa-Godoy, A., Ireland, J.J., Wortman, J.A., Ames, N.K. and Fogwell, R.L., 1981. Luteal function in heifers following destruction of ovarian follicles at three stages of diestrus. J. Anim. Sci., 53 (Suppl. 1): 366A.

Wilson, L., Jr., Butcher, R.L., Cenedella, R.J. and Inskeep, E.K., 1972. Effects of progesterone of endometrial prostaglandins in sheep. Prostaglandins, 1: 183-190.

Wilson, L., Jr. and Stanisic, D., 1981. Effect of progesterone and estradiol-17β on uterine PGF, PGE, TxB_2 and 6-keto-$PGF_{1\alpha}$ levels in the pregnant rat uterus. Biol. Reprod., 24 (Suppl. 1): 159A.

Wilson, L., Jr., Stanisic, D., Khan-Dawood, F. and Dawood, M.Y., 1982. Alterations in reproductive tissue prostaglandins E and F, 6-keto prostaglandin F1α and thromboxane B_2 with gestational age in the rat. Biol. Reprod., 27; 1207-1215.

Zelinski, M.B., Hirota, N.A., Keenan, E.J. and Stormshak, F., 1980. Influence of exogenous estradiol-17β on endometrial progesterone and estrogen receptors during the luteal phase of the ovine estrous cycle. Biol. Reprod., 23: 743-751.

Animal Reproduction Science, 7 (1984) 101—113
Elsevier Science Publishers B.V., Amsterdam — Printed in The Netherlands

OVARIAN OXYTOCIN AND LUTEAL FUNCTION IN THE EARLY PREGNANT SHEEP

E.L. Sheldrick and A.P.F. Flint

A.R.C. Institute of Animal Physiology, Babraham, Cambridge CB2 4AT, (UK).

ABSTRACT

Sheldrick, E.L. and Flint, A.P.F., 1984. Ovarian oxytocin and luteal function in the early pregnant sheep. Anim. Reprod. Sci., 7: 101-113.

INTRODUCTION

Although its exact significance remains to be assessed, there is a good deal of evidence implicating oxytocin in the control of luteal function, both at luteal regression and during the establishment of pregnancy. The luteolytic effects of administered oxytocin in the cow (Armstrong and Hansel, 1959; Hansel and Wagner, 1960), goat (Cooke and Knifton, 1981) and possibly the sheep (Milne, 1963), and the release of uterine prostaglandin $F_{2\alpha}$ ($PGF_{2\alpha}$) in response to oxytocin treatment (Sharma and Fitzpatrick, 1974; Mitchell et al., 1975; McCracken et al., 1981) point to a role for oxytocin in the control of luteolysis, and this is supported by the rise in uterine concentrations of the oxytocin receptor at the end of the non-pregnant cycle in the ewe (Roberts et al., 1976). Furthermore, episodes of release of oxytocin-neurophysin (Fairclough et al., 1980) or of oxytocin itself (Flint and Sheldrick, 1983a) accompany those of uterine $PGF_{2\alpha}$ at the end of the non-pregnant cycle. However, the best evidence of a role for oxytocin in causing luteal regression is probably that obtained from animals immunized against oxytocin, in which luteal regression is delayed (Sheldrick et al., 1980; Schams et al., 1982a); this suggests that endogenous oxytocin may be involved, and that it may act via the systemic circulation.

Factors causing luteal regression at the end of the luteal phase of a non-pregnant cycle are obviously relevant in respect to the establishment of preg-

0378-4320/84/$03.00

nancy, when luteal involution does not occur. It is therefore important, particularly in relation to the study of early embryonic growth, to identify the mechanisms by which such factors are prevented from acting in pregnant animals. The recent discovery that oxytocin is secreted by the corpus luteum itself has profound implications in the context of this process. We shall review the contribution made by the corpora lutea to circulating oxytocin concentrations in early pregnancy; mechanisms by which luteal oxytocin secretion may be controlled; and the consequences of the loss of luteal oxytocin which occurs at this time.

Luteal contribution to circulating oxytocin during the oestrous cycle and early pregnancy

Measurements of oxytocin in jugular venous plasma obtained once daily from non-pregnant sheep show that concentrations rise with the formation of the corpus luteum and fall at luteal regression (Fig. 1); similar results have been obtained with different breeds of domestic sheep using different radioimmunoassay methods (Sheldrick and Flint, 1981; Webb et al., 1981; Schams et al., 1982b). In non-pregnant sheep treated in midcycle with a luteolytic dose of cloprostenol (I.C.I. 80996, Estrumate) there is a transient increase in the circulating concentration of oxytocin, followed by a drop, as the corpora lutea regress (Flint and Sheldrick, 1983a).

These observations suggest that the ovary may secrete oxytocin. This is confirmed by measurements of oxytocin in ovarian venous and utero-ovarian venous blood (Flint and Sheldrick, 1983 a, b); on days 10-14 of the cycle concentrations of oxytocin in ovarian venous plasma exceed those in utero-ovarian venous and in femoral or carotid arterial plasma by factors of 5-8 and 10-100, respectively. The high concentration of oxytocin present in luteal tissue (Wathes and Swann, 1982; Flint and Sheldrick, 1982a) is consistent with its secretion by the corpus luteum, and this is confirmed in experiments in which oxytocin is measured in blood draining both luteal and non-luteal ovaries (Fig. 2).

Although it appears most likely that the corpus luteum contributes significantly to the circulating concentration of oxytocin, the proportion accounted for by luteal secretion is uncertain. During the luteal phase of the cycle, jugular venous oxytocin levels exceed those at oestrus by a factor of approximately 2, suggesting that the ovary may secrete 50% of oxytocin in the circulation; this

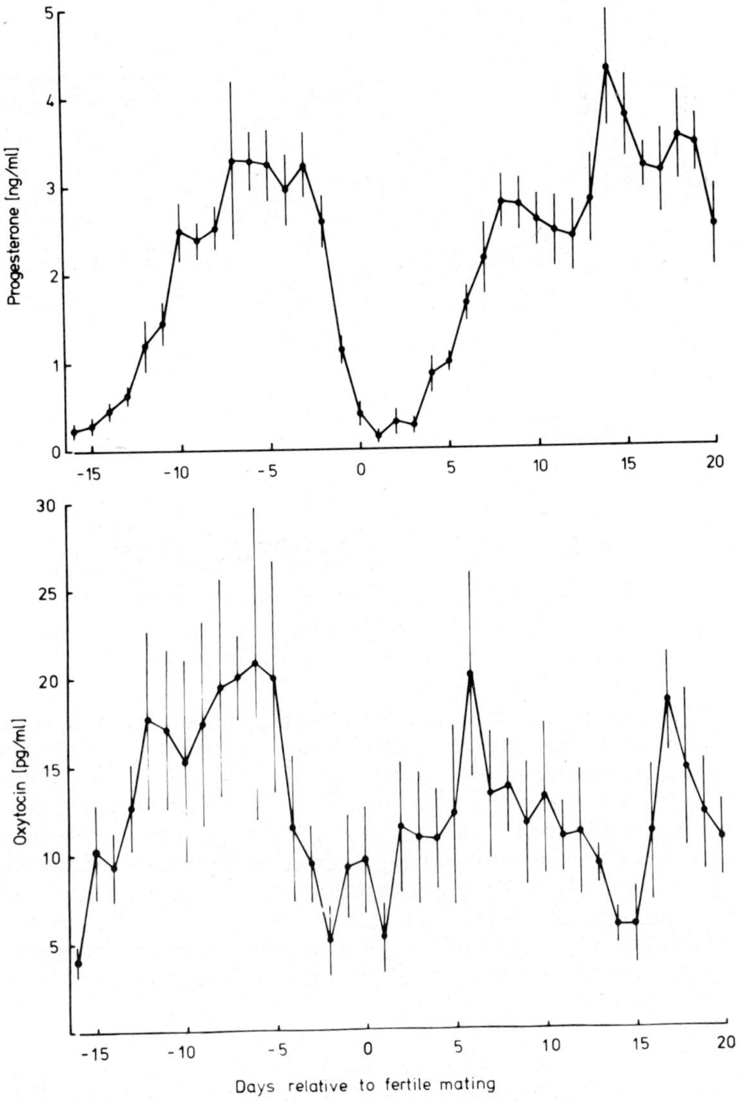

Fig. 1. Concentrations of progesterone and oxytocin in jugular venous plasma. Samples were obtained daily between 1400 and 1600 from 9 sheep during one infertile oestrous cycle (after mating to a vasectomized ram), and in 8 of these sheep after mating to an intact ram. Values are means ± SEM. All pregnant ewes delivered live lambs at normal term. (From Sheldrick and Flint, 1983a with permission).

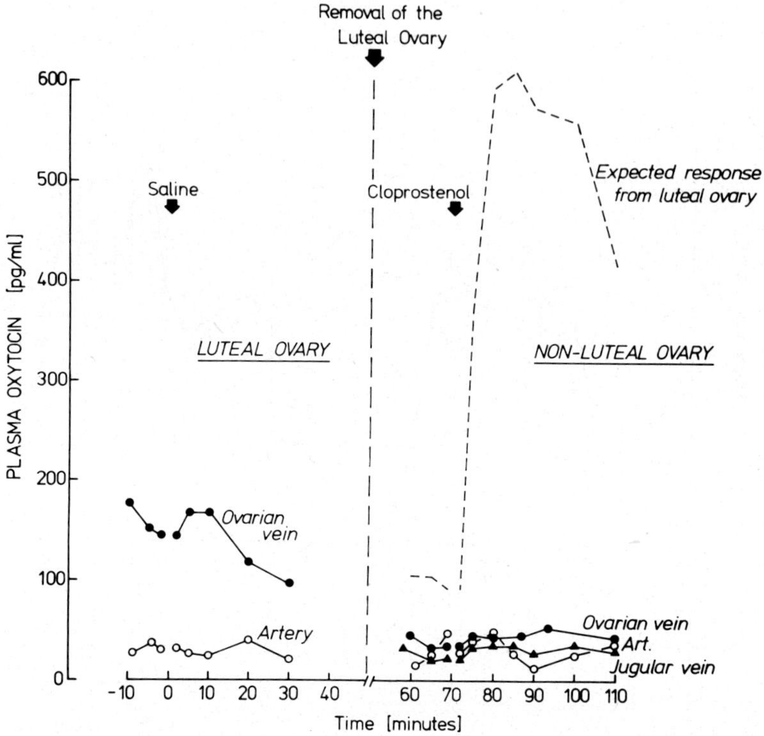

Fig. 2. Control experiment in which oxytocin release from a luteal ovary in response to saline, and from a non-luteal ovary in response to cloprostenol, was measured in an anaesthetized sheep on day 14 of the cycle. Catheters were inserted into both ovarian veins and into a femoral artery and a jugular vein. Samples were then collected from the vein draining the ovary with a corpus luteum, before and after administering a control injection of saline (left hand side of figure). The luteal ovary was subsequently removed, after which samples were collected from the opposite, non-luteal, ovary before and after administering cloprostenol (125 µg, i.m.). Levels of oxytocin normally present in the vein draining a luteal ovary before and after giving cloprostenol are shown as a dotted line.

would be consistent with the drop of approximately 50% observed in jugular venous levels after luteal regression induced in mid-cycle with cloprostenol (see above). Others, however, have found lower relative levels of oxytocin at oestrus, implying that a greater proportion may be secreted by the ovary (Schams et al., 1982b); this may in part be due to use of plasma to blank standard curves in the radioimmunoassay.

Determination of total blood production rates of oxytocin from metabolic clearance rates (MCR) and circulating concentrations allows an estimation to be made of the quantity of oxytocin entering the circulation; values obtained are of the order of 0.5 µg/h (MCR, 300 ml/min x jugular venous plasma concentration, 30 pg/ml; Flint and Sheldrick, 1983a). Production rates computed in this way exceed by approximately 10-fold ovarian secretion rates determined from arteriovenous concentration differences across the ovary and probable ovarian blood flow (ovarian secretion = (220-30) pg/ml x 4 ml/min = 0.046 µg/h). Possible reasons for this discrepancy have been dealt with elsewhere (Flint and Sheldrick, 1983a); however there is no clear reason for it, and more needs to be done to determine the proportion of circulating oxytocin accounted for by luteal secretion.

In early pregnancy circulating concentrations of oxytocin rise with the formation of the corpus luteum; levels reached are no lower than during a non-fertile cycle (Fig. 1). There is a significant fall in oxytocin on days 14 and 15 of pregnancy, and elevated progesterone levels at this time, followed by a return to raised oxytocin levels on days 16-18. In late pregnancy (8-32 days pre-partum) circulating concentrations of oxytocin are low (mean ± SEM, 6.2 ± 0.5 pg/ml, n = 115 samples obtained from the jugular vein). It is not certain at what stage of gestation circulating concentrations drop to those at term.

Identity of luteal oxytocin

Materials cross reacting in radioimmunoassays for oxytocin have been extracted from corpora lutea and identified by a variety of biological and physical techniques, including HPLC (Fig. 3), rat mammary pressure and uterine strip bioassays and gel filtration chromatography (Wathes and Swann, 1982). The immunoassay activity in ovarian venous plasma has also been identified by bioassay (Fig. 4). Further compelling evidence of identity is provided by the demonstration that the ovary secretes neurophysin I/II at a rate equal, mol:mol, to the rate of oxytocin secretion (Watkins et al., 1983; Fig.5). Arginine vasopressin is present in the human corpus luteum (Wathes et al., 1982), but there is no evidence for its secretion by the ovary in the sheep (Flint and Sheldrick, 1983b; Watkins et al., 1983). Thus although it has not been purified and sequenced, there is good evidence that the luteal peptide is authentic oxytocin.

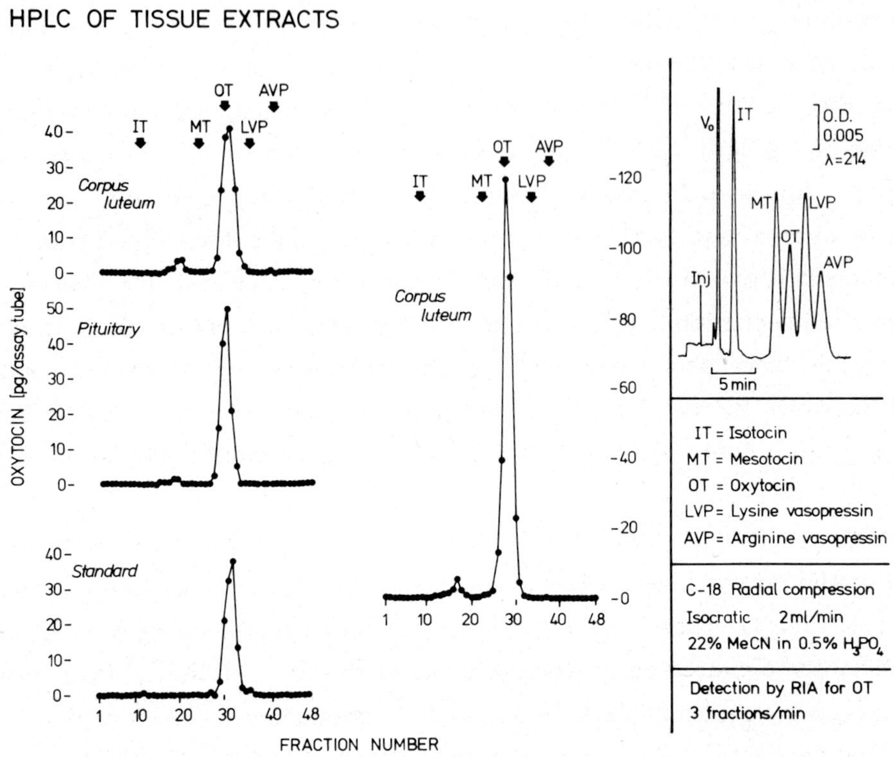

Fig. 3. Identification of oxytocin in corpus luteum and pituitary extracts using HPLC and radioimmunoassay. Acid extracts of corpora lutea and pituitary were concentrated by adsorption to and elution from octadecylsilica (ODS) cartridges (Sep-Paks; Water Associates) before application to HPLC (Waters Associates Radial-Pak ODS column run isocratically in 22% acetonitrile in 0.5% H_3PO_4, pH 2.1, 1.5 ml/min). Samples of effluent collected every 20 sec were assayed for oxytocin, after drying under N_2, by radioimmunoassay (Flint and Sheldrick, 1983a). Inset shows separation of standards; IT, isotocin ([Ser^4, Ile^8] oxytocin); MT, mesotocin ([Ile^8] oxytocin); OT, oxytocin; LVP, lysine vasopressin ([Phe^3, Lys^8] oxytocin); AVP, arginine vasopressin ([Phe^3, Arg^8] oxytocin). Data are shown for 2 extracts of corpora lutea from cycling sheep, one extract of posterior pituitary and a standard solution of oxytocin added to a blank extract.

Luteal oxytocin during early pregnancy

As shown in Fig. 6, concentrations of oxytocin are low in corpora lutea when their function is maintained beyond the expected time of luteal regression (Sheldrick and Flint, 1983 a, b). This applies whether luteal maintenance is caused by hysterectomy or pregnancy. The decline in luteal oxytocin concentration coincides closely with, or shortly follows, the time at which the corpora lutea would

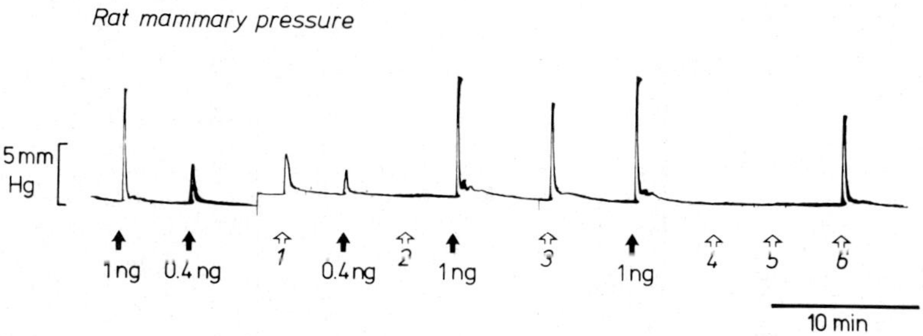

Fig. 4. Rat mammary pressure bioassay for oxytocin activity. Samples of ovarian venous plasma from ewes treated with cloprostenol were administered to a lactating rat via a jugular vein catheter, under pentobarbitone anaesthesia. Administration of oxytocin standards are shown by closed arrows. Open arrows indicate: 1, ovarian venous plasma from a sheep treated with cloprostenol (0.5 ml, equivalent to 0.5 ng oxytocin by radioimmunoassay); 2, carotid arterial plasma taken at the same time as sample 1 (equivalent to 25 pg); 3, ovarian venous plasma from a second sheep, after cloprostenol (0.48 ng); 4, jugular venous plasma taken at same time as sample 3 (15 pg); 5, ovarian venous plasma after administration of cloprostenol to a hysterectomized sheep (35 pg); 6, same sample as 5 after adding 1 ng oxytocin. (From Flint and Sheldrick (1983b) with permission).

normally regress. Luteal concentrations of oxytocin have also been shown to be low when luteal function is prolonged in the presence of a non-gravid uterus, by immunization against either $PGF_{2\alpha}$ or oxytocin (A.P.F. Flint and E.L. Sheldrick, unpublished observations). This drop in oxytocin does not appear to be due to the disappearance from the corpora lutea of the large luteal cells, which are responsible for oxytocin secretion (Rodgers et al., 1983), because these are detectable in pregnancy as well as in corpora lutea from hysterectomized sheep; at present therefore the mechanism causing the decline in luteal oxytocin under these conditions is unknown.

Prostaglandin E_2 and the secretion of luteal oxytocin

Measurement of circulating concentrations of oxytocin and progesterone during early gestation showed that oxytocin levels declined on days 14 and 15 after mating, as discussed above (Fig. 1). This decline in oxytocin, which was accompanied by a transient rise in progesterone level, may facilitate the prolongation of luteal function in view of the luteolytic actions of this peptide (see the Introduc-

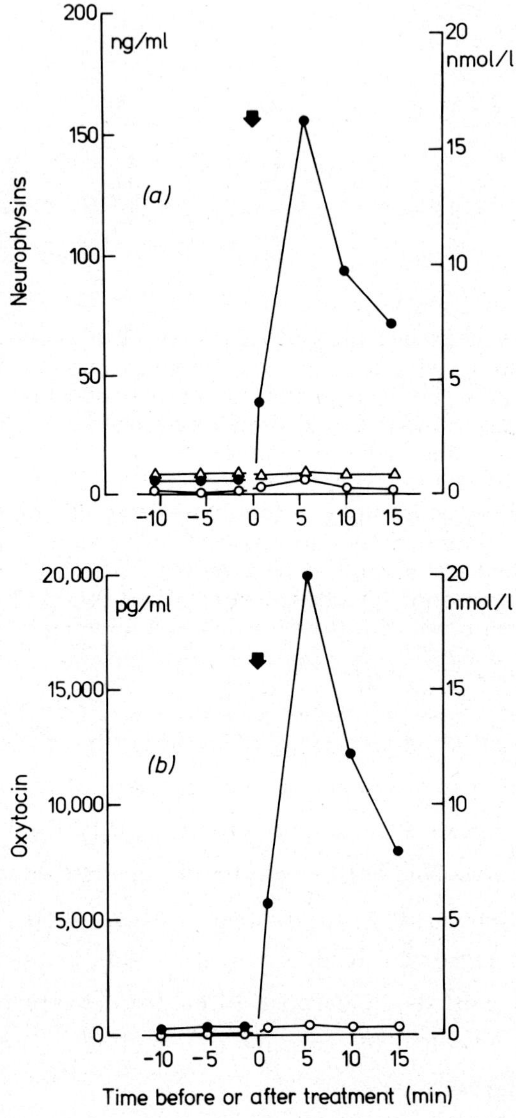

Fig. 5. Concentrations of oxytocin and neurophysins I/II and III in an anaesthetized non-pregnant sheep treated with cloprostenol (125 µg, i.m., at t = 0). Oxytocin and neurophysin I/II were measured in plasma from the ovarian vein (●) and femoral artery (○) and neurophysin III was determined in ovarian venous plasma (△), using specific radioimmunoassays. Mean (± SEM) molar ratio of oxytocin: neurophysin I/II secreted (i.e. corrected for levels recirculating in arterial plasma) was 1.17 ± 0.24. (From Watkins et al., 1983, with permission).

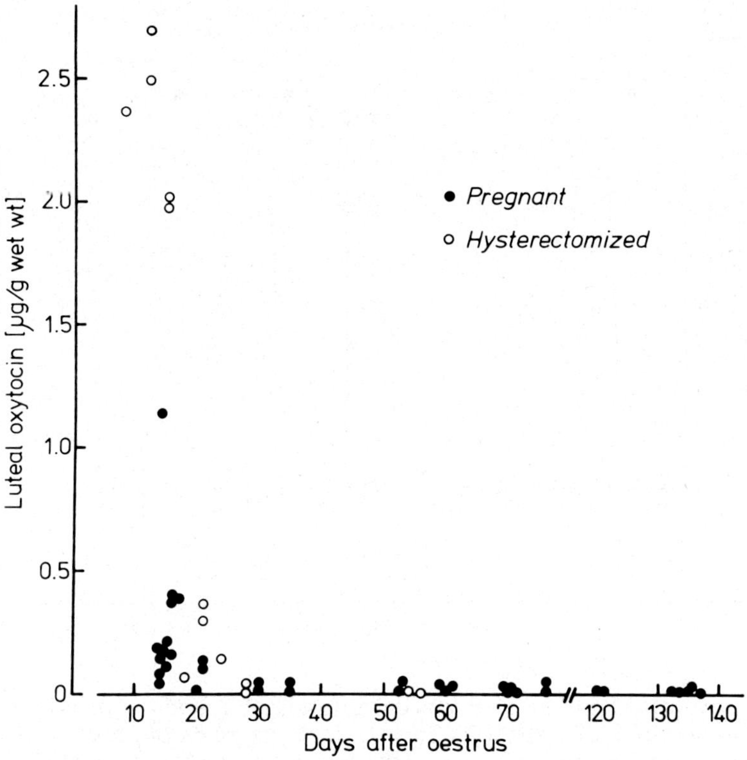

Fig. 6. Luteal concentrations of oxytocin measured by radioimmunoassay in pregnant and hysterectomized ewes. ●, Levels in pregnancy; ○, after hysterectomy. Maintenance of luteal function after hysterectomy was confirmed by progesterone assay. Abscissa indicates time since last oestrus; hysterectomy was performed on day 6 after oestrus.

tion). The hypothesis was tested, that uterine secretion of prostaglandin E_2 is involved in regulating luteal oxytocin, since concentrations of PGE_2 increase in early pregnancy (Lewis et al., 1978; Ellinwood et al., 1979) and this prostaglandin has been shown to be luteotrophic (Pratt et al., 1977). Furthermore, the stimulatory effect of the $PGF_{2\alpha}$ analogue, cloprostenol, on luteal secretion of oxytocin (Flint and Sheldrick, 1982b), raised the possibility that other prostaglandins may also influence this process. To test whether PGE_2 affected luteal oxytocin secretion, ewes were treated with the PGE_2 analogue, 15(S),15-methyl-PGE_2 (Fig. 7). When 15(S),15-methyl-PGE_2 was administered to 2 non-pregnant ewes on day 11 of the cycle, oxytocin secretion was stimulated; jugular venous plasma concen-

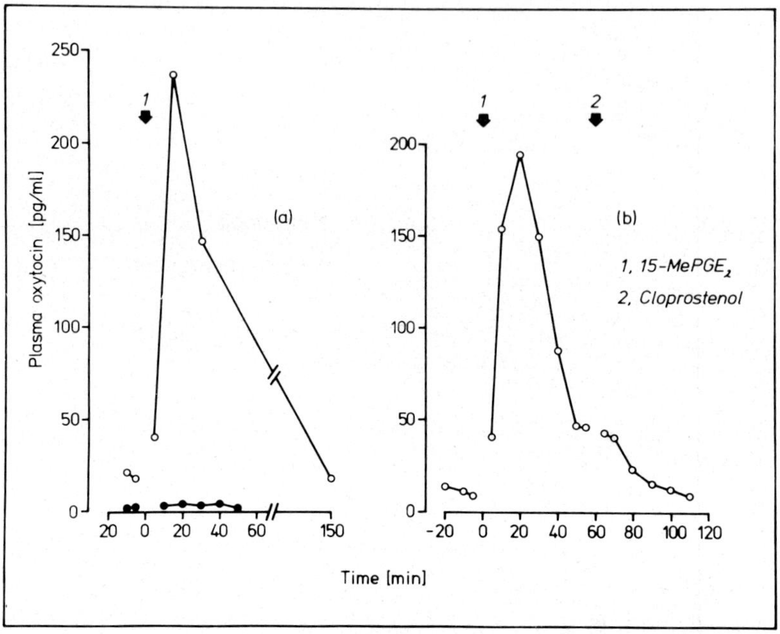

Fig. 7. Effects of 15(S),15-methylprostaglandin E_2 and cloprostenol on jugular venous concentrations of oxytocin. In (a), 2 intact ewes on day 11 of the oestrous cycle (○) or 2 ovariectomized ewes (●) were treated with 15(S),15-methylprostaglandin E_2 (125 µg, i.m.); in (b) 2 animals on days 9 and 11 of the cycle received 15(S),15-methylprostaglandin E_2 followed by cloprostenol (125 µg, i.m.). Times of administration of prostaglandins are indicated by arrows. Values are means for results from 2 animals in each case.

trations of oxytocin were significantly raised within 5 min. There was no release of oxytocin in ovariectomized animals (Fig. 7a). This effect resembled that observed with cloprostenol; however, in contrast to the action of the $PGF_{2\alpha}$ analogue, 15(S),15-methyl-PGE_2 did not cause luteal regression, and the intact ewes returned to oestrus on days 16 and 18 after the previous mating (to a vasectomized ram). In a further 2 animals treated with 15(S),15-methyl-PGE_2 followed by cloprostenol, peripheral oxytocin levels were raised after administration of the first prostaglandin, and no further release was caused by cloprostenol (Fig. 7b). These animals returned to oestrus 48 h after treatment. It appeared therefore, that ovarian stimulation by a PGE_2 analogue may cause at least temporary depletion of luteal oxytocin without inducing luteal regression. How long the corpora lutea remained depleted of oxytocin under these conditions is uncertain; such depletion may however be consistent with the temporary decline in circulating level found on days 14 and 15 of pregnancy.

Importance of luteal oxytocin in the establishment of pregnancy

As reviewed in the Introduction, several lines of evidence suggest that oxytocin is luteolytic, and the loss of luteal oxytocin in pregnancy and after hysterectomy, which is associated with luteal maintenance, is in keeping with its luteolytic effects. In experiments in which oxytocin has been administered to ewes between days 10 and 14 of pregnancy (25 I.U., s.c. twice per day) a return to oestrus occurred by day 20 (R.T. Duby and J.P. Tritschler II, pers. comm.) and this is consistent with a luteolytic effect of oxytocin at this time, either via loss of the embryo through increased uterine contractility, or by a more direct action on progesterone secretion.

The time at which oxytocin is lost from the corpus luteum may also be significant in relation to the mechanisms maintaining luteal function early in gestation. The antiluteolytic substance (named trophoblastin by Martal et al., 1979) which is present in embryo extracts on days 15 and 16 is lost from the embryo by day 21 (Rowson and Moor, 1967), as indicated by failure of extracts from day 21 embryos to prevent luteolysis when infused into the uterus of non-pregnant sheep. It is tempting to speculate that loss of embryonic antiluteolysin is tolerated because of the disappearance of luteal oxytocin and consequent reduction in 'luteolytic drive'.

We would propose therefore that there are at least 2 mechanisms operating to ensure maintenance of luteal function in early gestation in the sheep: the production of an embryonic antiluteolysin, which acts between approximately days 12 and 18; and a subsequent longer-lived decline in luteal oxytocin which becomes important after day 18. Superimposed on these mechanisms there may be embryonic secretion of PGE_2; this may be the cause of the reduced circulating levels of oxytocin on days 14 and 15, and lead to the transient rise in progesterone level observed at that time.

REFERENCES

Armstrong, D.T. and Hansel, W., 1959. Alteration of the bovine estrous cycle with oxytocin. J. Dairy Sci., 42: 533-542.

Cooke, R.G. and Knifton, A., 1981. Oxytocin-induced oestrus in the goat. Theriogenology, 16: 95-97.

Ellinwood, W.E., Nett, T.M. and Niswender, G.D., 1979. Maintenance of the corpus luteum of early pregnancy in the ewe. II. Prostaglandin secretion by the endometrium in vitro and in vivo. Biol. Reprod., 21: 845-856.

Fairclough, R.J., Moore, L.G., McGowan, L.T., Petersen, A.J., Smith, J.F., Tervit, H.R. and Watkins, W.B., 1980. Temporal relationships between plasma concentrations of 13,14-dihydro-15-keto-prostaglandin F and neurophysin I/II around luteolysis in sheep. Prostaglandins, 20: 199-208.

Flint, A.P.F. and Sheldrick, E.L., 1982a. Ovarian secretion of oxytocin in the sheep. J. Physiol. (Lond.), 330: 61P-62P.

Flint, A.P.F. and Sheldrick, E.L., 1982b. Ovarian secretion of oxytocin is stimulated by prostaglandin. Nature (Lond.), 297: 587-588.

Flint, A.P.F. and Sheldrick, E.L., 1983a. Evidence for a systemic role for ovarian oxytocin in luteal regression in sheep. J. Reprod. Fert., 67: 215-225.

Flint, A.P.F. and Sheldrick, E.L., 1983b. Secretion of oxytocin by the corpus luteum in sheep. In: B.A. Cross and G. Leng (Editors), The Neurohypophysis: Structure, Function and Control. Elsevier Biomedical Press, Amsterdam, (in press).

Hansel, W. and Wagner, W.C., 1960. Luteal inhibition in the bovine as a result of oxytocin injections, uterine dilatation, and intrauterine infusions of seminal and preputial fluids. J. Dairy Sci., 43: 796-805.

Lewis, G.S., Jenkins, P.E., Fogwell, R.L. and Inskeep, E.K., 1978. Concentrations of prostaglandins E_2 and $F_{2\alpha}$ and their relationship to luteal function in early pregnant ewes. J. Anim. Sci., 47: 1314-1323.

Martal, J., Lacroix, M.-C., Loudes, C., Saunier, M. and Wintenberger-Torres, S., 1979. Trophoblastin, an antiluteolytic protein present in early pregnancy in sheep. J. Reprod. Fert., 56: 63-73.

McCracken, J.A., Schramm, W., Barcikowski, B. and Wilson, L., 1981. The identification of prostaglandin $F_{2\alpha}$ as a uterine luteolytic hormone in the sheep and the endocrine control of its synthesis. Acta vet. scand., Suppl. 77: 71-88.

Milne, J.A., 1963. Effects of oxytocin on the oestrous cycle of the ewe. Aust. vet. J., 39: 51-52.

Mitchell, M.D., Flint, A.P.F. and Turnbull, A.C., 1975. Stimulation by oxytocin of prostaglandin F levels in uterine venous effluent in pregnant and puerperal sheep. Prostaglandins, 9: 47-56.

Pratt, B.R., Butcher, R.L. and Inskeep, E.K., 1977. Antiluteolytic effect of the conceptus and of PGE_2 in ewes. J. Anim. Sci., 46: 784-791.

Roberts, J.S., McCracken, J.A., Gavagan, J.E. and Soloff, M.S., 1976. Oxytocin stimulates release of prostaglandin $F_{2\alpha}$ from ovine endometrium in vitro: correlation with estrous cycle and oxytocin receptor binding. Endocrinology, 99: 1107-1114.

Rodgers, R.J., O'Shea, J.D., Findlay, J.K., Flint, A.P.F. and Sheldrick, E.L., 1983. Large luteal cells the source of luteal oxytocin in the sheep. Endocrinology (in press).

Rowson, L.E.A. and Moor, R.M., 1967. The influence of embryonic tissue homogenate infused into sheep uterus on the life-span of the corpus luteum. J. Reprod. Fert., 13: 511-516.

Schams, D., Prokopp. A. and Schmidt-Adamopoulou, B., 1982a. The effect of active immunization against oxytocin on ovarian cyclicity in ewes. Acta endocr., Suppl. 7: 246, (Abstract).

Schams, D., Lahlou-Kassi, A. and Glatzel, P., 1982b. Oxytocin concentrations in peripheral blood during the oestrous cycle and after ovariectomy in two breeds of sheep with low and high fecundity. J. Endocr., 92: 9-13.

Sharma, S.C. and Fitzpatrick, R.J., 1974. Effect of oestradiol-17β and oxytocin treatment on $PGF_{2\alpha}$ release in the anoestrous ewe. Prostaglandins, 6: 97-105.

Sheldrick, E.L., Mitchell, M.D. and Flint, A.P.F., 1980. Delayed luteal regression in ewes immunized against oxytocin. J. Reprod. Fert., 59: 37-42.

Sheldrick, E.L. and Flint, A.P.F., 1981. Circulating concentrations of oxytocin during the estrous cycle and early pregnancy in sheep. Prostaglandins, 22: 631-636.

Sheldrick, E.L. and Flint, A.P.F., 1983a. Luteal concentrations of oxytocin decline during early pregnancy in the ewe. J. Reprod. Fert., 68: 477-480.

Sheldrick, E.L. and Flint, A.P.F., 1983b. Regression of the corpora lutea in sheep in response to cloprostenol is not affected by loss of luteal oxytocin after hysterectomy. J. Reprod. Fert., 68: 155-160.

Wathes, D.C. and Swann, R.W., 1982. Is oxytocin an ovarian hormone? Nature (Lond.), 297: 225-227.

Wathes, D.C., Swann, R.W., Pickering, B.T., Porter, D.G., Hull, M.G.R. and Drife, J.O., 1982. Neurohypophysial hormones in the human ovary. The Lancet, August 21, 410-412.

Watkins, W.B., Moore, L.G., Flint, A.P.F. and Sheldrick, E.L., 1983. Secretion of neurophysin I/II and oxytocin by the ovary in sheep. Peptides (in press).

Webb, R., Mitchell, M.D., Falconer, J. and Robinson, J.S., 1981. Temporal relationships between peripheral plasma concentrations of oxytocin, progesterone and 13,14-dihydro-15-keto-prostaglandin $F_{2\alpha}$ during the oestrous cycle and early pregnancy in the ewe. Prostaglandins, 22: 443-453.

Animal Reproduction Science, 7 (1984) 115—132
Elsevier Science Publishers B.V., Amsterdam — Printed in The Netherlands

EXOCRINE VERSUS ENDOCRINE SECRETION OF PROSTAGLANDIN $F_{2\alpha}$ IN THE CONTROL OF PREGNANCY IN SWINE

F.W. Bazer*, S. R. Marengo*, R.D. Geisert* and W.W. Thatcher

Departments of Animal Science* and Dairy Science, University of Florida, Gainesville, FL 32611, (USA).

ABSTRACT

Bazer, F.W., Marengo, S.R., Geisert, R.D. and Thatcher, W.W., 1984. Exocrine versus endocrine secretion of prostaglandin $F_{2\alpha}$ in the control of pregnancy in swine. Anim. Reprod. Sci., 7: 115-132.

INTRODUCTION

In the pig, restrictions are placed upon the extent to which energy is expended for support of endometrial secretory activity. In the absence of fertilized ova and blastocysts developing in synchrony with the uterine endometrium, the period of endometrial secretory activity is limited primarily to the mid to late luteal phase of the oestrous cycle. Endometrial secretory activity is dependent upon maintenance of functional corpora lutea (CL). The life span of the CL, in turn, is limited during the oestrous cycle by endometrial production of a uterine luteolytic factor which is assumed to be prostaglandin $F_{2\alpha}$ ($PGF_{2\alpha}$). Release of $PGF_{2\alpha}$ from non-pregnant uterine endometrium results in morphological regression of CL, progesterone secretion ceases and, therefore, progesterone dependent endometrial secretory activity is terminated in the late luteal phase of the oestrous cycle.

In pigs having normally developing blastocysts, endometrial-blastocyst interactions begin to occur by day 11 of pregnancy. Blastocysts signal their presence, presumably through oestrogen production, which results in CL maintenance and continued endometrial development and secretory activity. Species such as the pig, that have central or fusion implantation (Schlafke and Enders, 1975), depend upon endometrial histotroph (Amoroso, 1952) for a major portion of pregnancy.

0378-4320/84/$03.00

This chapter will discuss evidence for the theory that oestrogens, of blastocyst origin, allow for continued secretion of $PGF_{2\alpha}$ and other components of endometrial histotroph into the uterine lumen, i.e. in an exocrine direction, and prevents their release toward the uterine vascular bed, i.e. in an endocrine direction. This mechanism appears to be essential for the establishment and maintenance of pregnancy in pigs.

The oestrous cycle

Sexually mature female pigs have recurring oestrous cycles of 18 to 21 days throughout the year. Behavioral oestrus lasts for 24-72 h and ovulation occurs 36-42 h after onset of oestrus. Corpora lutea are well formed by day 4 or 5 of the oestrous cycle and progesterone secretion increase from that time to maximum production between days 12 and 14 of the oestrous cycle (Guthrie et al., 1972). Luteal regression begins on about day 15 in non-pregnant females and plasma progesterone concentrations decline rapidly to basal levels (1 ng/ml or less) by days 17 to 18, leading to recurrent oestrous cycles unless interrupted by pregnancy.

The pig uterine endometrium has been identified as the source of a substance (uterine luteolysin) which causes morphological regression of CL and cessation of progesterone secretion. Loeb (1923) first noted that hysterectomy of guinea pigs during the luteal phase of the oestrous cycle allowed prolonged CL maintenance. This was later demonstrated in pigs by Spies et al. (1958) and Du Mesnil Du Buisson and Dauzier (1959). Anderson et al. (1961) found that bilateral hysterectomy in the early to mid luteal phase of the oestrous cycle resulted in CL maintenance for periods equal to or longer than the 114 days of a normal pregnancy. Destruction of the endometrial epithelium and congenital absence of the uterine endometrium also leads to prolonged CL maintenance (Anderson et al., 1969).

Prostaglandin $F_{2\alpha}$ ($PGF_{2\alpha}$) acts as a luteolytic agent when injected into the uterine lumen or intramuscularly in swine on day 12 or later in the oestrous cycle (see Bazer et al., 1982). The observations that pig CL were refractory to $PGF_{2\alpha}$ until day 12 is consistent with reports that the CL are "autonomous" with respect to lack of a need for pituitary support for the first 12 days of the oestrous cycle. Henderson and McNatty (1975) suggested that pig CL may remain refractory to $PGF_{2\alpha}$ until LH begins to dissociate from luteal cell membrane receptors. At this time, on or about day 12, conformational changes within the luteal cell membrane may facilitate $PGF_{2\alpha}$ binding. The $PGF_{2\alpha}$ in turn, is then proposed to inactivate the adenyl cyclase system to inhibit progesterone secretion and activate lysosomal enzymes to cause morphological regression of the luteal cells. Guthrie and Rexroad (1981) reported that a single injection of hCG on day 12 of the cycle temporarily blocked luteolysis. However, daily gonadotropin replacement therapy will not prevent the return to oestrus (Du Mesnil Du Buisson, 1966).

Endometrial extracts from days 13 to 17 of the oestrous cycle and day 19 of pregnancy exert a luteolytic effect in unilaterally pregnant gilts having induced CL (Christenson and Day, 1972). Patek and Watson (1976) reported that PGF production in vitro was greater for endometrial tissues from the mid and late luteal phase, compared with tissue obtained earlier in the oestrous cycle, and that this production could be blocked by indomethacin. Watson and Patek (1979) reported similar results in that $PGF_{2\alpha}$ production which was similar for endometrium from the mid-luteal stage of the oestrous cycle and early pregnancy. Guthrie et al. (1978) also demonstrated in vitro PGF production by pig endometrium representing days 7-8 and 15-16 of the oestrous cycle, but they did not detect a difference in PGF production between the two stages. Prostaglandin F is produced in vitro by pig endometrium taken from either day 8, 12, 14, 16 or 18 of the oestrous cycle (Guthrie and Rexroad, 1981), but only production values for days 16 and 18 were significantly greater than those for days 8, 12 and 14.

In support of these in vitro studies, other research has been directed toward determining concentrations of immunoreactive PGF in utero-ovarian vein plasma. Gleeson et al. (1974), Killian et al. (1976), Moeljono et al. (1977) and Frank et al. (1977) all concluded that utero-ovarian vein plasma PGF concentrations were greater during the period of expected luteolysis in non-pregnant gilts. Halton and First (unpublished data) reported that intrauterine infusion of indomethacin resulted in a prolonged luteal phase of the oestrous cycle in the pig. This effect of indomethacin could be overcome by intramuscular injection of $PGF_{2\alpha}$ which indicated that treatment had no effect on CL sensitivity to the luteolytic effect of exogenous $PGF_{2\alpha}$.

Collectively, available data suggest that endometrial production of PGF is greater during the period of expected luteolysis and that exogenous $PGF_{2\alpha}$ is luteolytic in gilts by day 12 after onset of oestrus.

Pregnancy, corpus luteum, oestrogens

In pregnant gilts, embryos move from the oviducts and into the uterine horns at about the 4-cell stage, i.e. about 60-72 h after onset of oestrus. Embryos reach the blastocyst stage by day 5 and emerge from the zona pellucida (hatching) between days 6 and 7. Blastocysts expand from about 0.5-1 mm diameter at hatching, to 2-6 mm diameter on day 10 and then elongate rapidly to a thread-like organism 700-1000 mm in length by days 14-16 of pregnancy (Perry and Rowlands, 1962; Anderson, 1978; Geisert et al., 1982b). Perry et al. (1973) first demonstrated conversion of tritium-labelled dehydroepiandrosterone, androstenedione, progesterone and oestrone-sulphate to unconjugated oestradiol and oestrone by pig blastocysts. These findings have been extended and confirmed by several investigators (see Gadsby et al., 1980). Geisert et al. (1982b) estimated that the transition period from large spherical blastocysts to the filamentous form represents the time that close apposition is achieved between blastocysts and endometrium (Crombie, 1972) and oestrogen production begins (Geisert et al., 1982a).

In order to closely examine the relationship between stage of blastocyst development and oestrogen production, uterine flushings were obtained from pregnant and non-pregnant gilts between days 10.5 and 12 after onset of oestrus. Uterine flushings were obtained from one uterine horn from all females on day 10.5 and the second uterine horn was flushed on either day 11, 11.5 or 12 of the oestrous cycle. In examining the data, it was obvious that total recoverable oestrogens in uterine flushings were closely associated with stage of blastocyst development. Data in Table I are, therefore, organized and summarized relative to the stage of blastocyst development. Total recoverable oestrone (E_1), oestradiol (E_2) and oestrone sulphate (E_1S) were greater in flushings having tubular and filamentous blastocysts. However, there was no discernible change in total recoverable E_1, E_2 or E_1S in uterine flushings from days 10.5, 11, 11.5 and 12 of the oestrous cycle. It should be noted that total recoverable E_2 is greater than E_1 in uterine flushings containing filamentous blastocysts. Zavy et al. (1980) also found total recoverable E_2 to be greater than E_1 in uterine flushings recovered on days 6, 10 and 12 of pregnancy, but total E_1 was greater than E_2 in uterine flushings collected on days 14, 15, 16 and 18 of gestation. Observations on total recoverable E_1, E_2 and E_1S, in uterine flushings from day 14 of pregnancy indi-

Table I. Changes in oestrogens, calcium, protein and prostaglandins $F_{2\alpha}$ and E_2 ($\bar{x} \pm SEM$) in uterine flushings from pregnant and non-pregnant gilts

Item	Non-pregnant Gilts, day [a]					Pregnant Gilts (Blastocyst size, mm/day)[b,c]				
						5.0	5.0-8.8	Tubular	Filamentous	
	10.5	11.0	11.5	12.0	14	10.5	10.5-11.0	11.0-12.0	11.5-12.0	14
Total E_1, ng	0.4 ± 0.1	0.4 ± 0.1	0.1 ± 0.04	0.4 ± 0.01	0.2 ± 0.03	0.5 ± 0.1	0.5 ± 0.1	0.8 ± 0.2	2.8 ± 1.0	0.6 ± 0.1
Total E_2, ng	0.4 ± 0.1	0.5 ± 0.3	0.2 ± 0.04	0.4 ± 0.2	0.2 ± 0.1	0.7 ± 0.3	0.6 ± 0.1	2.3 ± 0.5	4.4 ± 1.6	0.4 ± 0.1
Total E_1S, ng	0.9 ± 0.3	0.9 ± 0.5	0.6 ± 0.1	0.5 ± 0.1	1.6 ± 0.8	0.8 ± 0.2	1.3 ± 0.4	3.4 ± 0.8	2.2 ± 0.3	0.8 ± 0.2
Total Ca^{++}, mg	0.2 ± 0.1	0.5 ± 0.1	0.2 ± 0.1	0.4 ± 0.1	0.5 ± 0.2	0.1 ± 0.1	0.6 ± 0.2	1.5 ± 0.2	0.8 ± 0.2	0.1 ± 0.1
Total protein, mg	13 ± 3	8 ± 2	16 ± 6	13 ± 2	55 ± 10	13 ± 3	11 ± 2	31 ± 9	35 ± 4	69 ± 12
Total PGF, ng	2 ± 0.2	2 ± 0.2	3 ± 1	3 ± 1	4 ± 2	1 ± 0.4	3 ± 0.3	6 ± 1	27 ± 13	108 ± 23
Total PGE_2, ng	1 ± 0.4	1 ± 0.3	1 ± 0.3	2 ± 0.4	3 ± 0.2	3 ± 1	16 ± 6	24 ± 6	66 ± 19	197 ± 2

[a] For non-pregnant gilts day trends were significant for total protein ($P < 0.01$) and PGE_2 ($P < 0.05$).
[b] For pregnant gilts trends related to blastocyst size were significant ($P < 0.01$) for total E_2, E_1S, protein, calcium, PGF and PGE_2.
[c] Day by status (pregnant vs non-pregnant) interactions were significant for total E_2 and E_1S ($P < 0.05$), total protein ($P < 0.10$) and total calcium, PGF and PGE_2 ($P < 0.01$).

cate a decrease for all of these oestrogens relative to values found on day 12 (early filamentous blastocysts). This implies that a transient increase in oestrogen production accompanies the transition from the spherical to early filamentous stages of pig blastocyst development and then declines to days 14 to 16.

Assuming that free oestrogens provide the signal for maternal recognition of pregnancy in pigs and that this signal is provided on days 11 and 12 of pregnancy, injection of exogenous oestrogens should allow for continued CL maintenance (luteostasis) and uterine function. It has been demonstrated that injection of oestrogen into gilts between days 11 and 15 of the oestrous cycle does result in luteal maintenance (see Bazer et al., 1982). Since prolonged CL maintenance (pseudopregnancy) can be induced in gilts by administration of 5 mg oestradiol-valerate (E_2V) daily on days 11-15 of the oestrous cycle, Frank et al. (1977) studied utero-ovarian vein plasma PGF and peripheral plasma progestin concentrations in E_2V-treated and control gilts. Utero-ovarian vein plasma PGF concentrations were greater between days 12 and 17 than on either day 10 and 11 or 18 of the oestrous cycle for control pigs. Plasma progesterone concentrations decreased rapidly between days 13 and 15 and the average interoestrus interval for the control gilts was 19 ± 0.6 ($\bar{x}$ ± SEM) days. Mean utero-ovarian vein plasma PGF concentrations, number of PGF peaks (concentrations greater than $\bar{x}$ ± 2 S.D. for each gilt) and frequency of PGF peaks were all lower for E_2V-treated gilts than for control gilts. The mean interoestrus interval was 146.5 ± 74.8 days for E_2V-treated gilts.

Gleeson et al. (1974) first reported a temporal relationship between elevated PGF concentrations in utero-ovarian vein plasma of pigs and declining plasma progesterone concentrations associated with CL regression. Moeljono et al. (1977) later compared utero-ovarian vein plasma PGF and peripheral plasma progestins in non-pregnant and pregnant pigs. Utero-ovarian vein PGF concentrations were elevated ($P<0.01$) between days 13 and 17 of the oestrous cycle when plasma progestin concentrations were declining rapidly. For pregnant gilts, however, there were no significant changes in utero-ovarian vein PGF concentrations between days 12 and 25 of pregnancy. Killian et al. (1976) compared utero-ovarian vein plasma concentrations of PGF and progesterone between days 10 and 18 of the oestrous cycle and pregnancy. For non-pregnant gilts, a correlation coefficient of -0.96 ($P<0.01$) was reported for the temporal relationship between plasma progesterone and PGF concentrations in non-pregnant gilts. For pregnant gilts, plasma PGF concentrations were not affected by day of pregnancy and a correlation coefficient of only 0.02 was found between plasma progesterone and PGF concentrations. Collectively, data of Gleeson et al. (1974), Killian et al. (1976), Moeljono et al. (1977) and Frank et al. (1977) indicate a strong temporal relationship between elevated utero-ovarian vein plasma PGF concentrations and decreasing plasma progesterone concentrations associated with luteolysis. However, plasma progesterone and PGF were not related in pregnant gilts (Killian et al., 1976; Moeljono et al., 1977).

Subsequent to reports of Killian et al. (1976) and Moeljono et al. (1977), Ford and Christenson (1979) observed that uterine arterial blood flow increases 2- to 4-fold from day 11 (50 ml/min) to days 12 and 13 (150-200 ml/min) of pregnancy. By day 14 of pregnancy, uterine blood flow decreases slightly to 100-150 ml/min, but remains higher than for comparable days of the oestrous cycle. These data raise the question of whether or not utero-ovarian vein concentrations of $PGF_{2\alpha}$ are lower between days 12 and 16 of pregnancy because of the dilution effect of increased uterine blood flow. Data are not available which compare rate of PGF release from the pregnant and non-pregnant uterus. However, results of studies of the PGF metabolite, 15-keto-13,14-dihydro-$PGF_{2\alpha}$ (PGFM) in peripheral plasma of non-pregnant and pregnant gilts have been reported (Shille et al., 1979; Terqui et al., 1979; Guthrie and Rexroad, 1981). In these studies, samples were obtained from peripheral veins and PGFM concentrations reflect sampling from a plasma pool not affected by changes in uterine blood flow. Data from these studies support the conclusions of Kilian et al. (1976) and Moeljono et al. (1977) that there is increased PGF release from the uterus during the period of luteal regression. However, only basal prostaglandin metabolite concentrations were detected during comparable periods of pregnancy, except for occasional transient increases.

Collectively, these data indicate that there is greater release of PGF into the utero-ovarian vein during the period of luteolysis in non-pregnant pigs. However, release of PGF into the utero-ovarian vein of pregnant pigs, remains at a 'basal' level and CL maintenance results. However, there are opposing views. Ford et al. (1982a) attempt to convince one that the brief periods of elevated peripheral plasma PGFM reported by Shille et al. (1979) for pregnant gilts either on day 11,

12 to 13 or 12 are inconsistent with the concept of reduced secretion of $PGF_{2\alpha}$ into the uterine venous drainage. Their arguments are not well founded since they suggest that days 12 and 13 critical to maternal recognition of pregnancy, but they only consider this phenomenon with respect to CL function. As discussed earlier, pig CL are just becoming susceptible to luteolysis by day 12 and would likely be unaffected by transient increases in $PGF_{2\alpha}$ in the utero-ovarian vein blood. Ford et al. (1982a) do make the point that $PGF_{2\alpha}$ released may be diluted in utero-ovarian vein blood due to an increase in rate of uterine blood flow between days 11 and 13 of pregnancy. Ford et al. (1982b) also suggest that oestrogens of conceptus origin may be responsible for increased progesterone output by the CL. Oestrogen is known to reduce catecholamine concentrations in adrenergic nerves of the rat uterus (McKercher et al., 1973) and to stimulate mobilization of free calcium in the uterine endometrium (Geisert et al., 1982a). Both catecholamines and calcium are stimulatory to progesterone production by porcine luteal tissue (Veldhuis and Klase, 1982) and there is evidence that oestrogen may directly stimulate progesterone production by porcine granulosa cells (Goldenberg et al., 1972).

Hunter and Poyser (1982) compared $PGF_{2\alpha}$ in uterine vein plasma of pregnant and non-pregnant gilts between days 1 and 21 after onset of oestrus. All samples were collected from anesthetized gilts by either direct venipuncture or at 15 min intervals for 4 to 6 h after having inserted and fixed an indwelling catheter into the major uterine vein. They concluded that there was little difference in uterine vein plasma $PGF_{2\alpha}$ concentrations between non-pregnant and pregnant gilts and suggested that a "luteotrophic" agent of conceptus origin may be required for CL maintenance. Evidence for a porcine conceptus luteotrophin is not available; however, CL maintenance can be extended by injecting PMSG on day 14 (Hunter, 1964) or hCG on day 12 (Guthrie and Rexroad, 1981).

Data available from several laboratories indicate that $PGF_{2\alpha}$ in utero-ovarian vein plasma and PGFM in peripheral plasma is lower in pregnant than in non-pregnant gilts. These experiments raise the question of whether or not PGF production by the uterine endometrium is altered during pregnancy. An experiment was designed to examine total recoverable PGF in uterine flushings representing days 8, 10, 12, 14, 15, 16 and 18 of the oestrous cycle and pregnancy. Total recoverable PGF was greater ($P < 0.01$) in uterine flushings from pregnant gilts than from non-pregnant gilts (Zavy et al., 1980). For non-pregnant gilts, total recoverable PGF ($\bar{x}$ ± SEM) of 465 ± 38 ng on day 18 was maximum, compared with 22,688 ± 1772 ng on day 18 of pregnancy. These data indicate that PGF synthesis and secretion is not being inhibited during pregnancy and suggest that PGF of endometrial and/or conceptus origin is sequestered within the uterine lumen of pregnant gilts.

In a subsesquent experiment, Frank et al. (1978) examined total recoverable PGF in uterine flushings recovered on days 11, 13, 15, 17 and 19 of the oestrous cycle and E_2V-induced pseudopregnancy. Total recoverable PGF per uterine horn ($\bar{x}$ ± SEM) was maximal (210.2 ± 58.7 ng) on day 17 of the oestrous cycle. For pseudopregnant gilts, however, there was a linear increase in total recoverable PGF from 1.9 ± 0.1 ng on day 11 to 5113 ± 1735 ng on day 19. These data indicate that PGF secretion is not inhibited in pseudopregnant pigs. Rather, instead of PGF entering the utero-ovarian venous drainage, as appears to occur in non-pregnant gilts, PGF is sequestered within the uterine lumen as previously discussed for pregnant gilts. It should be noted that total recoverable PGF was only 5113 ng per uterine horn or about 10,226 ng for both horns, as compared to 22,688 ng from both uterine horns of pregnant gilts. This may reflect contributions of PGF by pig blastocysts as reported by Lewis and Waterman (1982).

Uteroferrin, histotroph

Uteroferrin is the purple colored progesterone-induced, intra-uterine glycoprotein of pig uterine secretions (histotroph) which appears to be involved in transport of iron from the uterine endometrium to the porcine conceptus (see review by Roberts and Bazer, 1980). Uteroferrin is synthesized and secreted by surface and glandular epithelium of pig uterine endometrium (Chen et al., 1975). In non-pregnant gilts, uteroferrin is secreted into the lumen of the uterine glands between days 9 and 13 of the oestrous cycle. Beginning on day 14, however, uteroferrin begins to become localized within the uterine endometrial stroma surrounding the basement membrane of uterine glands. It is possible that uteroferrin enters the interstitial fluid and/or vascular system within the endometrial stroma for transport to the spleen or liver for degradation and conservation of the iron. In contrast to the changing pattern of localization of uteroferrin in non-pregnant gilts, uteroferrin was not found in the endometrial stroma at any of the stages of pregnancy studied, i.e., days 6, 8, 12, 14, 16, 18, 30, 50, 70 and 90 of gestation. Uteroferrin is always localized within the epithelial cells and lumen of uterine glands and within the placental areolae after day 30 of pregnancy. Therefore, in the pregnant animal, uteroferrin, one component of uterine histotroph, continues to be secreted into the uterine lumen where it is available to provide a nutrient source to the developing conceptus for a major portion of gestation. These observations on uteroferrin localization by Chen et al. (1975) provided the basis for the question of whether or not oestrogens might prevent CL regression in the pig by affecting the direction of secretion of PGF from the uterine endometrium.

Theory of maternal recognition of pregnancy in the pig

In non-pregnant pigs it is proposed that progesterone enhances and/or induces PGF synthesis by the uterine endometrium and secretion is associated with elevated plasma oestradiol concentrations between days 12 and 18 of the oestrous cycle (Zavy et al., 1980). This oestrogen is likely to be of ovarian origin and is not present in adequate concentration in the endometrium to elicit a pseudopregnancy response. Secretion of PGF during the mid to late luteal phase of the oestrous cycle is primarily in an endocrine direction, i.e., toward the uterine venous drainage. This is based on the observations that utero-ovarian vein PGF concentrations are greatest between days 12 and 18 of the oestrous cycle when events leading to luteolysis are initiated. It has also been pointed out that there is little PGF accumulation within the uterine lumen during the oestrous cycle.

In pregnant pigs, it is also assumed that progesterone enhances and/or induces PGF synthesis by the uterine endometrium and that secretion is enhanced by markedly elevated oestrogen concentrations due to blastocyst production of oestrogens. However, oestrogens produced by pig blastocysts as early as day 11 (tubular and filamentous blastocysts) of pregnancy, alter the direction of movement of PGF so that its secretion remains in an exocrine direction, i.e., toward the uterine lumen. Maintenance of PGF secretion in an exocrine direction would prevent PGF from entering the uterine venous drainage and exerting a luteolytic effect on the CL. Of comparable significance is the fact that maintenance of histotroph secretion in an exocrine direction is essential for nourishment of the pig conceptus.

The proposed theory requires that mechanisms exist for altering the direction of movement of PGF from the epithelial cells of the uterine endometrium where it is assumed to be produced, and that PGF can be compartmentalized and sequestered when the uterus is under appropriate hormonal stimulation. It is known that $PGF_{2\alpha}$ can be compartmentalized and sequestered within the uterine fluids of pregnant and pseudopregnant pigs, allantoic fluid of pigs, uterine fluids of unilaterally pregnant ewes, uterine fluid and allantoic fluid of cows and within the uterine lumen of cows with pyometra (see Bazer et al., 1982). Data are not available to explain the mechanism whereby prostaglandins may be sequestered and compartmentalized within the uterus and conceptus.

In a recent study, we conducted two experiments. The first determined the effect of E_2V on movement of $PGF_{2\alpha}$, injected acutely into the uterine lumen, into the vasculature and the second experiment evaluated effects of PGE_2 and PGE_1 on movement of exogenous $PGF_{2\alpha}$ from the uterine lumen into the vasculature.

The first experiment was a 2 x 2 factorial involving 16 gilts assigned to the following groups (pre-treatment + uterine lumen injection): sesame oil + saline; sesame oil + $PGF_{2\alpha}$; E_2V + saline; or E_2V + $PGF_{2\alpha}$. On day 14 catheters were inserted about 10 cm into the lumen of each uterine horn via a small incision in the oviduct, each catheter was secured to the oviduct and the free ends exteriorized. Catheters were also placed into each radial vein. The E_2V-treated gilts received 5 mg E_2V on days 11 through 14 and control gilts received 0.5 ml sesame oil on each of those days. During the experiment, blood samples were drawn from one radial vein so that the plasma could be used to determine PGFM as an indirect measure of $PGF_{2\alpha}$ movement from the uterine lumen into the vasculature. During surgery blood samples were collected at the time the radial veins were catheterized, when the incision was made to begin the mid-ventral laparotomy, when the cervix was ligated and when the catheters were placed into the uterine horns (surgical period). The reproductive tract was returned to the abdominal cavity and blood samples were taken at 15 min intervals over the next 60 min or equilibration period. At the end of the equilibration period, there was an acute infusion into each uterine horn of either 5 ml of 0.9% saline or 50 µg $PGF_{2\alpha}$ in 5 ml saline. The first infusion was followed immediately with an additional 5 ml saline. The total infusion time was 30 to 50 sec. Blood samples were taken at 5 min intervals for the first 45 min after $PGF_{2\alpha}$ infusion and at 15 min intervals thereafter until 120 min. During the surgery and equilibration periods PGFM ($\bar{x} \pm$ SEM) was lower ($P < 0.05$) for E_2V gilts (518 ± 143 pg/ml) than sesame oil treated gilts (1998 ± 143 pg/ml). PGFM values were lower ($P < 0.01$) during the equilibration period (984 ± 112 pg/ml) than the surgical period (1531 ± 125 pg/ml) irrespective of treatment. However, PGFM concentrations were not different between these two periods for E_2V gilts (456 ± 158 vs 597 ± 179 pg/ml), but were different for control gilts (1512 ± 160 vs 2483 ± 1177 pg/ml). During the post-uterine $PGF_{2\alpha}$ injection interval, PGFM concentrations were also lower for E_2V-treated gilts irrespective of which uterine injection they received (E_2V + saline and E_2V + $PGF_{2\alpha}$, 438 ± 124 pg/ml vs sesame oil + saline and sesame oil + $PGF_{2\alpha}$, 842 ± 124 pg/ml). Least squares regression analyses indicated that plasma PGFM concentrations were lower immediately post-injection for E_2V gilts (1059 ± 215 pg/ml) than sesame oil gilts (1801 ± 222 pg/ml) and the slope of the post-uterine PGFM concentration changes was less for E_2V gilts.

These data indicate that E_2V-treated gilts apparently release less $PGF_{2\alpha}$ in response to physical handling of the reproductive tract during the surgical period and the rate of movement of exogenous $PGF_{2\alpha}$ from the uterine lumen was slower for E_2V-treated gilts. These results support the theory that an oestrogen effect reduces the movement of $PGF_{2\alpha}$ in an endocrine direction.

In pregnant pigs, there is a temporal relationship between blastocyst elongation and increasing quantities of PGE_2 and $PGF_{2\alpha}$, as well as, oestrogens in uterine flushings (Geisert et al., 1982a). Therefore, the second experiment was conducted to determine if PGE_2 and/or PGE_1 affected the movement of exogenous $PGF_{2\alpha}$ from the uterine lumen into the vasculature. The movement of $PGF_{2\alpha}$ into the vasculature was measured indirectly by determining peripheral plasma concentrations of PGFM. Four gilts were assigned to each of two 4 x 4 latin squares with gilts in one latin square being treated with Banamine (flunixine meglumine, donated by Schering Corp., Kenilworth, NJ). Results were essentially identical for Banamine-treated and saline injected gilts; therefore, only data from Banamine-treated gilts will be discussed.

On day 14 of the oestrous cycle Banamine was injected i.m. at 125 mg at 180 min and 45 min prior to the first $PGF_{2\alpha}$ infusion and at 135 min intervals thereafter. Intra-uterine infusions of treatment solutions and collection of blood samples from the saphenous artery utilized catheters which had been introduced on day 13 with the gilts under general anesthesia. All PGE treatments consisted of a total of either 50 μg PGE_1, 50 μg PGE_2 or 25 μg PGE_1 + 25 μg PGE2 in 5 ml phosphate buffered saline (PBS) infused into the uterine lumen of each horn 15 min prior to infusion of 25 μg $PGF_{2\alpha}$ in 0.9% saline. Blood samples were collected every 5 min from time (T) 0 min to 15 min (Pre-PGE) and from 15 min to 30 min (Post-PGE). The $PGF_{2\alpha}$ was infused at T30 min and blood samples collected every 2 min to T50 and then every 15 min to T75 min. Each gilt received each treatment sequence, i.e., PGE_1-$PGF_{2\alpha}$, PGE_2-$PGF_{2\alpha}$, PGE_1 + PGE_2-$PGF_{2\alpha}$ and PBS-$PGF_{2\alpha}$. Concentrations of PGFM ($\bar{x}$ ± SEM) were higher ($P < 0.10$) post-$PGF_{2\alpha}$ (793 ± 24 pg/ml) than pre-PGE (458 ± 49 pg/ml) or post-PGE (455 ± 50 pg/ml).

Treatment sequences involving PGE_1, PGE_2 and PGE_1 + PGE_2 had higher ($P < 0.05$) overall mean PGFM concentrations than that following PBS (820 ± 48, 885 ± 48 and 1019 ± 48 pg/ml vs 449 ± 48 pg/ml). In the PGE treatment sequences, PGFM concentrations began to increase ($P < 0.01$) immediately post-PGE infusion, while this did not occur until the post-$PGF_{2\alpha}$ period if a PBS infusion preceded the $PGF_{2\alpha}$ infusion. These data do not support the hypothesis that PGE_2 is involved in the mechanism whereby oestrogen causes $PGF_{2\alpha}$ to be sequestered within the uterine lumen.

Schneider et al. (1982) also failed to extend the interoestrus interval of gilts by infusing either 200 or 300 μg PGE_2 into the uterine lumen every 6 h each day between days 12 and 19 of the oestrous cycle. Data from our study suggest that PGE_2 may increase uterine blood flow (see Ford, 1982) and thereby increase the rate of transport of $PGF_{2\alpha}$ from the uterine lumen into the vasculature.

Significance of prostaglandins of the E and F series in the pregnant uterus

The accumulation of secretory vesicles in the apical portion of uterine endometrial epithelial cells occurs prior to about day 12 of pregnancy in pigs (Geisert et al., 1982a) and termination of delayed implantation in roe deer (Aitken, 1979). Release of these secretory vesicles is associated with onset of production of oestrogens by the blastocysts and an increase in free Ca^{++} in uterine flushings. Oestrogens have been shown to influence the movement of Ca^{++} in cells of rat endometrium (Pietras and Szego, 1977) and Ca^{++} is associated with release of secretory vesicles in cells of the adrenal medulla and cortex, salivary glands, pancreas, kidney and nerve cells (see Rubin and Laychock, 1978). It has been suggested that Ca^{++} activates phospholipase A2 to enhance release of arachidonic acid and lysophospholipids which are necessary for vesicular and plasma membrane fusion during exocytosis and endocytosis (Pollard et al., 1981). Rubin and Laychock (1978) also indicated an increase in both $PGF_{2\alpha}$ and PGE_2 during ACTH stimulation of the adrenal cortex. As summarized in table I, a close temporal association exists between initiation of oestrogen production, and increases in Ca^{++}, protein, $PGF_{2\alpha}$ and PGE2 in uterine flushings of pigs between days 10 and 12 of pregnancy. The process of exocytosis by uterine epithelial cells and endocytosis by trophectoderm (Geisert et al., 1982b) may require that phospholipase A_2 activate the arachidonic acid cascade to promote membrane fusion necessary for these events to occur (Fig. 1). This raises the question of whether or not production of prostaglandins of the E and F series is a by-product of oestrogen mediated endometrial epithelial secretory activity. It is also known that pig blastocysts produce $PGF_{2\alpha}$ and PGE_2 (Lewis and Waterman, 1982).

Prostaglandins exert an array of biochemical, physiological and endocrinological effects on cellular processes which may be critical to the establishment and maintenance of pregnancy. These effects include those on uterine blood flow (Ford, 1982), vascular permeability (Kennedy, 1980), fluid and electrolyte transport (Benos and Biggers, 1981), cellular proliferation (MacManus and Whitefield, 1974), steroid biosynthesis (Batta, 1975) and immunoprotection of the conceptus (Goodwin and Webb, 1980).

ACKNOWLEDGEMENTS

Research results from our laboratories reported herein have been supported by NIH Grants HD08560 and HD10436 and U.S.D.A. Cooperative Agreement 58-7B30-0-255. We thank Mrs. Cheryl Combs for preparation of this manuscript.

REFERENCES

Aitken, R.J., 1979. The hormonal control of implantation. In: J. Whelan (Editor), Maternal Recognition of Pregnancy. Excerpta Medica, Amsterdan, pp. 53-84.
Amoroso, E.C., 1952. Placentation. In: A.S. Parkes (Editor), Marshall's Physiology of Reproduction, Little, Brown and Co., Boston, pp. 127-311.

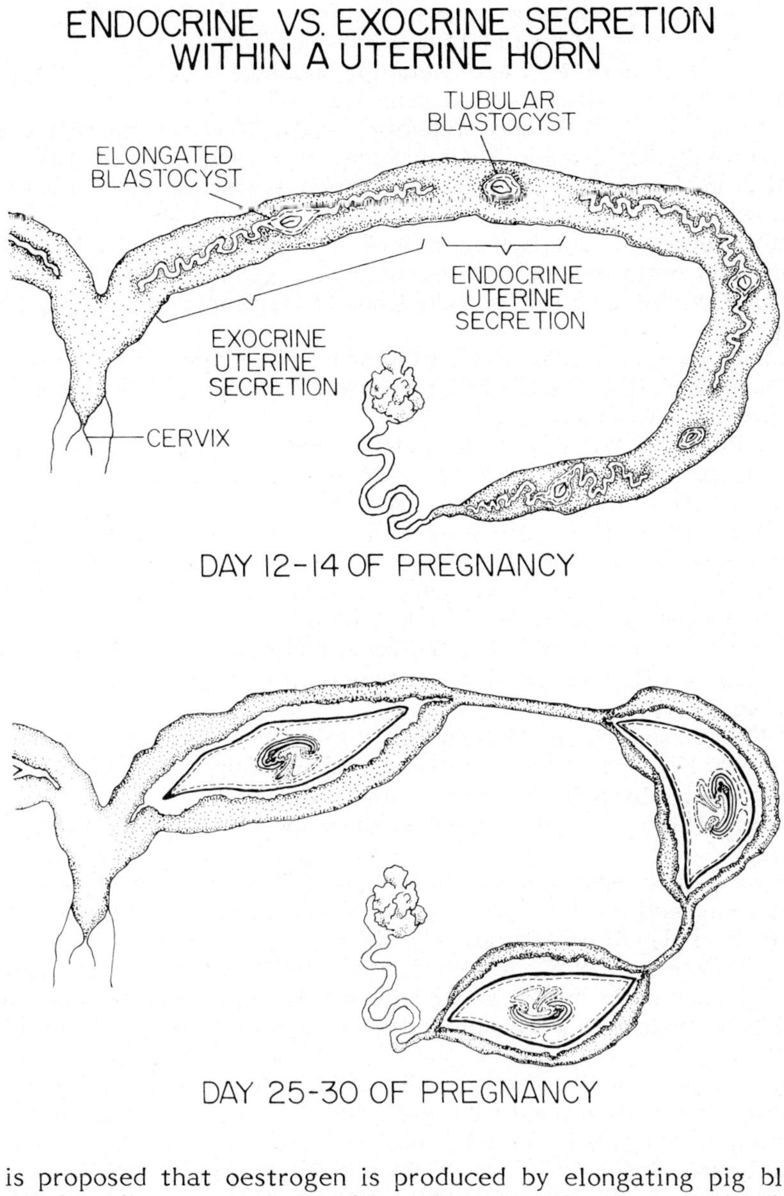

Fig.1. It is proposed that oestrogen is produced by elongating pig blastocysts which act on the adjacent uterine endometrium to maintain secretion in an exocrine direction. This prevents release of $PGF_{2\alpha}$ in an endocrine direction and ensures continued secretion of histotroph into the uterine lumen to support conceptus development. Endometrium associated with blastocysts not elongating normally and not producing oestrogens at the appropriate time (days 11-12) would not be stimulated to maintain secretions in an exocrine direction. Rather, $PGF_{2\alpha}$ and histotroph would move in an endocrine direction resulting in failure of the uterus to secrete and sequester $PGF_{2\alpha}$ and histotroph necessary for conceptus development. This may explain why conceptus tissues adjacent to an area of unoccupied uterine endometrium do not grow into that area as depicted at day 25-30 of gestation.

Anderson, L.L., 1978. Growth, protein content and distribution of early pig embryos. Anat. Rec., 190: 143-154.

Anderson. L.L., Butcher, R.L. and Melampy, R.M., 1961. Subtotal hysterectomy and ovarian function in gilts. Endocrinology, 69: 571-580.

Anderson. L.L., Bland, K.P. and Melampy, R.M., 1969. Comparative aspects of uterine-luteal relationships. Recent Prog. Horm. Res., 25: 57-104.

Batta, S.K., 1975. Effect of prostaglandins on steroid biosynthesis. J. Ster. Biochem., 6: 1075-1080.

Bazer, F.W., Geisert, R.D., Thatcher, W.W. and Roberts, R.M., 1982. The establishment and maintenance of pregnancy. In: D.S.A. Cole and G.R. Foxcraft (Editors), Control of Pig Reproduction. Butterworth Scientific, London, pp. 227-252.

Benos, D.J. and Biggers, J.D., 1981. Blastocyst fluid formation. In: L. Mastoianni and J.D. Biggers (Editors), Fertilization and Embryonic Development In Vitro. Plenum Press, New York, pp. 285-297.

Chen, T.T., Bazer, F.W., Gebhardt, B.M. and Roberts, R.M., 1975. Uterine secretion in mammals; Synthesis and placental transport of a purple acid phosphatase in pigs. Biol. Reprod., 13: 304-313.

Crombie, P.R., 1972. The morphology and ultrastructure of the pig's placenta throughout pregnancy. Ph.D. Dissertation. University of Cambridge.

Christenson, R.K. and Day, B.N., 1972. Luteolytic effects of endometrial extracts in the pig. J. Anim. Sci., 34: 620-626.

Du Mesnil Du Buisson, F., 1966. Contribution a l'etude de maintien du corps jaune de la truie, Ph.D. Dissertation. Institut National de la Recherche Agronomique, Nouzilly, France.

Du Mesnil Du Buisson, F. and Dauzier, L., 1959. Controle mutuel de l'uterus et de l'ovaire chez la truie. Annls. Zootech., Serie D, Suppl. 149-151.

Ford, S.P., 1982. Control of uterine and ovarian blood flow throughout the estrous cycle and pregnancy of ewes, sows and cows. J. Anim. Sci., 55, Suppl. II: 32-42.

Ford, S.P. and Christenson, R.K., 1979. Blood flow to uteri of sows during the estrous cycle and early pregnancy: Local effect of the conceptus on the uterine blood supply. Biol. Reprod., 21: 617-624.

Ford, S.P., Christenson, R.K. and Ford, J.J., 1982a. Uterine blood flow and uterine arterial, venous and lumenal concentrations of estrogens on days 11, 13 and 15 after estrus in pregnant and nonpregnant sows. J. Reprod. Fert., 64: 185-190.

Ford, S.P., Magness, R.R., Farley, D.B. and Van Orden, D.E., 1982b. Local and systemic, effects of intrauterine estradiol-17β on luteal function of nonpregnant sows. J. Anim. Sci., 55: 657-664.

Frank, M., Bazer, F.W., Thatcher, W.W. and Wilcox, C.J., 1977. A study of prostaglandin $F_{2\alpha}$ as the luteolysin in swine: III. Effects of estradiol valerate on prostaglandin F, progestins, estrone, and estradiol concentrations in the utero-ovarian vein of nonpregnant gilts. Prostaglandins, 14: 1183-1196.

Frank, M., Bazer, F.W., Thatcher, W.W. and Wilcox, C.J., 1978. A study of prostaglandin $F_{2\alpha}$ as the luteolysin in swine: IV. An explanation for the luteotrophic effect of estradiol. Prostaglandins, 15: 151-160.

Gadsby, J.E., Heap, R.B. and Burton, R.D., 1980. Oestrogen production by blastocyst and early embryonic tissue of various species. J. Reprod. Fert., 50: 409-417.

Geisert, R.D., Renegar, R.H., Thatcher, W.W., Roberts, R.M. and Bazer, F.W., 1982a. Establishment of pregnancy in the pig. I. Interrelationships between preimplantation development of the pig blastocyst and uterine endometrial secretions. Biol. Reprod., 27: 925-939.

Geisert, R.D., Brookbank, J.W., Roberts, R.M. and Bazer, F.W., 1982b. Establishment of pregnancy in the pig. II. Cellular remodelling of the porcine blastocyst during elongation on Day 12 of pregnancy. Biol. Reprod., 27: 941-955.

Gleeson, A.R., Thorburn, G.D. and Cox, R.J., 1974. Prostaglandin F concentrations in the utero-ovarian vein plasma of the sow during the late luteal phase of the estrous cycle. Prostaglandins, 5: 521-530.

Goldenberg, R.L., Bridson, W.E. and Kohler, P.B., 1972. Estrogen dimutation of progesterone synthesis by porcine granulosa cells in culture. Biochem. Biophys. Res. Commun., 48: 101-106.

Goodwin, J.S. and Webb, D.R., 1980. Regulation of the immune response by prostaglandins. Clin. Immunol. Immunopath., 15: 106-122.

Guthrie, H.D. and Rexroad, C.E., Jr., 1980. Progesterone secretion and prostaglandin F release in vitro by endometrial and luteal tissue of cyclic pigs. J. Reprod. Fert., 60: 157-163.

Guthrie, H.D., Henricks, D.M. and Handlin, D.L., 1972. Plasma estrogen, progesterone and luteinizing hormone prior to estrus and during early pregnancy in pigs. Endocrinology, 91: 675-679.

Guthrie, H.D., Rexroad, C.E., Jr. and Bolt, D.J., 1978. In vitro synthesis of endometrial tissue from the pig. Prostaglandins, 16: 433-440.

Guthrie, H.D. and Rexroad, C.E., 1981. Endometrial prostaglandin F release in vitro and plasma 13,14-dihydro-15-keto-prostaglandin $F_{2\alpha}$ in pigs with luteolysis blocked by pregnancy, estradiol benzoate or human chorionic gonadotropin. J. Anim. Sci., 52: 330-337.

Henderson, K.M. and McNatty, K.P., 1975. A biochemical hypothesis to explain the mechanism of luteal regression. Prostaglandins, 9: 779-797.

Hunter, R.H.F., 1964. Superovulation and fertility in the pig. Anim. Prod., 6: 189-194.

Hunter, R.H.F. and Poyser, N.L., 1982. Uterine secretion of prostaglandin $F_{2\alpha}$ in anesthetized pigs during the estrous cycle and early pregnancy. Reprod. Nutr. Develop., 22: 1013-1023.

Kennedy, T.G., 1980. Estrogen and uterine sensitization for the decidual cell reaction: role of prostaglandins. Biol. Reprod., 23: 955-962.

Killian, D.B., Davis, D.L. and Day, B.N., 1976. Plasma PGF and hormonal changes during the estrous cycle and early pregnancy in the gilt. Proc. Int. Pig Vet. Soc., Ames, Iowa, pp. D.1.

Lewis, G.S. and Waterman, R.A., 1982. Arachidonic acid metabolism by porcine blastocysts and endometrium. Biol. Reprod., 26, Suppl. 1: 64A. (Abstract).

Loeb, L., 1923. The effect of extirpation of the uterus on the life and function of the corpus luteum in the guinea pig. Proc. Soc. Exp. Biol. Med., 20: 441-464.

MacManus, J.P. and Whitfield, J.F., 1974. Cyclic AMP, prostaglandins and the control of cell proliferation. Prostaglandins, 6: 475-487.

McKercher, T.C., Van Orden, L.S., Bhatnagar, R.K. and Burke, J.P., 1973. Estrogen-induced biogenic amine reduction in the rat uterus. J. Pharmacol. Exp. Therap., 185: 514-517.

Moeljono, M.P.E., Thatcher, W.W., Bazer, F.W., Frank, M., Owens, L.J. and Wilcox, C.J., 1977. A study of prostaglandin $F_{2\alpha}$ as the luteolysin in swine: II. Characterization and comparison of prostaglandin F, estrogen and progestin concentrations in utero-ovarian vein plasma of nonpregnant gilts. Prostaglandins, 14: 543-555.

Patek, C.E. and Watson, J., 1976. Prostaglandin F and progesterone secretion by porcine endometrium and corpus luteum in vitro. Prostaglandins, 12: 97-111.

Perry J.S. and Rowlands, I.W., 1962. Early pregnancy in the pig. J. Reprod. Fert., 4: 175-188.

Perry, J.S., Heap, R.B. and Amoroso, E.C., 1973. Steroid hormone production by pig blastocysts. Nature, 245: 45-47.

Pietras, R.J. and Szego, C.M., 1977. Specific binding sites for estrogen at the outer surface of isolated endometrial cells. Nature, 265: 69-72.

Pollard, H.B., Pazoles, C.J. and Creutz, C.E., 1981. Mechanism of calcium action and release of vesicle-bound hormones during exocytosis. Rec. Prog. Horm. Res., 37: 299-315.

Roberts, R.M. and Bazer, F.W., 1980. The properties, function and hormonal control of synthesis of uteroferrin, the purple protein of the pig uterus. In: M. Beato (Editor), Steroid Induced Uterine Proteins, Elsevier North Holland Biomedical, Amsterdam, pp. 133-149.

Rubin, R.P. and Laychock, S.G., 1978. Prostaglandins and calcium-membrane interactions in secretory glands. Annals NY. Acad. Sci., 307: 377-390.

Schlafke, S. and Enders, A.C., 1975. Cellular basis of interaction between trophoblast and uterus at implantation. Biol. Reprod., 12: 41-65.

Schneider, T.M., Tilton, J.E., Okrasa, S., Jeng, M., Weigl, R. and Williams, G.L., 1982. Effect of intra-uterine infusions of prostaglandin E_2 (PGE_2) on luteal function in nonpregnant gilts. J. Anim. Sci., Suppl. 1: 388. (Abstract).

Shille, V.M., Karlbom, I., Einarsson, S., Larsson, K., Kindahl, H. and Edqvist, L.-E., 1979. Concentrations of progesterone and 15-keto-13,14-dihydroprostaglandin $F_{2\alpha}$ in peripheral plasma during the estrous cycle and early pregnancy in gilts. Zentbl. Vet. Med. A., 26: 169-181.

Spies, H.G., Zimmerman, D.R., Self, H.L. and Casida, L.E., 1958. Influence of hysterectomy and exogenous progesterone on size and progesterone content of the corpora lutea in gilts. J. Anim. Sci., 17: 1234. (Abstract).

Terqui, M., Martinat-Botte, F. and Thatcher, W.W., 1979. Early diagnosis of pregnancy in sows: Determination of $PGF_{2\alpha}$ levels on the 14th and 15th days after insemination. Journees de la recherche porcine en France, pp. 365-371.

Veldhuis, J.D. and Klase, P.A., 1982. Calcium ions modulate hormonally stimulated progesterone production in isolated ovarian cells. Biochem. J., 202: 381-386.

Watson, J. and Patek, C.E., 1979. Steroid and prostaglandin secretion by the corpus luteum, endometrium and embryos of cyclic and pregnant gilts. J. Endocr., 82: 425-428.

Zavy, M.T., Bazer, F.W., Thatcher, W.W. and Wilcox, C.J., 1980. A study of prostaglandin $F_{2\alpha}$ as the luteolysin in swine: V. Comparison of prostaglandin F, progestins, estrone and estradiol in uterine flushings from pregnant and nonpregnant gilts. Prostaglandins, 20: 837-851.

Animal Reproduction Science, 7 (1984) 133—148
Elsevier Science Publishers B.V., Amsterdam — Printed in The Netherlands

LEVELS OF PROSTAGLANDIN $F_{2\alpha}$ METABOLITES IN BLOOD AND URINE DURING EARLY PREGNANCY

H. Kindahl, S. Basu, G. Fredriksson, A. Goff[1], A. Kunavongkrit[2] and L.-E. Edqvist*

From the Departments of Obstetrics and Gynaecology and of Clinical Chemistry*, College of Veterinary Medicine, Swedish University of Agricultural Sciences, S-750 07 Uppsala, (Sweden).

ABSTRACT

Kindahl, H., Basu, S., Fredriksson, G., Goff, A., Kunavongkrit, A. and Edqvist, L.-E., 1984. Levels of prostaglandin $F_{2\alpha}$ metabolites in blood and urine during early pregnancy. Anim. Reprod. Sci., 7: 133-148.

INTRODUCTION

The quantitative estimation of prostaglandin $F_{2\alpha}$ ($PGF_{2\alpha}$) release requires reliable assay parameters. To monitor release of $PGF_{2\alpha}$ from the uterus several approaches are possible. The first attempts to evaluate the biosynthesis and release of $PGF_{2\alpha}$ in this organ were made by collecting plasma from its venous drainage for $PGF_{2\alpha}$ analysis. Data on $PGF_{2\alpha}$ release during the oestrous cycle and early pregnancy using such an approach is available from several species (for a review, see Horton and Poyser, 1976). Collection of such plasma samples requires surgical intervention and insertion of chronical catheters with subsequent risks of disturbing reproductive physiological events. The administration of large doses of heparin systemically to maintain patency of such catheters is very often employed, but should be avoided since heparin is lipolytic and the released

---

[1] On leave from the Centre de Recherche en Reproduction Animale, Faculté de Médecine Vétérinaire, Université de Montréal, Québec, Canada.

[2] On leave from Faculty of Veterinary Medicine, Chulalongkorn University, Bangkok, Thailand.

0378-4320/84/$03.00

fatty acids can disturb the radioimmunological measurements of prostaglandins (cf. Metz et al., 1979).

From studies on the metabolism of $PGF_{2\alpha}$ in the human (Granström, 1972) it became apparent that another approach to monitor $PGF_{2\alpha}$ release is to analyze the main initially formed metabolite of $PGF_{2\alpha}$, 15-keto-13,14-dihydro-$PGF_{2\alpha}$ (15-ketodihydro-$PGF_{2\alpha}$), in peripheral blood plasma. This approach avoids surgical cannulation of the utero-ovarian or uterine veins and permits long-term studies on reproductive events in undisturbed subjects (Peterson et al., 1975; Kindahl et al., 1976a). To obtain a reliable picture of $PGF_{2\alpha}$ release, blood samples must be collected very frequently, otherwise substantial peaks can be missed (Kindahl et al., 1981).

More detailed metabolic studies in the human have revealed that $PGF_{2\alpha}$ is initially metabolized to 15-ketodihydro-$PGF_{2\alpha}$; however the pattern of metabolites appearing with time in the peripheral circulation gradually began to resemble the spectrum of metabolites in the urine (Granström et al., 1982). The major metabolite in human plasma a few minutes after bolus i.v. injection of $PGF_{2\alpha}$ is an 11-ketotetranor PGF metabolite, 5α,7α-dihydroxy-11-ketotetranorprostane-1,16-dioic acid, which is also the major metabolite in human urine (Granström and Samuelsson, 1971). This more degraded metabolite is longer-lived and can be detected in peripheral plasma several hours after the exogenous injection or endogenous release of $PGF_{2\alpha}$ into the blood stream, while the initially formed 15-ketodihydro-$PGF_{2\alpha}$ disappears rapidly from the circulation (Granström et al., 1982). Large differences have been found in circulating prostaglandin metabolites in different animal species (Granström and Kindahl, 1982). However, in all studied species, 15-ketodihydro-$PGF_{2\alpha}$ is the initially formed metabolite, but is gradually replaced by a number of more degraded products, the profiles of which are relatively typical for each species. Since these so-called 11-ketotetranor PGF metabolites also appear in large quantities in urine, an alternative approach to studying $PGF_{2\alpha}$ release is to measure these metabolites in urine (Granström and Kindahl, 1976). This latter approach allows determination of $PGF_{2\alpha}$ release in totally undisturbed subjects for very long periods.

In studies on prostaglandin release it is thus important to investigate the prostaglandin metabolism in the species under study in order to make an adjustment between body fluid, sample frequency and assay parameter.

This paper examines some aspects on metabolism of PGF$_{2\alpha}$ and levels of PGF$_{2\alpha}$ metabolites during luteolysis and early pregnancy in the cow, guinea pig, mare, sow and doe.

Cow

[9β-^{3}H]-Prostaglandin F$_{2\alpha}$ was injected in male calves and a heifer to study the appearance of metabolites in plasma and urine (Kindahl, 1980; Granström and Kindahl, 1982). The dominating product in plasma was always found to be 15-ketodihydro-PGF$_{2\alpha}$. With time a few more degraded metabolites appeared in plasma in lower quantities. The tentative identities of these metabolites suggest four different tetranor metabolites (Granström and Kindahl, 1982). In the calves as much as 80% of the injected radioactivity appeared in urine in about 6 h (Fig. 1). Three major tetranor (C_{16}) metabolites were identified: 5α,7α,11-trihydroxytetranorprostane-1,16-dioic acid (main metabolite), 5α,7α-dihydroxy-11-ketotetranorprostane-1,16-dioic acid, 5α,7α,16-trihydroxy-11-ketotetranorprostanoic acid; and two minor tetranor, ω-dinor (C_{14}) metabolites: 5α,7α,11-trihydroxy(tetranor,ω-dinor)prostane-1,14-dioic acid and 5α,7α-dihydroxy-11-keto(tetranor,ω-dinor)prostane-1,14-dioic acid (Kindahl, 1980).

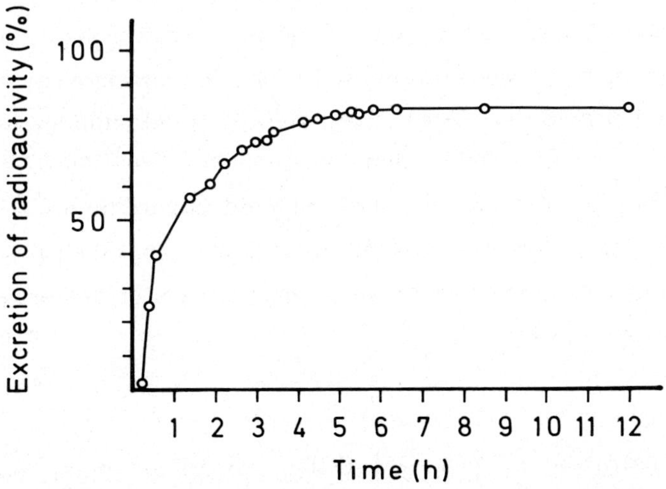

Fig. 1. Excretion of radioactivity in the urine after intravenous injection of 1 mg [9β-^{3}H]-PGF$_{2\alpha}$ (specific activity, 35 Ci/mol) in a male calf.

It could thus be expected that 15-ketodihydro-$PGF_{2\alpha}$ should be a suitable parameter of $PGF_{2\alpha}$ release in cattle, which is in contrast to the human (Granström et al., 1982). This concept was further verified by exogenous injection of unlabelled $PGF_{2\alpha}$ and measurement of 15-ketodihydro-$PGF_{2\alpha}$ and 11-keto-tetranor PGF metabolites by separate radioimmunoassays. From this kinetic experiment it was evident that the 15-keto metabolite always exceeded the levels of the more degraded C_{16} metabolites (Granström and Kindahl, 1982). A similar relationship between the two types of metabolites was also seen during luteolysis and 15-ketodihydro-$PGF_{2\alpha}$ proved to be the most reliable parameter of $PGF_{2\alpha}$ release in blood. It is, however, important that samples are collected frequently, otherwise a false idea of $PGF_{2\alpha}$ release could be formed (Kindahl et al., 1981).

The 15-ketodihydro-$PGF_{2\alpha}$ levels in a heifer followed by blood samples every third hour during luteolysis and the first 24 days of pregnancy are depicted in Fig. 2 (Kindahl et al., 1976a). During luteolysis $PGF_{2\alpha}$ is released as pulses of short duration during 2-3 days (Fig. 2). During the same period of early pregnancy the levels of 15-ketodihydro-$PGF_{2\alpha}$ are different and the luteolytic pulse release pattern is abolished. Also in other studies of early pregnant cows, although employing less frequent sampling schedules, we have been unable to record elevated prostaglandin levels (Kindahl et al., 1982a). One important difference between cycling and early pregnant animals is the lack of a pulsatile prostaglandin release pattern in the latter condition. This probably plays an important role in rescuing the corpus luteum (CL) function during early pregnancy.

As mentioned previously, large quantities of prostaglandin metabolites are excreted in the urine of cattle. Some metabolites are identified as 11-keto-tetranor PGF metabolites and can be analyzed according to a previously described method (Granström and Kindahl, 1976). This approach has been used for long-term studies during oestrous cycles, embryo transfer and pregnancy (Harvey et al., 1984, this publication).

Guinea pig

After administration of [9β-^{3}H]-$PGF_{2\alpha}$ to guinea pigs, 5α,7α-dihydroxy-11-ketotetranorprostanoic acid was found to be the dominating metabolite in urine (Granström and Samuelsson, 1969). When plasma samples were collected, low levels of radioactivity were seen, and most of this radioactivity was due to

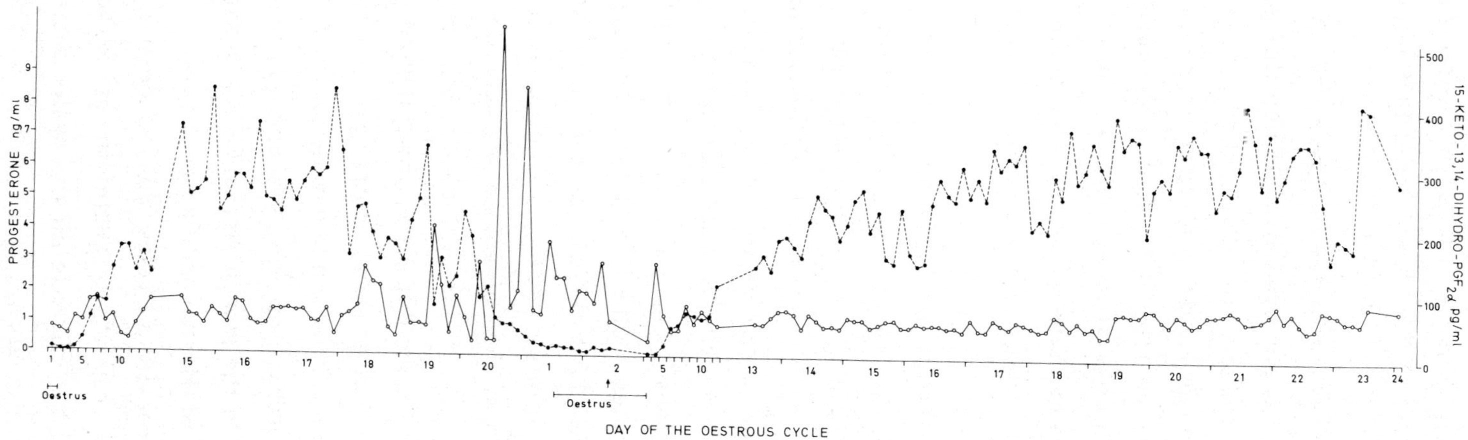

Fig. 2. Peripheral plasma levels of 15-ketodihydro-$PGF_{2\alpha}$ (○——○) and progesterone (●- - -●) during one oestrous cycle and early pregnancy in a heifer. Arrow denotes time of mating.

triated water (Granström and Kindahl, 1982). This latter finding indicates that the guinea pig converted part of the administered $PGF_{2\alpha}$ into PGE metabolites with the consequent loss of the 9β-^3H label.

The lipophilic material was dominated by 15-ketodihydro-$PGF_{2\alpha}$ early after injection of radioactive $PGF_{2\alpha}$ and even after 40 min this metabolite dominated the spectrum. More polar products also appeared with time and the above mentioned metabolite, $5\alpha,7\alpha$-dihydroxy-11-ketotetranorprostanoic acid, was rather prominent after 15-40 min, but later decreased in abundance. Some more polar metabolites were also seen in the latter samples, but they could not be identified due to insufficient amounts of material. At 90 min, which was the last time of sampling, no single product dominated the picture, but relatively large amounts of 15-ketodihydro-$PGF_{2\alpha}$ were seen.

Judging from this experiment, several possibilities could be considered when the best plasma parameter for monitoring $PGF_{2\alpha}$ release is sought. The best should be 15-ketodihydro-$PGF_{2\alpha}$ or its tetranor counterpart. However, the guinea pig has a relatively rapid clearance of $PGF_{2\alpha}$ or metabolites from the circulation, resulting in low levels of total metabolites in the blood stream. Furthermore collection of frequent blood samples is not easy without disturbing the animals. For these reasons it is possible that plasma measurement should be avoided entirely in this species. Collection of total daily urine and estimation of the major metabolite, $5\alpha,7\alpha$-dihydroxy-11-ketotetranorprostanoic acid, may be a better alternative for monitoring $PGF_{2\alpha}$ prodction in the guinea pig. This approach has been used in studies of guinea pig reproduction (Granström and Kindahl, 1976). The collection of urine in this species is simple. The animals are kept in individual cages and urine and feces are collected separately. The urine could be collected at room temperature, centrifuged and diluted prior to analysis. Long-term studies can easily be made with analysis of metabolites of $PGF_{2\alpha}$ in the urine. One major drawback of urinary analysis is that no pulsatile $PGF_{2\alpha}$ release can be monitored. In the guinea pig, however, pulsatile release of $PGF_{2\alpha}$ has never been demonstrated (Horton and Poyser, 1976).

A total number of 18 oestrous cycles from 11 oestrous female guinea pigs have been followed with respect to excretion of $5\alpha,7\alpha$-dihydroxy-11-keto-tetranorprostanoic acid in the urine. In all the cycles a marked increase of the metabolite levels was seen around days 11-13 of the cycle, and the highest levels were generally seen during the two last days of the cycle and the first day of the next cycle (Fig. 3) (Granström and Kindahl, 1976).

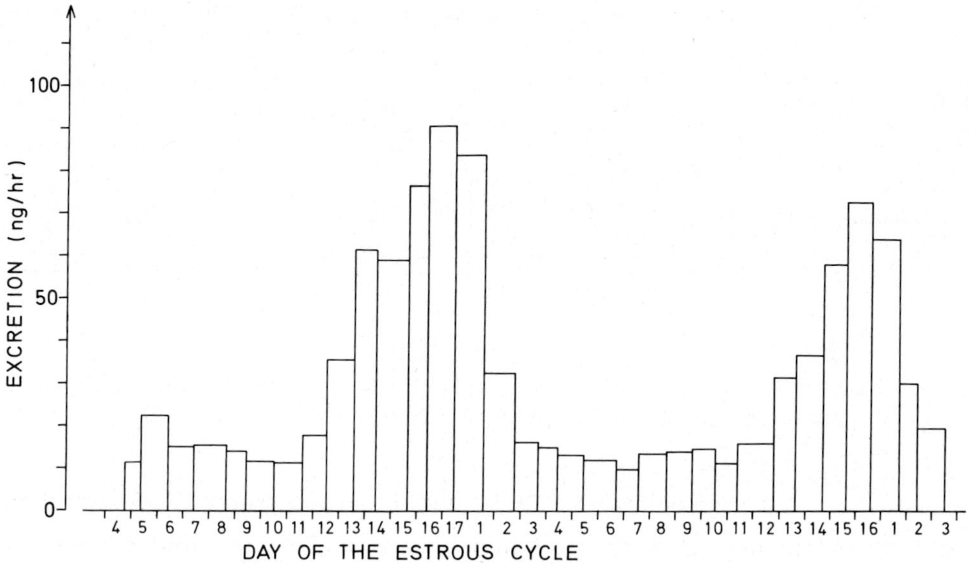

Fig. 3. Excretion of 5α,7α-dihydroxy-11-ketotetranorprostanoic acid in the urine during 2 consecutive oestrous cycles in a guinea pig.

Three animals were mated and became pregnant. No increase in the urinary metabolite levels was seen during days 11-13 as in the case with the non-pregnant animals (Fig. 4). A slight increase in the metabolite levels started around day 25 after mating, and high levels comparable to those normally occurring during luteolysis were obtained around day 35. At this period the CL activity decreases (Kovacic, 1970). The animal shown in Fig. 4 was subjected to hysterectomy on day 33 after mating and the level of the urinary metabolite rapidly decreased to a basal level, indicating that the increased production was caused by the uterus or its content.

Mare

No qualitative metabolic studies have been performed in the horse, and profiles of metabolites in plasma or urine are not known. However, appearance and

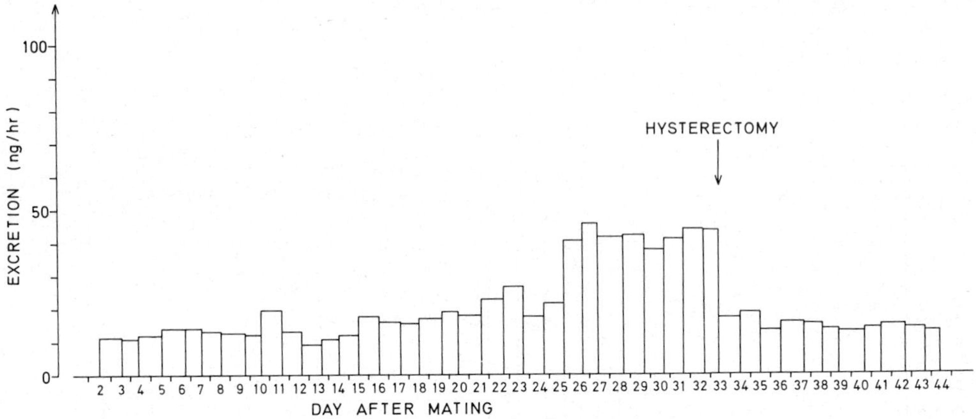

Fig. 4. Excretion of 5α,7α-dihydroxy-11-ketotetranorprostanoic acid in the urine during early pregnancy and after hysterectomy in a guinea pig.

disappearance of 15-ketodihydro-PGF$_{2\alpha}$ and 11-ketotetranor PGF metabolites were followed in plasma with radioimmunoassays after i.v. injection of 1 mg unlabelled PGF$_{2\alpha}$ (Fig. 5) (Goff et al., 1983). As can be seen from the figure, large quantities of 15-ketodihydro-PGF$_{2\alpha}$ are formed immediately after injection of PGF$_{2\alpha}$ (from basal levels of around 25 pmol/l to about 27,000 pmol/l 2 min after the injection). On the other hand rather high basal levels of the 11-ketotetranor PGF metabolites are seen (around 300 pmol/l) reaching a maximum of about 3,000 pmol/l at 25 min. The formation and elimination of both types of metabolites resemble the picture seen in the ewe (Granström and Kindahl, 1982). It is thus likely that both types of metabolites and especially the longer-lived 11-ketotetranor PGF metabolites are suitable parameters to monitor PGF$_{2\alpha}$ synthesis and release in the mare.

Earlier studies concerning quantitative estimations of PGF$_{2\alpha}$ release have employed analysis of PGF$_{2\alpha}$ in the uterine venous plasma (Douglas and Ginther, 1976) or 15-ketodihydro-PGF$_{2\alpha}$ in the peripheral circulation (Kindahl et al., 1976b; Neely et al., 1979; Allen and Pashen, 1981; Kindahl et al., 1982b). During luteolysis a pulsatile release pattern of 15-ketodihydro-PGF$_{2\alpha}$ has been seen concomitantly with the decrease in progesterone levels. This pattern resembles that seen in the cow. During early pregnancy no peak pattern of 15-keto-

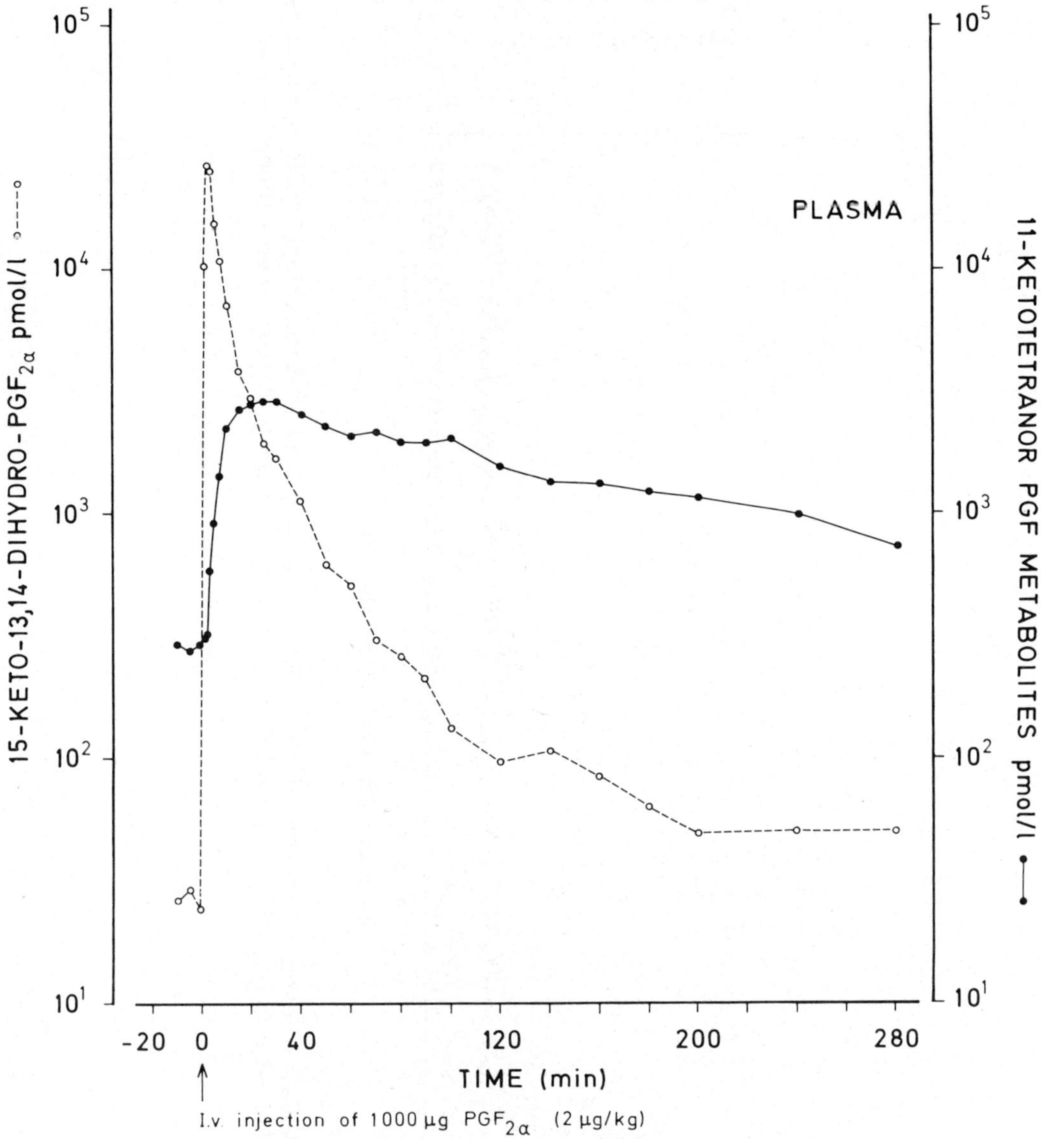

Fig. 5. Kinetics of appearance and disappearance of $PGF_{2\alpha}$ metabolites in the equine circulation after intravenous injection of unlabelled $PGF_{2\alpha}$.

dihydro-$PGF_{2\alpha}$ is seen and progesterone levels are maintained (Kindahl et al., 1982b). To obtain a more reliable picture of the events during luteolysis and early pregnancy a comparison between the two types of $PGF_{2\alpha}$ metabolites has been performed (Fig. 6). It is obvious that both metabolites reflect $PGF_{2\alpha}$ release in a similar way. However, by measuring the more degraded metabolite,

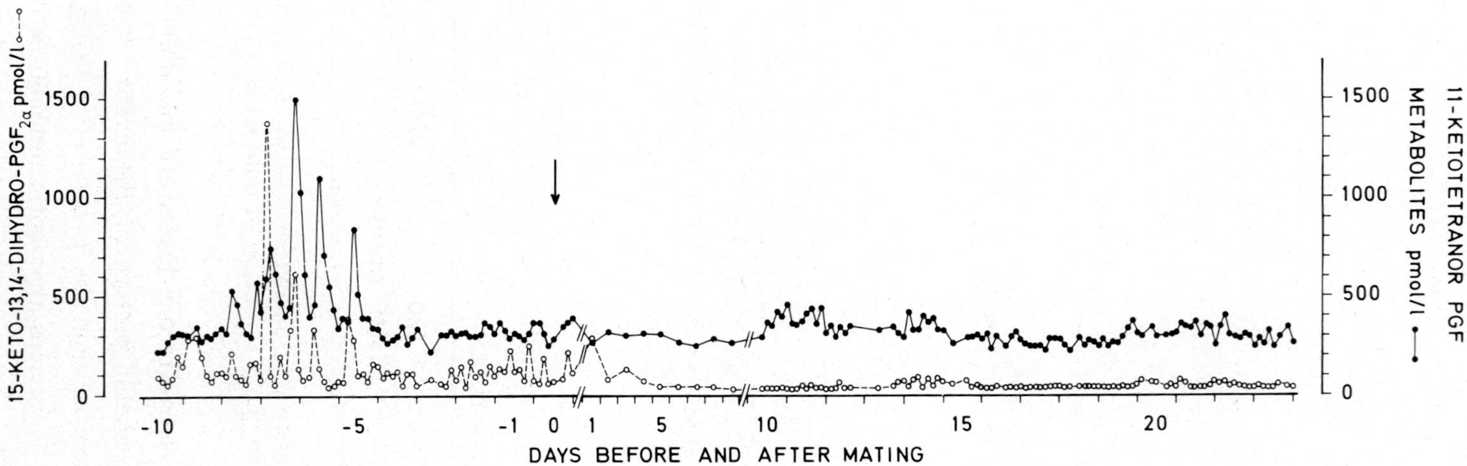

Fig. 6. Peripheral blood plasma levels of 15-ketodihydro-PGF$_{2\alpha}$ (○- - -○) and 11-ketotetranor PGF metabolites (●——●) in a Standardbred mare during the late luteal phase and early pregnancy. Arrow denotes time of mating.

more sustained peaks are seen during luteolysis. Because of the higher basal levels and the higher precision of the assay, more reliable determinations of the 11-ketotetranor PGF metabolites can be made than of the 15-ketodihydro-$PGF_{2\alpha}$. The low basal levels of 15-ketodihydro-$PGF_{2\alpha}$ are around the limit of detection although 0.5 ml plasma is used. By using the assay for 11-ketotetranor PGF metabolites a reliable estimation of the basal levels can be made in 50 µl plasma.

Large quantities of 11-ketotetranor PGF metabolites also appear in urine after administration of unlabelled $PGF_{2\alpha}$ (Goff et al., 1983). Thus long-term studies of $PGF_{2\alpha}$ release can be performed by collecting urine for analysis of 11-ketotetranor PGF metabolites.

Sow

In this species no detailed metabolic studies have been performed. Kunavongkrit et al. (1983) followed the appearance and disappearance of 15-ketodihydro-$PGF_{2\alpha}$ after both i.v. (100 µg) and i.m. (12.5 mg) injection of unlabelled $PGF_{2\alpha}$ (Fig. 7). From that study it is obvious that 15-ketodihydro-$PGF_{2\alpha}$ is a reliable parameter of $PGF_{2\alpha}$ release. A similar study, but with less frequent sampling has been performed earlier (Shille et al., 1979). In the latter study $PGF_{2\alpha}$ release was followed both during luteolysis and early pregnancy. The sow seems to have the longest period of $PGF_{2\alpha}$ release during the oestrous cycle (6 days as compared to 4-5 days for the guinea pig and about 3 days for the cow and mare) (Fig. 8, upper panel). During early pregnancy the peak release of $PGF_{2\alpha}$ is abolished, but transient increases in the levels are seen (Fig. 8, lower panel). This is in agreement with other studies (for a review, see Bazer et al., 1984, this publication).

Doe

Peripheral plasma levels of 15-ketodihydro-$PGF_{2\alpha}$ and 11-ketotetranor PGF metabolites were measured during normal oestrous cycle and early pregnancy in 6 goats (Fig. 9) (Fredriksson et al., 1983). The does were followed with blood samples twice daily until day 12 of the oestrous cycle and thereafter 8 times daily until the onset of oestrus. The does were mated at oestrus and the blood sampling protocol was repeated until day 28 of pregnancy. All 6 does became pregnant.

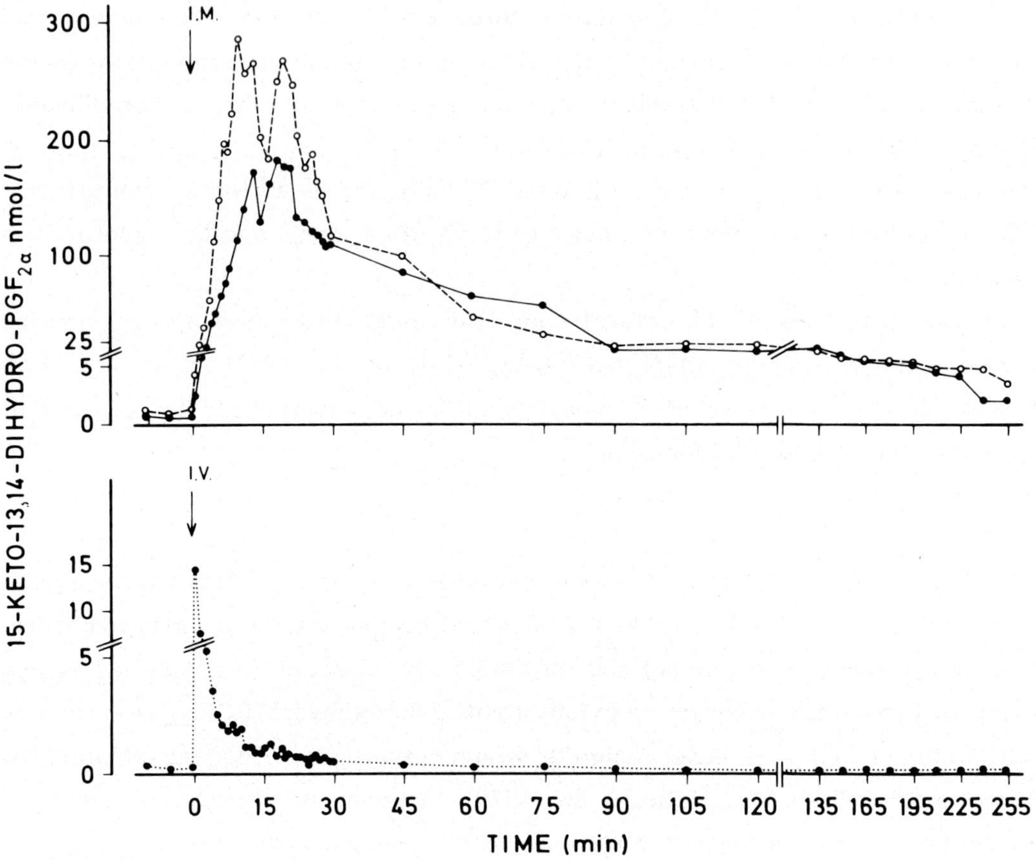

Fig. 7. Peripheral plasma levels of 15-keto-13,14-dihydro-$PGF_{2\alpha}$ before and after intramuscular (i.m.) injection of $PGF_{2\alpha}$ in two sows (upper panel); before and after intravenous (i.v.) injection of $PGF_{2\alpha}$ in one sow (lower panel).

Basal peripheral plasma levels of the two prostaglandin metabolites change during the oestrous cycle. A transient increase is seen from day 12 to 16, which is about twice the level recorded during early luteal phase. From about day 15 to day 17 the does had elevated baseline concentrations and repeated surge releases of prostaglandins. The pulsatile release continued until onset of oestrus. Luteolysis as indicated by decreasing progesterone levels occurred at day 16-18. During pregnancy, baseline levels of the two prostaglandin metabolites are of the same order of magnitude as during the oestrous cycle, and an increase is seen around day 12-14. However, during pregnancy the levels are maintained on these

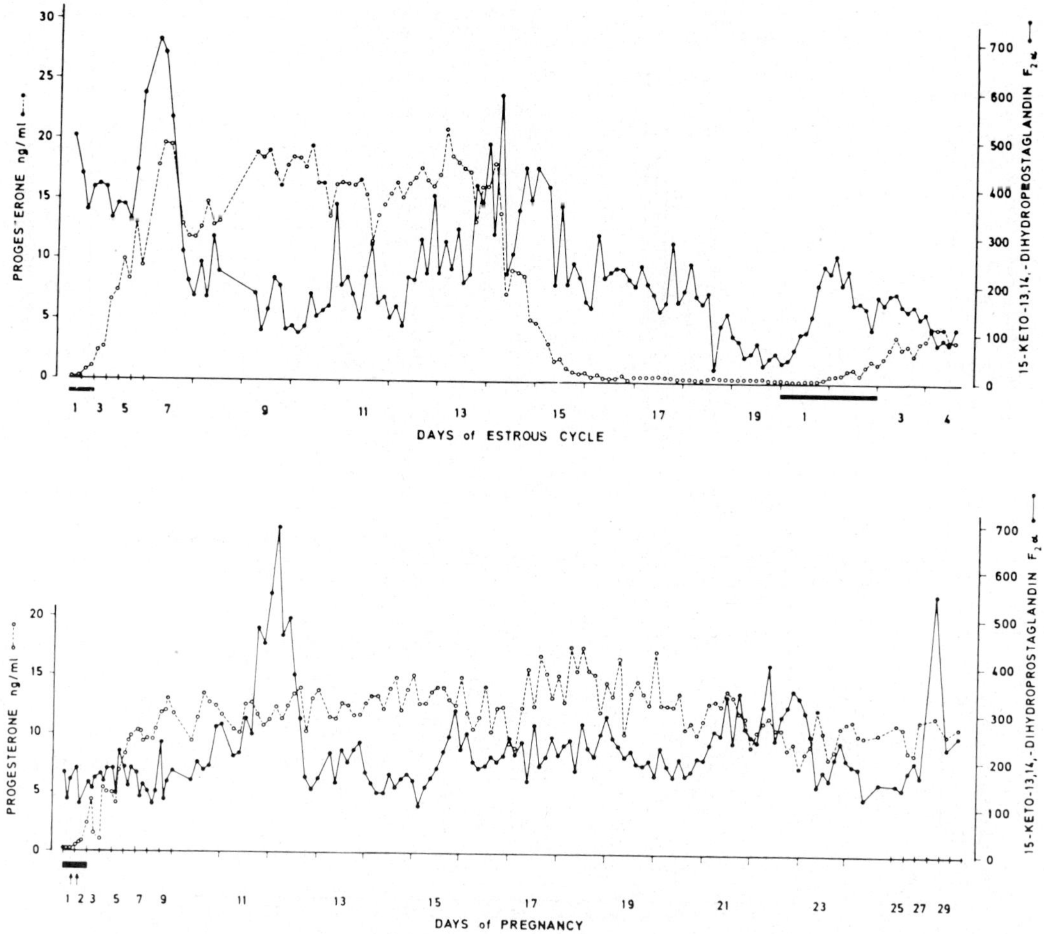

Fig. 8. Peripheral plasma levels of 15-ketodihydro-$PGF_{2\alpha}$ (●——●) and progesterone (○- - -○) during the oestrous cycle (upper panel) and early pregnancy (lower panel) in a gilt. Standing heat is indicated by the horizontal black bar. Arrows denote time of inseminations.

higher levels throughout the observation period. With the exception of a few sporadic peaks, no repeated surge release of prostaglandin metabolites was recorded and high progesterone levels were maintained. Homeida and Cooke (1982) reported a constant baseline of 15-ketodihydro-$PGF_{2\alpha}$ during the oestrous cycle and early pregnancy, with pulsatile release of $PGF_{2\alpha}$ during luteolysis. The discrepancy in the baseline elevations may be due to differences both in the sensitivity for the assay of the prostaglandin metabolite and in sampling frequencies.

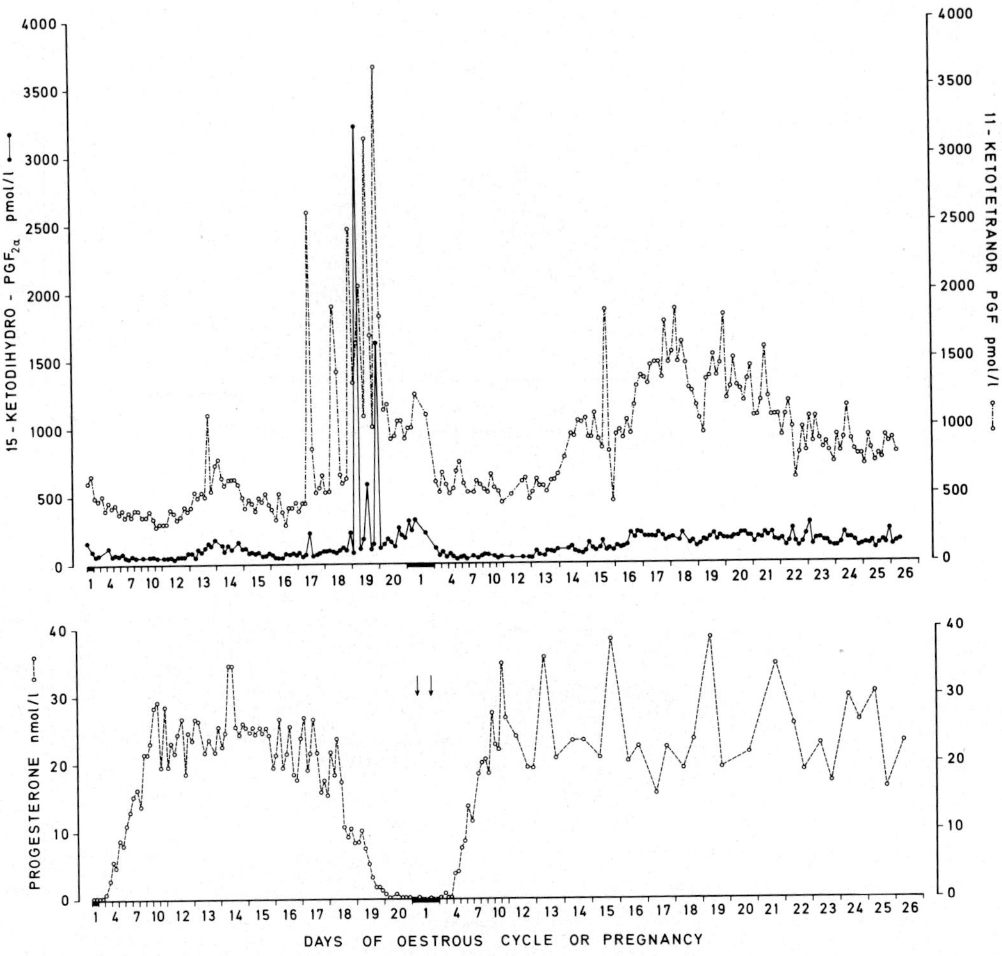

Fig. 9. Peripheral blood plasma levels of 15-ketodihydro-$PGF_{2\alpha}$ (●——●) and 11-ketotetranor PGF metabolites (○-·-·-·-○) (upper panel), and progesterone (○- - -○) (lower panel) in a goat. Horizontal black bar denotes the time of oestrus, and the solid arrows indicate time of mating.

The two prostaglandin metabolites behaved in most cases similarly (Fig. 9). Due to the more sustained circulation time of the shorter metabolites they were more easily recorded with the blood sampling regime employed. In some cases the assay of 11-ketotetranor PGF metabolites record a peak where 15-keto-dihydro-$PGF_{2\alpha}$ levels are only marginally increased or not increased at all. Furthermore the basal levels and the high surge levels are much higher for the

11-ketotetranor PGF metabolites than for the 15-ketodihydro-$PGF_{2\alpha}$ levels. Due to the slower formation of the peak levels of the shorter metabolites these peaks occur somewhat later than the corresponding peaks for 15-ketodihydro-$PGF_{2\alpha}$. Thus the measurement of the 11-ketotetranor PGF metabolites seems to reflect the $PGF_{2\alpha}$ release more accurately than the measurement of 15-keto dihydro $PGF_{2\alpha}$. More information is of course obtained by measuring both types of metabolites.

ACKNOWLEDGEMENT

The authors are indebted to the Swedish Council for Forestry and Agricultural Research for financial support.

REFERENCES

Allen, W.R. and Pashen, R.L., 1981. The role of prostaglandins during parturition in the mare. Acta vet. scand., Suppl. 77: 279-298.

Bazer, F.W., Marengo, S.R., Geisert, R.D. and Thatcher, W.W., 1984. Exocrine versus endocrine secretion of prostaglandin $F_{2\alpha}$ in the control of pregnancy in swine. Anim. Reprod. Sci., Suppl. 7: 31-55.

Douglas, R.H. and Ginther, O.J., 1976. Concentrations of prostaglandin F in the uterine venous plasma of anesthetized mares during the estrous cycle and early pregnancy. Prostaglandins, 11: 251-260.

Fredriksson, G., Kindahl, H. and Edqvist, L.-E., 1983. 11-Ketotetranor PGF metabolites, a suitable indicator for measuring prostaglandin release during the normal oestrous cycle and early pregnancy in the goat. Submitted for publication.

Goff, A., Basu, S. and Kindahl, H., 1983. Measurement of 11-ketotetranor PGF metabolites: A novel approach to judging prostaglandin $F_{2\alpha}$ release in the mare. Submitted for publication.

Granström, E., 1972. On the metabolism of prostaglandin $F_{2\alpha}$ in female subjects. Structures of two metabolites in blood. Eur. J. Biochem., 27: 462-469.

Granström, E. and Samuelsson, B., 1969. The structure of the main urinary metabolite of prostaglandin $F_{2\alpha}$ in the guinea pig. Eur. J. Biochem., 10: 411-418.

Granström, E. and Samuelsson, B., 1971. On the metabolism of prostaglandin $F_{2\alpha}$ in female subjects. J. Biol. Chem., 246: 5254-5263.

Granström, E. and Kindahl, H., 1976. Radioimmunoassay for urinary metabolites of prostaglandin $F_{2\alpha}$. Prostaglandins, 12: 759-783.

Granström, E. and Kindahl, H., 1982. Species differences in circulating prostaglandin metabolites - relevance for the assay of prostaglandin release. Biochem. Biophys. Acta, 713: 555-569.

Granström, E., Kindahl, H. and Swahn, M.L., 1982. Profiles of prostaglandin metabolites in the human circulation - identification of late-appearing, long-lived products. Biochem. Biophys., Acta, 713: 46-60.

Harvey, D., Basu, S., Betteridge, K.J., Goff, A.K. and Kindahl, H., 1984. The influence of pregnancy on PGF$_{2\alpha}$ secretion in cattle. II. Urinary levels of 11-ketotetranor PGF metabolites and plasma progesterone concentration during the oestrous cycle and early pregnancy. Anim. Reprod. Sci., Suppl. 7: 219-236.

Homeida, A.M. and Cooke, R.G., 1982. Peripheral plasma concentrations of 13,14-dihydro-15-keto-prostaglandin F$_{2\alpha}$ and progesterone around luteolysis and during early pregnancy in the goat. Prostaglandins, 24: 313-321.

Horton, E.W. and Poyser, N.L., 1976. Uterine luteolytic hormone: A physiological role for prostaglandin F$_{2\alpha}$. Physiol. Rev., 56: 595-651.

Kindahl, H., 1980. Prostaglandin biosynthesis and metabolism. J. Amer. Vet. Med. Assoc., 176: 1173-1177.

Kindahl, H., Edqvist, L.-E., Bane, A. and Granström, E., 1976a. Blood levels of progesterone and 15-keto-13,14-dihydro-prostaglandin F$_{2\alpha}$ during the normal oestrous cycle and early pregnancy in heifers. Acta endocr. (Kbh)., 82: 134-149.

Kindahl, H., Granström, E., Edqvist, L.-E., Neely, D., Hughes, J. and Stabenfeldt, G., 1976b. The advantages of measuring a prostaglandin F$_{2\alpha}$ metabolite in peripheral blood in studies of the physiological role of prostaglandin release during luteolysis in domestic animals. Proc. 8th Int. Congr. Anim. Reprod. & A.I., Kracow, pp. 145-148.

Kindahl, H., Lindell, J.-O. and Edqvist, L.-E., 1981. Release of prostaglandin F$_{2\alpha}$ during the oestrous cycle. Acta vet. scand., Suppl. 77: 143-158.

Kindahl, H., Edqvist, L.-E., Larsson, K. and Malmqvist, Å., 1982a. Influence of prostaglandins on ovarian function post partum. In: Factors influencing fertility in the postpartum cow. Current Topics in Veterinary Medicine and Animal Science, Vol. 20 (editors H. Karg and E. Schallenberger), Matinus Nijhoff Publishers, The Hague, pp. 173-196.

Kindahl, H., Knudsen, O., Madej, A. and Edqvist, L.-E., 1982b. Progesterone, prostaglandin F$_{2\alpha}$, PMSG and oestrone sulphate during early pregnancy in the mare. J. Reprod. Fert., Suppl. 32: 353-359.

Kovacic, N.M.I., 1970. Endocrinology of reproduction. In: Reproduction and Breeding Techniques for Laboratory Animals, (editor E.S.E. Hafez), Lea & Febiger, Philadelphia, pp. 5-27.

Kunavongkrit, A., Kindahl, H., Einarsson, S. and Edqvist, L.-E., 1983. Clinical and endocrinological studies in primiparous zero-weaned sows. 1. Clinical and morphological findings with special reference to the effect of PGF$_{2\alpha}$ treatment. Zbl. Vet. Med. A (in press).

Metz, S.A, Rice, M.G. and Robertson, R.P., 1979. Applications and limitations of measurement of 15-keto,13,14-dihydro prostaglandin E$_2$ in human blood by radioimmunoassay. Prostaglandins, 17: 839-861.

Neely, D.P., Kindahl, H., Stabenfeldt, G.H., Edqvist, L.-E. and Hughes, J.P., 1979. Prostaglandin release patterns in the mare: Physiological, pathophysiological, and therapeutic responses. J. Reprod. Fert., Suppl 27: 181-189.

Peterson, A.J., Fairclough, R.J., Payne, E. and Smith, J.F., 1975. Hormonal changes around bovine luteolysis. Prostaglandins, 10: 675-681.

Shille, V.M., Karlbom, I., Einarsson, S., Larsson, K., Kindahl, H. and Edqvist, L.-E., 1979. Concentrations of progesterone and 15-keto-13,14-dihydroprostaglandin F$_{2\alpha}$ in peripheral plasma during the estrous cycle and early pregnancy in gilts. Zbl. Vet. Med. A, 26: 169-181.

Animal Reproduction Science, 7 (1984) 149—176
Elsevier Science Publishers B.V., Amsterdam — Printed in The Netherlands

PROSTAGLANDIN DYNAMICS ASSOCIATED WITH DEVELOPMENT OF THE BOVINE CONCEPTUS

W.W. Thatcher[1], D. Wolfenson[1,a], J.S. Curl[1], L.E. Rico[1], J.J. Knickerbocker[1], F.W. Bazer[2] and M. Drost[3]

Departments of Dairy[1] and Animal[2] Sciences, Department of Reproduction[3], University of Florida, Gainesville, Florida 32611, (USA).

ABSTRACT

Thatcher, W.W., Wolfenson, D., Curl, J.S., Rico, L.E., Knickerbocker, J.J., Bazer, F.W. and Drost, M., 1984. Prostaglandin dynamics associated with development of the bovine conceptus. Anim. Reprod. Sci., 7: 149-176.

Oestradiol injection induces an acute increase in uterine production of $PGF_{2\alpha}$ that can be monitored in the peripheral circulation of cattle by measurements of 15-keto-13,14-dihydro-$PGF_{2\alpha}$ (PGFM). Spikes of PGFM were detected prior to luteolysis. An oestradiol induced increase in PGFM concentrations is reduced in pregnant cows at days 18 and 20 compared to day 18 cyclic cows. Evidence is presented that supports a local venoarterial pathway for exchange of $PGF_{2\alpha}$ from the uterus to the ovary. Lower concentrations of $PGF_{2\alpha}$ in the ovarian artery of pregnant cows also are consistent with a conceptus induced antiluteolytic effect. Endometrial tissues from day 17 pregnant cows synthesized, in vitro, less $PGF_{2\alpha}$ than day 17 cyclic cows. Tissues from uterine endometrium, ovarian vein and ovarian artery of day 17 pregnant cows also accumulated less $PGF_{2\alpha}$, in vitro, than day 17 cyclic cows. Pregnancy at 17 days does not appear to influence transport of exogenously administered $PGF_{2\alpha}$ from the uterine lumen. Such agents as oestradiol and PGE_2 enhance $PGF_{2\alpha}$ movement from the uterine lumen. The early pregnant cow appears to have a diminished endometrial secretion of $PGF_{2\alpha}$ that favors luteal maintenance. Conceptus induced alterations in tissue permeability and conceptus control of delivery systems from the uterus also may influence balance of luteolytic and luteotrophic substances delivered to the ovary for luteal maintenance.

---

[a] Present Address: Department of Animal Science, Faculty of Agricultural, The Hebrew University of Jerusalem, Rehovot 76-100, (Israel).

0378-4320/84/$03.00

INTRODUCTION

Survival of the embryo and associated maintenance of the corpus luteum (CL) are major biological limitations that contribute to the high rate of embryonic mortality that occurs in cattle (15 %; Hawk, 1979) and extends the interval between births. Furthermore, such normal levels of pregnancy wastage are exacerbated during periods of thermal stress due to high temperatures and humidity. An understanding of the basic mechanisms involved with processes of pregnancy recognition may provide the basis to improve reproductive efficiency in the future.

Purpose of this presentation is to describe the potential systems of control between the developing conceptus and endometrium that lead to maintenance of the CL in cattle. Major emphasis will be placed on the dynamics and control of endometrial-conceptus secretion and transport of prostaglandins.

In vivo release of prostaglandin $F_{2\alpha}$ during the oestrous cycle and early pregnancy

The sexually mature bovine female exhibits recurring oestrous cycles of 20 to 21 days which are not affected appreciably by season of year. Essential to recurring oestrous cycles is regression of the CL which appears to be regulated by follicular components of the ovary via the uterus. Importance of follicular function is exemplified by the observations (Villa-Godoy et al., 1981) that destruction of bovine follicles (electrocauterization of visible follicles and X-irradiation of both ovaries) on days 9, 12 or 15 post-oestrus resulted in heavier CL at day 24 post-oestrus compared to control animals (5.5 g vs 1.11 g). Luteolytic action of follicles appears to be exerted after day 15, and results support the concept that follicles initiate luteal regression in cattle. Although follicles may regulate luteolysis, the ultimate effector component appears to be secreted via the uterus. The uterus has been shown to control lifespan of the bovine CL (Wiltbank and Casida, 1956; Malven and Hansel, 1964) since hysterectomy leads to prolonged CL maintenance ($\cong$150 days).

Free and conjugated oestrogens shorten life span of the CL in cattle if administered during the luteal phase of the oestrous cycle (Greenstein et al., 1958; Wiltbank et al., 1961; Eley et al., 1979), but oestradiol (E_2) was not effective in hysterectomized heifers (Brunner et al., 1969). Luteolytic activity of oestrogen is thought to be mediated through stimulation of uterine prostaglandin $F_{2\alpha}$ ($PGF_{2\alpha}$) synthesis and release. Luteolytic activity of exogenous $PGF_{2\alpha}$ in cattle has been documented extensively (Hansel et al., 1973; Hafs et al., 1974; Lauderdale, 1974; Thatcher and Chenault, 1976). Elevated levels of $PGF_{2\alpha}$ in uterine venous drainage (Nancarrow et al., 1973; Shemesh and Hansel, 1975), uterine tissue (Shemesh and Hansel, 1975) and uterine flushings (Lamothe et al., 1977; Bartol et al., 1981a) coincide with the transient rise in follicular E2 prior to ovulation in cyclic cattle (Nancarrow et al., 1973; Chenault et al., 1975). Peterson et al. (1975) and Kindahl et al. (1976) reported elevations in peripheral plasma 15-keto-13,14-dihydro-$PGF_{2\alpha}$ (PGFM), the inactive metabolite of $PGF_{2\alpha}$, which were temporally associated with CL regression in cyclic heifers. Fairclough et al. (1981) demonstrated that luteolysis can be halted by passive immunization with $PGF_{2\alpha}$ antiplasma on day 16 of the oestrous cycle. Luteolytic action of E_2 in heifers was blocked when indomethacin was administered within the uterus of heifers between days 14 and 21 of the cycle, but not when it was administered systemically. Collectively, these observations support the concept that follicular oestrogen secretion regulates onset on luteolysis via a uterine induced secretion of $PGF_{2\alpha}$.

Oestradiol (3 mg) given intravenously on day 13 of the oestrous cycle to five heifers caused a rise in PGFM concentrations that began at 3 h, reached a peak by 6 h, and returned to basal concentrations by 9 h post-E_2 administration (Thatcher et al., 1979). However, in heifers (n=5) receiving only the vehicle solution (6 ml of ethanol: saline [1:1]), no change in PGFM was detected for an acute 11 h period comprised of a 30 min sampling sequence (Fig. 1). An additional study indicated that this increase in peripheral plasma PGFM concentration was associated with an elevation of $PGF_{2\alpha}$ in uterine flushings collected at 6 h after injection of E_2 (Bartol et al., 1981b).

Synthesis and metabolism of $PGF_{2\alpha}$ by the uterus, in response to E_2 administration, is documented in Fig. 2 (Knickerbocker et al., 1982). On day 17 of the oestrous cycle, seven dairy cows were surgically prepared and fitted, ipsilateral to the CL, with: an electromagnetic flow transducer around the uterine artery (UA; n=5); catheters within the ovarian vein (OV; n=7) via a uterine branch of the ovarian vein, uterine branch of the ovarian artery (UBOA; n=5) and facial artery (FA; n=7). On day 18, blood samples were collected at 30 min intervals for 1 h

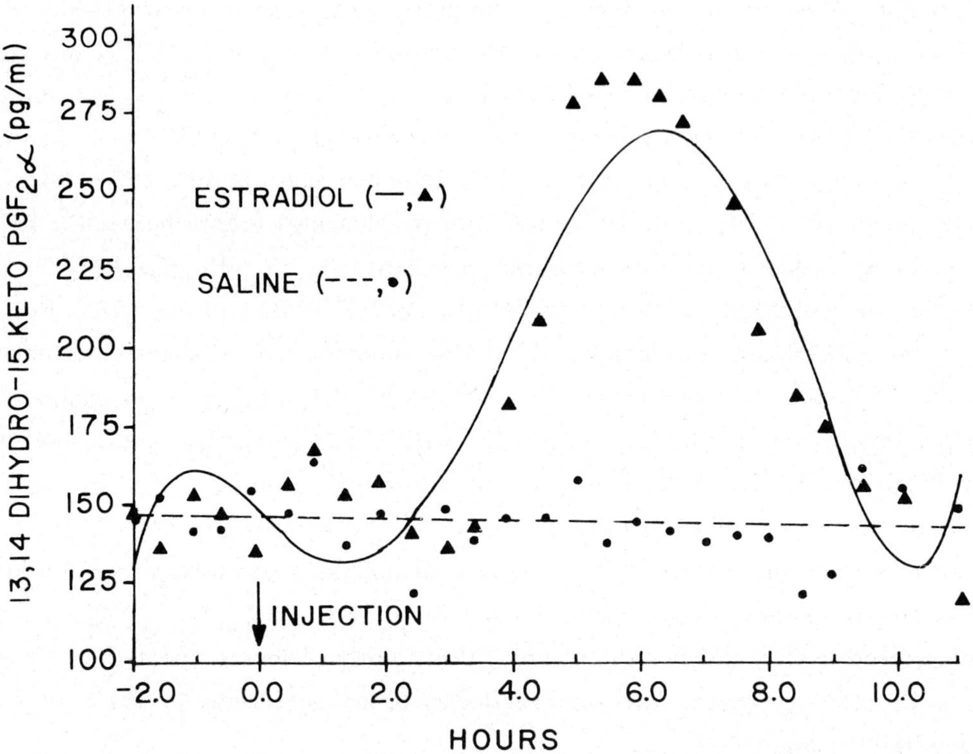

Fig. 1. Regressions and time means of plasma 15-keto-13,14-dihydro-$PGF_{2\alpha}$ concentrations in heifers injected with oestradiol (3 mg) or saline.

prior to injection of E_2 (3 mg) and 12 h post-E_2. Uterine blood flow was monitored continuously and plasma samples analyzed for $PGF_{2\alpha}$ and PGFM. Exact locations of catheters in reproductive tracts were verified post-slaughter. Maximal production of uterine $PGF_{2\alpha}$ (0.93 ± 0.25 μg/min, Fig. 2), PGFM (0.35 ± 0.08 μg/min) and total prostaglandin ($PGF_{2\alpha}$ and PGFM; 1.26 ± 0.32 μg/min) were observed at 6 h post-E_2. Peak prostaglandin responses followed maximal increase in uterine blood flow at 3 h. Results indicated that synthesis of $PGF_{2\alpha}$ and partial metabolism to PGFM occur within the uterus following injection of E_2. Peripheral concentrations of $PGF_{2\alpha}$ were basal during the pre- and post-E_2 phases (Fig. 2) and were not correlated with any uterine prostaglandin response. In contrast, measurement of the initial metabolite of $PGF_{2\alpha}$, PGFM, in peripheral plasma was related to uterine $PGF_{2\alpha}$ production (r=0.66; Fig. 2). Results

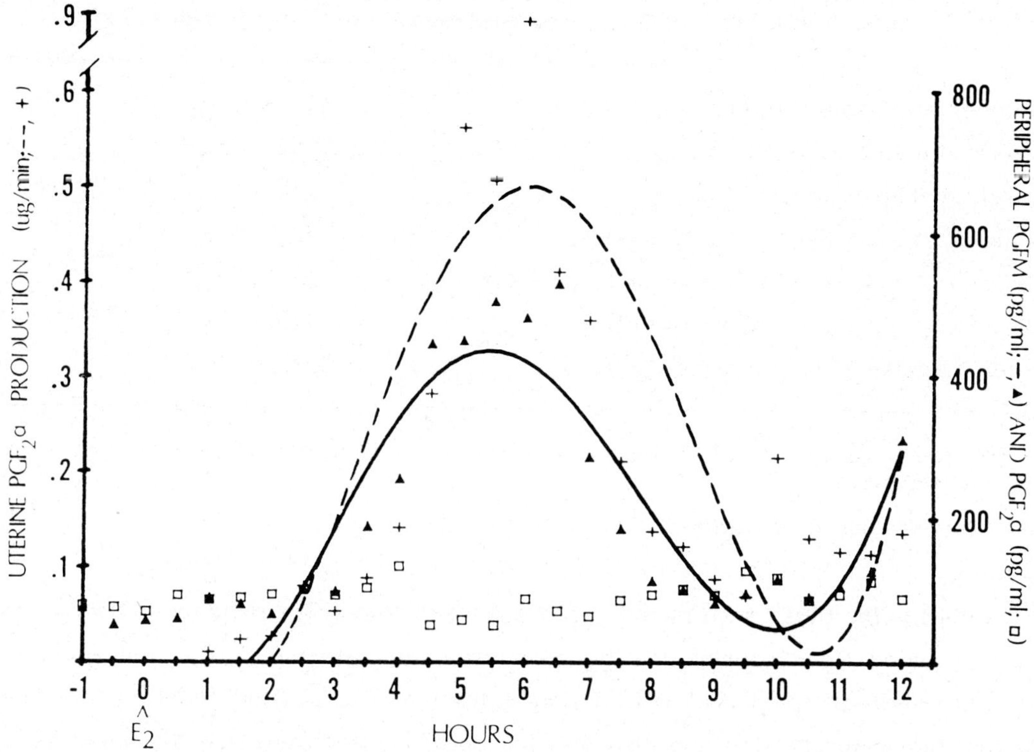

Fig. 2. Regressions and time means of prostaglandin responses in cows injected with oestradiol (3 mg).

support use of peripheral PGFM as an index of uterine prostaglandin production in cattle (Kindahl et al., 1976; Kindahl et al., 1981).

Injection of E_2 (3 mg) on day 13 of the oestrous cycle (n=5 heifers) caused luteolysis (progesterone concentration < 1 ng/ml) to occur earlier following injection (96.9 ± 10.6 < 153.6 ± 17.7 h; P < 0.05) than in control heifers (n=5; Thatcher et al., 1979). However, there was no acute change in progesterone concentrations in association with the 6 h induced peak of PGFM. Dynamics of PGFM concentrations in peripheral plasma relative to luteolysis for all 10 heifers are depicted in Table I. Of the 39 samples classified as PGFM spikes (6.4% of 606 samples; $\bar{x}$ of 323 pg/ml), 21 occurred before the time of luteolysis ($\bar{x}$ = 18.9 h prior). This is supportive of the concept that PGFM measurements and their increases are temporally associated with luteolysis on a within cow basis (e.g. Fig. 3).

Table I. Characteristics of 15-keto-13,14-dihydro-$PGF_{2\alpha}$ concentrations in peripheral plasma of 10 cows[a] during the oestrous cycle (day 13 to oestrus)

Number of plasma samples	606
Overall mean ± SEM (pg/ml)	129.2 ± 3.49
Total number of spikes[b] (%)	39 (6.4%)
Number spikes (%) prior to luteolysis	21 (53%)
Time (h; $\bar{X}$ ± SEM) of spikes prior to luteolysis	18.9 ± 3.86
Mean basal concentration[b] (pg/ml)	104.6 ± 16.4
Mean magnitude of all spikes[b] (pg/ml)	323.0 ± 50.0

[a] Treated with oestradiol (3 mg; n=5) or saline (n=5) on day 13; sampled every 3 h until oestrus

[b] Responses based on method of Christian et al., 1978

Collectively, these results indicated that exogenous E_2 induced increases in uterine blood flow, uterine $PGF_{2\alpha}$ production and uterine $PGF_{2\alpha}$ metabolism. Measurement of peripheral PGFM concentration was a good index of uterine prostaglandin production and also was temporally associated with the process of luteolysis. The E_2 injection scheme was utilized as an experimental model to investigate the interaction between conceptus and uterus at day 18 and 20 of pregnancy (Rico et al., 1981). Crossbred beef cows (n=17), fitted with jugular catheters, were injected i.v. with E_2 (3 mg) on day 18 of the cycle (n=6) or day 18 postbreeding (n=11). Blood samples were collected at 30 min intervals from 2 h preinjection to 10 h postinjection. On day 19, uteri of bred cows were flushed nonsurgically to confirm presence of a blastocyst (9 of 11 were pregnant; comparisons made of 9 pregnant versus cyclic animals). Plasma concentrations of PGFM and progesterone were measured by radioimmunoassay. Pregnant and cyclic treatments did not differ in preinjection concentrations of progesterone (8.8 ± 1.1 versus 6.8 ± 0.93 ng/ml) and PGFM (71.7 ± 10.9 versus 54.3 ± 11.5 pg/ml). Mean concentrations of PGFM differed (Pregnant < Cyclic; $P < 0.05$) for the 10 h postinjection phase between Pregnant and Cyclic treatments (120.3 versus 190.0 pg/ml). Overall response curve indicated that PGFM increased by 4.5 h, peaked at $\cong$ 6.5 h, and returned to basal levels by 9.5 h. However, treatment response curves differed (Pregnant < Cyclic; $P < 0.01$; Fig. 4). Mean peak concentrations

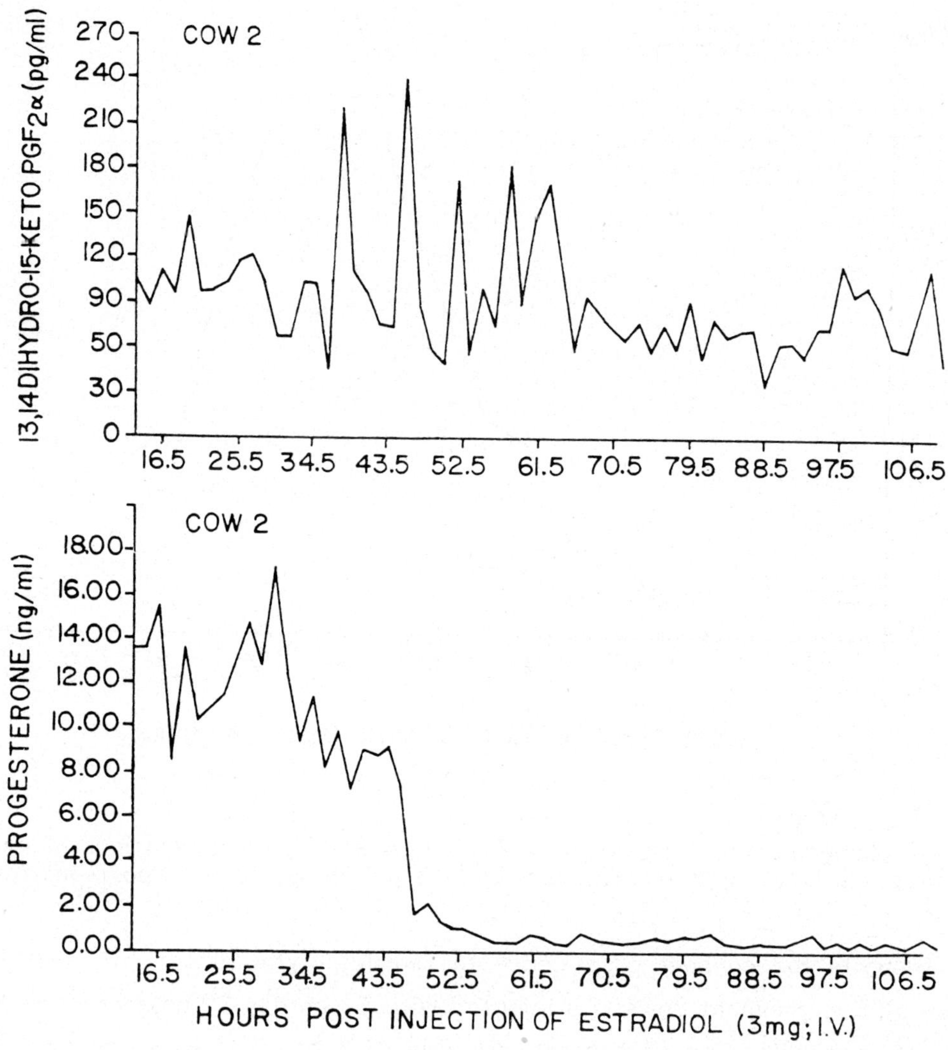

Fig. 3. Temporal patterns of 15-keto-13,14-dihydro-$PGF_{2\alpha}$ and progesterone concentrations in peripheral plasma of a cow following an injection of oestradiol (3 mg).

of PGFM were Pregnant = 362.8 ± 84.3 versus Cyclic = 724.3 ± 136.8 pg/ml ($P < 0.01$). Three of nine pregnant cows failed to have an E_2 induced increase in PGFM. Rate of increase in PGFM was less for Pregnant cows when comparing Pregnant and Cyclic response curves. Results suggest, that at day 18, presence of a conceptus alters the temporal pattern of $PGF_{2\alpha}$ secretion in response to E_2

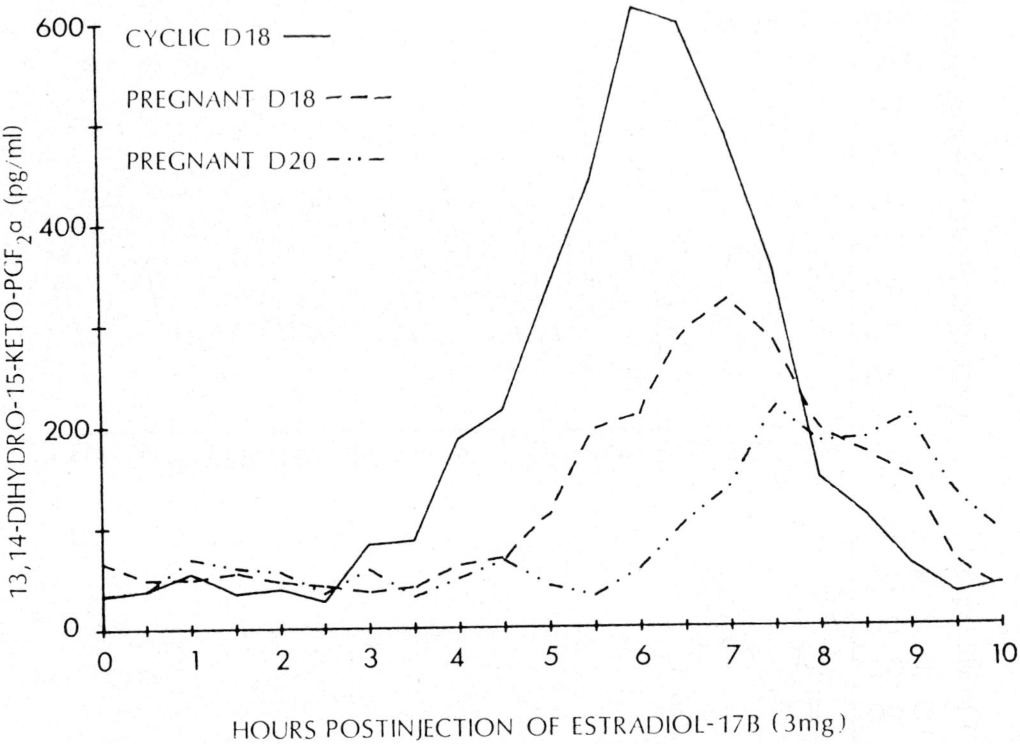

Fig. 4. Time means of 15-keto-13,14-dihydro-PGF$_{2\alpha}$ concentrations in cyclic and pregnant cows injected with oestradiol (3 mg) on day 18 or 20 post-oestrus.

and supports the concept of a conceptus antiluteolytic effect. An additional group of five pregnant cows was injected with E_2 on day 20 postbreeding. Pretreatment progesterone concentrations were 11.7 ± 1.5 ng/ml; embryos were recovered, nonsurgically, from all cows on day 21. Mean concentration of PGFM for the 10 h postinjection phase at day 20 was 86.7 ± 49.5 pg/ml. Peak concentrations of PGFM were 218.7 ± 91.5, but two of five cows did not respond to E_2 injections. It is clear from Fig. 4 that presence of the conceptus attenuates the PGFM response. As the conceptus continues to develop and occupies additional surface area of the endometrium (e.g. day 18 to day 20), the PGFM response to E_2 is further diminished. Variation among pregnant cows within day 18 or day 20, regarding ability to express an attenuated response, may be related to conceptus quality. For example, one pregnant cow at day 18 had a PGFM response characteristic of the cyclic group. The recovered conceptus may not have been viable.

Increase in PGFM for day 18 cyclic cows would represent a uterine (i.e. endometrial production) $PGF_{2\alpha}$ response since a conceptus is absent. By deduction, it is likely that endometrial tissue of pregnant cows has a diminished ability to secrete $PGF_{2\alpha}$ due to an antiluteolytic effect exerted by the conceptus.

The bovine conceptus can alter CL function by day 17 postmating, since the interoestrus interval was extended when conceptuses were removed from cows on day 17 (25 ± 1.2 days) compared to either non-mated controls (21 days) or when conceptuses were removed on day 15 (20.2 ± 0.8 days; Northey and French, 1980). Furthermore, these workers demonstrated that intra-uterine infusions of embryonic homogenates during days 15 through 17 extended the interoestrus interval and delayed the decline in progesterone associated with CL regression. An important point was that progesterone was not stimulated, but the decline delayed. An antiluteolytic mechanism may involve some process, initiated by the conceptus, that stabilizes and/or reduces endometrial secretion of the luteolysin ($PGF_{2\alpha}$). Kindahl et al. (1976) reported that there was an absence of elevated concentrations of PGFM in peripheral plasma from one pregnant heifer compared to non-pregnant heifers in which pulses of PGFM were temporally associated with luteolysis. A similar trend also has been reported recently for buffaloes (Batra and Pandey, 1983). However, in the latter study a limited sequential sampling sequence of 6 h intervals may not provide a reliable characterization of the $PGF_{2\alpha}$ release pattern (Kindahl et al., 1981). The PGFM response (Fig. 4) to E_2 injection of cyclic (day 18) versus pregnant (day 18 and 20) cows does support an antiluteolytic effect exerted by the conceptus.

An elegant series of surgical preparations clearly demonstrated that luteolysis in cattle is exerted through a local venoarterial pathway involving the veins which drain the uterus and the ipsilateral ovarian artery (Mapletoft et al., 1975; Ginther, 1981). A main uterine branch of the ovarian vein and ipsilateral ovarian artery is an adequate uterine outlet and ovarian inlet, respectively, for countercurrent exchange of a luteolytic agent. Results described above indicate that the bovine uterine luteolysin is likely to be $PGF_{2\alpha}$. Elevated endometrial and uterine vein plasma $PGF_{2\alpha}$ concentrations between days 15 and 20 of the oestrous cycle (Nancarrow et al., 1973; Shemesh and Hansel, 1975) and preferential transfer of $PGF_{2\alpha}$ from uterine vein to ovarian artery following intra-uterine deposition of $PGF_{2\alpha}$ (Hixon and Hansel, 1974) support the concept. In a subsequent experiment (Shemesh and Hansel, 1975), evidence for a preferential transfer of endogenous $PGF_{2\alpha}$ from uterine vein to ovarian artery, between 15 to 20 days of the oestrous cycle, was not supported. However, only single blood samples were collected at surgery, and a more chronically sustained sampling period within animal at the time of luteolysis is needed to test the "countercurrent" concept.

We have examined recently the temporal pattern of $PGF_{2\alpha}$ concentrations in ovarian arterial blood ipsilateral to the ovary bearing the CL during the oestrous cycle or early pregnancy (days 17 to 20; Wolfenson and Thatcher, 1983a). Objective of the experiment was to see if there was a localized increase in $PGF2\alpha$ within ovarian arterial blood, and if such a localized elevation differed between cyclic and pregnant cattle. A total of 26 cattle were fitted with polyvinyl catheters that were inserted (surgery on days 15 to 17) into the uterine branch of the ovarian artery (UBOA), a peripheral facial artery (PA, catheter tip inserted into the internal carotid artery), and in some preparations a branch of the uterine branch of ovarian vein (UBOV) was catherized (tip of catheter advanced to ovarian vein, OV). Catheters were secured within vessels by ligation at the point of catheter entry which prevented blood flow past this point. Catheter fixed within the UBOA represented a blood sample from the ovarian artery (OA); possible exchange of $PGF_{2\alpha}$ would occur between the OV and the OA below the point of confluence with the UBOV (point of confluence refers to junction where ovarian branch of OV and UBOV merge). Beginning on the day following surgery, six sequential blood samples were collected at 30 to 45 min intervals twice daily between 0700 to 1200 h and 1700 to 2200 h. Due to non-functional catheters and mistaken identification of UBOA, a total of 8 cyclic and 6 pregnant cows were sampled successfully. Cyclic cows were not inseminated, and pregnancy was confirmed in bred cows by nonsurgical embryo flushes at the end of the experiment. Average length of the sampling period was 2.0 days (range 1 to 4 days). A total of 818 blood samples was collected, and this included 216 concurrent OA and PA arterial pairs. Placement of catheters were confirmed at slaughter.

An overall vessel effect (OA > PA, $P < 0.01$) and a vessel by pregnancy status interaction ($P < 0.10$) were detected for $PGF_{2\alpha}$ concentrations (OA > PA): cyclic, $208 \pm 15 > 115 \pm 14$ pg/ml; pregnant, $173 \pm 17 > 122 \pm 17$ pg/ml. Since concentrations of $PGF_{2\alpha}$ were higher in the OA, statistical analyses were conducted on the series of paired samples whereby variation in net concentrations (OA minus PA, pg/ml) was evaluated. A status x day interaction was detected ($P < 0.01$, Table II). Net concentrations of $PGF_{2\alpha}$ (OA-PA) within plasma of the OA increased in the cyclic group on days 19 and 20, whereas concentrations were maintained at a stable level between days 17 to 20 in pregnant cattle. A spike analysis of $PGF_{2\alpha}$ concentrations in the OA of pregnant and cyclic cows revealed no difference in frequency of spikes, but $PGF_{2\alpha}$ concentrations of sample spikes were greater for cyclic cows (n=12, 725 ± 189 pg/ml > 397 ± 196 pg/ml, n=9).

Table II. Least squares means for difference in $PGF_{2\alpha}$ (pg/ml) between ovarian-peripheral arteries (OA-PA) of pregnant (n=5) and cyclic (n=6) cows

Day	Pregnant*	Cyclic*
17	29	38
18	136	1
19	70	167
20	54	161
Mean (sample number)	72 (84)	92 (114)

*$P < 0.01$ Status x Day interaction

The positive difference (OA - PA) in $PGF_{2\alpha}$ concentrations is suggestive of a local venoarterial countercurrent exchange system for $PGF_{2\alpha}$. This was further assessed for the cyclic group of cows. Mean concentrations of $PGF_{2\alpha}$ (pg/ml) for the OV (n=194), OA (n=141) and PA (n=211) were: 562 ± 149.9, 228 ± 19.2, and 106 ± 7.7 pg/ml, respectively. A sub-sample of 82 sample times was evaluated where plasma was collected concurrently from the OV, OA and PA (Table III). Net concentration of $PGF_{2\alpha}$ in the OA (OA - PA) was +85.0 pg/ml which was 10% of the mean $PGF_{2\alpha}$ concentration in the OV (810 pg/ml). This percentage is a potential measure of exchange efficiency between uterine and ovarian vascular beds. Additional evidence for a functional role of the local increase of $PGF_{2\alpha}$ in the OA is illustrated in Fig. 5. The transitory decrease in peripheral progesterone

Table III. Least squares means for $PGF_{2\alpha}$ (pg/ml) in concurrent samples of ovarian vein (OV), ovarian artery (OA) and peripheral artery (PA) of cows (n=8) during the oestrous cycle

Day	OV	OA	PA	(OA-PA)*	(OA-PA)% / OV	N
17	215	126	94	32	14.9	25
18	335	54	50	4	1.2	18
19	1095	272	104	168	15.3	24
20	1595	243	108	135	8.5	15
Mean	810	174	89	85	10.0	82

* = $P < 0.01$

concentrations associated with luteolysis was associated with temporal increases of $PGF_{2\alpha}$ concentrations in the ovarian vein and with net concentrations of $PGF_{2\alpha}$ (OA - PA) in the OA of a cyclic cow. Collectively, results from this experiment are supportive of a local venoarterial pathway for exchange of $PGF_{2\alpha}$ from the uterus to the ovary. Lower concentrations of $PGF_{2\alpha}$ in the OA of pregnant cows is consistent with evidence for a conceptus induced antiluteolytic effect.

In vitro production, metabolism and uptake of prostaglandins by bovine uterine endometrial tissues

The previous in vivo experiments suggest that cyclic cattle enter into a luteolytic pathway, whereas pregnant cattle enter into an embryotrophic pathway of which luteal maintenance is an essential component. Critical to understanding this process in cattle is whether endometrial production of prostaglandins differs between pregnant and cyclic animals at the time when the process of pregnancy recognition has been initiated (e.g. day 17, Nortley and French, 1980). Denovo and maximal denovo synthesis of $PGF_{2\alpha}$, PGFM and PGE_2, from endometrial tissues of cyclic (n=5) and pregnant (n=5) cattle, has been examined at day 17 post-oestrus (Curl et al., 1983). Endometrium ipsilateral to the CL bearing

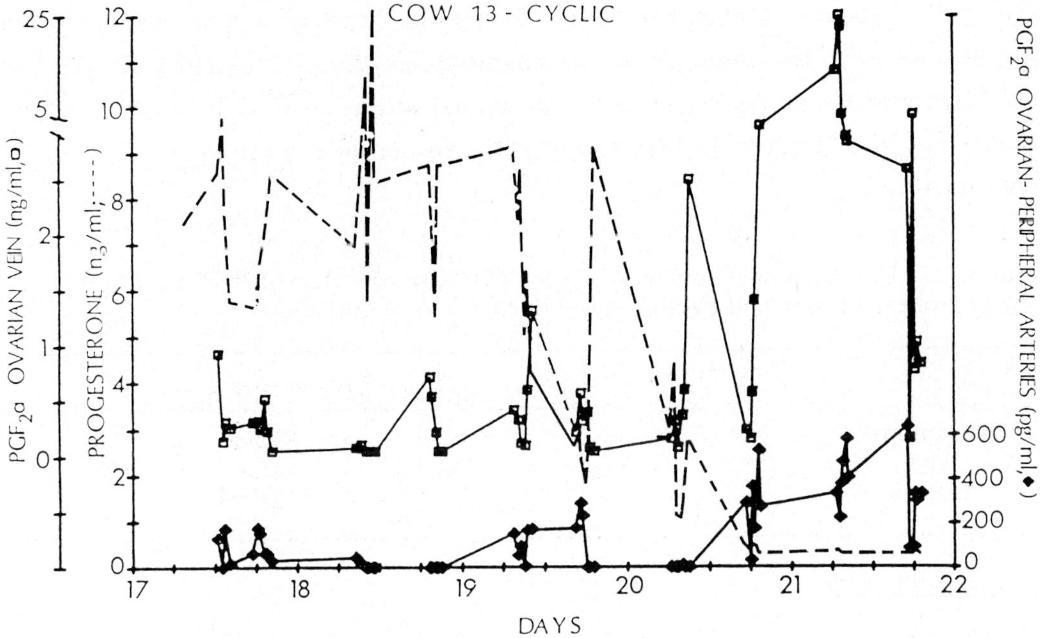

Fig. 5. Association of peripheral progesterone concentrations during luteolysis with temporal patterns of $PGF_{2\alpha}$ concentrations in the ovarian vein and ovarian artery of a cyclic cow.

ovary was dissected and minced. Tissue (500 mg, in duplicate) was incubated in 15 ml minimum essential media for 24 h at 37C under 50% N_2, 45% O_2, 5% CO_2. Endometrium was incubated either alone (A), with 200 µg arachidonic acid (B), or with 1.5 mg flunixin meglumine (FM). $PGF_{2\alpha}$, PGFM and PGE_2 were measured by RIA in incubation media and in endometrial tissue frozen immediately after dissection (initial tissue content). Releasable pool (RP) of PG was that sequestered in endometrium and released into media during incubation with FM. Denovo synthesis of PG was A minus RP; maximal denovo synthesis was B minus RP. No differences were detected in initial endometrial tissue content of prostaglandins for cyclic and pregnant animals. Overall means for $PGF_{2\alpha}$, PGFM and PGE_2 were: 4.26, 1.45 and 17.91 ng/500 mg, respectively. Addition of arachidonic acid to endometrium increased maximal denovo synthesis of all prostaglandins compared to denovo synthesis in which no added arachidonic acid substrate was added (Table IV). Endometrium from cyclic cattle produced more $PGF_{2\alpha}$ than pregnant cattle when evaluating either denovo ($P < 0.01$) or maximal denovo ($P < 0.01$)

synthetic responses (Table IV). PGFM responses between cyclic and pregnant animals were of the same order of magnitude (cyclic > pregnant) but did not approach statistical significance due to animal variation ($P \cong 0.10$). In contrast there was no difference in PGE_2 production responses between cyclic and pregnant cattle.

Table IV. Least squares means of endometrial production of $PGF_{2\alpha}$, PGFM[a] and PGE_2 from pregnant and cyclic cows[b] during a 24 h culture

Treatment[c] and Status	Response (ng/24 h/500 mg)			
	$PGF_{2\alpha}$	PGFM	$PGF_{2\alpha}$ and PGFM	PGE_2
Realeasable Pool				
Pregnant	19.89^{1}	64.41^{1}	84.30^{1}	24.93^{1}
Cyclic	18.84^{1}	74.65^{1}	93.80^{1}	19.71^{1}
Denovo Synthesis[e]				
Pregnant	78.86^{2}	88.11^{2}	166.97^{2}	243.05^{2}
Cyclic	207.59^{3}	197.95^{2}	405.54^{3}	262.71^{2}
Maximal Denovo Synthesis[d,e]				
Pregnant	125.05^{4}	224.44^{3}	349.48^{4}	537.28^{3}
Cyclic	268.34^{5}	456.82^{3}	725.16^{5}	517.07^{3}

[a] PGFM = 15-keto-13,14-dihydro-$PGF_{2\alpha}$

[b] Day 17 post-oestrus; 5 cows per group

[c] Two ANOVA per response variable; first on releasable pool, second on Denovo and Maximal; Means within each analysis with different column superscripts differ $P < 0.01$

[d] Cultured with 200 µg arachidonic acid

[e] Denovo and Maximal Synthesis responses adjusted for releasable pool

It is clearly evident that the bovine endometrium is capable of metabolizing $PGF_{2\alpha}$ (Table IV; Lewis et al., 1982) to PGFM (15-keto-13,14-dihydro-$PGF_{2\alpha}$). Thus the 15-prostaglandin dehydrogenase and Δ^{13} - reductase enzyme are present in the bovine endometrium. Local metabolism and possible differences in metabolism between cyclic and pregnant cattle may indeed be important since it has been reported recently that the $PGF_{2\alpha}$ metabolite, 13,14-dihydro-$PGF_{2\alpha}$, is luteolytic when administered to heifers (Milvae and Hansel, 1983). The present experiment indicates that processes of early pregnancy recognition appeared to have stabilized the endometrium such that in vitro production of $PGF_{2\alpha}$ is less than in cyclic cattle.

Relative to dynamics of prostaglandin production in early pregnancy, it is important to know what prostaglandin products are produced by the conceptus in order to understand conceptus-endometrial-ovarian interrelationships. Bovine conceptuses recovered on days 16 and 19 of pregnancy produced considerably more $PGF_{2\alpha}$ and PGE_2 per milligram of tissue than did endometrial slices recovered from uterine horns either ipsilateral or contralateral to the CL (Lewis et al., 1982). For example, day 16 conceptuses produced 66 and 284 ng of PGE_2 and $PGF_{2\alpha}$/mg of tissue compared to only 0.6 and 2.1 ng of PGE_2 and $PGF_{2\alpha}$/mg of endometrial tissue. In contrast, endometrial and conceptus tissues produced approximately equal amounts of PGFM. In fact, endometrial slices were more capable of metabolizing $PGF_{2\alpha}$ than were conceptuses (Lewis et al., 1982). This study, as well as that of Shemesh et al. (1979), indicate that bovine conceptuses produce $PGF_{2\alpha}$ and PGE_2; production increased with day of pregnancy. To some degree, these prostaglandins appear to be sequestered and/or compartmentalized within the uterine lumen. For example, $PGF_{2\alpha}$ and PGE_2 contents of uterine flushings from pregnant cattle were greater than cyclic or non-pregnant cattle (Bartol et al., 1981a; Lewis et al., 1982).

The following series of observations led us to test whether there was a difference in $PGF_{2\alpha}$ uptake by reproductive tissues of pregnant versus cyclic cattle at day 17. Pregnant endometrial tissue synthesized less $PGF_{2\alpha}$ in vitro, and pregnant cattle had an attenuated PGFM response in vivo to an injection of E_2. This diminished response occurred even though it is known that conceptuses themselves synthesized copious quantities of $PGF_{2\alpha}$, and a considerable pool of $PGF_{2\alpha}$ in uterine flushings is bathing the underlying endometrial tissues. It was hypothesized that uptake of $PGF_{2\alpha}$ by endometrium and vascular components, leading to the ultimate delivery of $PGF_{2\alpha}$ to the CL leading ovary, may be less in pregnant cattle (Curl and Thatcher, 1983).

Cows were slaughtered and reproductive tracts recovered at 17 days post-oestrus (n=6; cyclic) or of pregnancy (n=6) as confirmed by presence of a conceptus. Endometrial tissues ipsilateral (EI) to the CL bearing ovary were sectioned and minced. Ovarian artery (OA) and vein (OV), ipsilateral to the CL bearing ovary, were dissected from connective tissue, washed and minced. Endometrium (200 mg in triplicate) and vessels (80 mg in duplicate) were incubated in 25 ml flasks with 5 ml minimum essential media for 1 h at 37C under an atmosphere of 50% N_2, 45% O_2 and 5% CO_2. Media was changed, and all tissues received 10 ng $[9\beta\text{-}^3H]$-$PGF_{2\alpha}$ and 0.1 µCi $[^{14}C]$-inulin (carboxyl-^{14}C). Treatments were: EI received 0, 10, 40, 100, 1000 or 5000 ng of radioinert $PGF_{2\alpha}$; OA and OV received either 0, 100 or 1000 ng of radioinert $PGF_{2\alpha}$. Flasks were regassed and incubations continued for 4 h. Tissues were solubilized in 1 ml NCS, and DPM of $[^3H]$-$PGF_{2\alpha}$ and $[^{14}C]$-inulin determined. Uptake of $PGF_{2\alpha}$ (ng $PGF_{2\alpha}$/ml cell water) was calculated after correction for intercellular space. Overall $PGF_{2\alpha}$ uptake of EI was less ($P < 0.025$) for pregnant than cyclic animals ($62.1 < 96.9$ ng/ml) and regressions of uptake differed ($P < 0.001$): Pregnant, EI ($\hat{Y} = 5.47 + 55.937X$) versus Cyclic, EI ($\hat{Y} = 1.86 + 91.989X$) were $\hat{Y}$=ng $PGF_{2\alpha}$/ml cell water and X=ng $PGF_{2\alpha}$ added. A three way interaction [pregnancy status by vessel by dose of $PGF_{2\alpha}$ (0, 100, 1000 ng)] was detected ($P < 0.01$) for vascular uptake of $PGF_{2\alpha}$. The interaction is illustrated in Fig. 6. In cyclic cattle, mean uptake (U) and slope (b) responses were greater for the OA (U=31.8 ng/ml, b=81.25) and OV (U=51.1 ng/ml, b=133.4) compared to pregnant cattle, OA (U=16.7 ng/ml, b=46.8) and OV (U=11.7 ng/ml, b=32.2). Not only were the levels of uptake (slopes) less for pregnant cattle, but the rank in slopes of uptake responses changed between cyclic (OV > OA) and pregnant (OA > OV) animals (Fig. 6).

Several salient points can be gleaned from these uptake responses of endometrial and vascular tissues. $PGF_{2\alpha}$ was accumulated in these tissues, however, uptake was not saturable in endometrial (0 to 5000 ng) or vascular (OV and OA; 0 to 1000 ng) tissues of cyclic and pregnant cattle. This suggests that a specific receptor-dependent binding system was not functioning. Differences in uptake (cyclic > pregnant) may be due to a decrease in tissue permeability to $PGF_{2\alpha}$

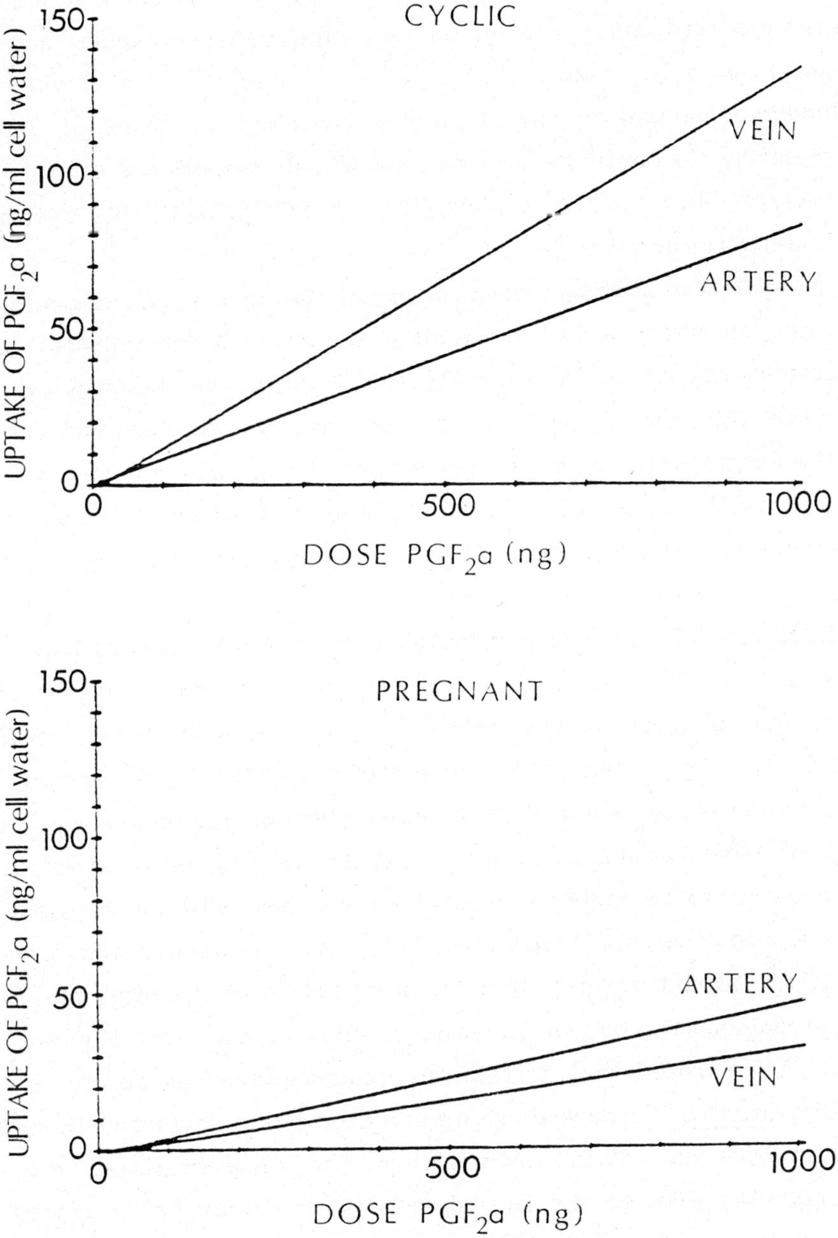

Fig. 6. Regressions of in vitro $PGF_{2\alpha}$ uptakes by ovarian veins and ovarian arteries of pregnant and cyclic cows at day 17 post-oestrus.

that was associated with pregnancy or presence of a conceptus. Riehl et al. (1983) also reported that endometrial epithelial cells of rabbits accumulated $PGF_{2\alpha}$ and PGE_2 in a manner that was not saturable, and that magnitude of prostaglandin accumulation varied among physiological states of oestrus and pseudopregnancy. In contrast, smooth muscle cell suspensions from the rabbit oviduct have distinct specific binding sites for $PGF_{2\alpha}$ and PGE_2 that are saturable (Riehl and Harper, 1981).

It is possible that $PGF_{2\alpha}$ transport within the bovine endometrium, ovarian vein and ovarian artery is by simple diffusion across a concentration gradient. Such a system may occur from the OV to OA during the oestrous cycle where there is a 4.6-fold difference in $PGF_{2\alpha}$ concentration between the OV and OA (Table III). Permeability of these tissues to $PGF_{2\alpha}$ may be decreased during pregnancy (Fig. 6), and the concentration gradient also decreased due to a diminished endometrial production of $PGF_{2\alpha}$ by the pregnant endometrium (Table IV, Fig. 4).

Localized steroid synthesis and metabolism by the conceptus may influence tissue permeability. Shemesh et al. (1979) reported that bovine blastocysts at 13, 15 and 16 days of age produced progesterone, some testosterone, and limited amounts of E_2. Eley et al. (1983) demonstrated that androstenedione can be converted to limited amounts of E_2 by bovine blastocysts recovered on days 15, 16 and 17 post-oestrus. Furthermore, extensive metabolism of androstenedione and progesterone to 5β-reduced metabolites also has been documented (Eley et al., 1983; Chenault et al., 1980). A striking observation was the predominant activity of the 5β-reductase system in conceptus tissue, whereas the 5α-reductase was predominant in endometrial tissues (Eley et al., 1983). Kubli-Garfias et al. (1979) demonstrated that certain 5β-reduced pregnanes suppressed uterine contractility in vitro. Perhaps these metabolites alter membrane function, stabilize target cells such as the endometrium, and influence tissue permeability. An additional function, of the 5β-reductase system may be to reduce steroid substrate availability for subsequent aromatization to oestrogens. Such metabolism would lower oestrogen production by the bovine conceptus (e.g. compared to porcine conceptuses) and be of a biological advantage since E_2 is luteolytic in cattle. Additional studies on the biological functions of the 5β-reduced steroid metabolites in relation to pregnancy recognition are warranted.

Effect of the bovine conceptus and conceptus produced products (PGE_2 and E_2) on delivery of exogenous $PGF_{2\alpha}$ from the uterine lumen

Three experiments were completed to examine whether delivery of exogenous $PGF_{2\alpha}$ from the uterine lumen was altered by early pregnancy or by agents such as E_2 or PGE_2 that are produced by the bovine conceptus. Objectives of experiment one were to determine whether $PGF_{2\alpha}$ and/or its metabolite PGFM may be transported from the uterine lumen, and the extent that such transport is altered by early pregnancy (Williams et al., 1980). On day 17 post-oestrus or post-mating, heifers were given intra-uterine injections of saline (2 pregnant, 2 non-pregnant) or 200 µg $PGF_{2\alpha}$ (7 pregnant, 6 non-pregnant) through cannulae implanted surgically into the lumen of the uterine horn ipsilateral to the CL bearing ovary. Jugular blood samples were collected at 5 to 10 min intervals prior to and during laparotomy, and for 90 min following the intra-uterine injection. Plasma was subsequently assayed for PGFM. At the end of the experiment, a second laparatomy was performed to confirm proper cannula placement, and uterine horns flushed to determine if blastocysts were present in mated heifers. Peripheral concentrations of PGFM were higher in pregnant compared to non-pregnant heifers during the presurgery (68 ± 26 versus 24 ± 26 pg/ml; $P < 0.025$) and surgery (186 ± 47 versus 65 ± 17 pg/ml; $P < 0.05$) periods. Pregnancy status did not alter the mean concentrations of PGFM (pregnant, 554 ± 70 pg/ml; non-pregnant, 422 ± 81 pg/ml) or the half-life of its decline in concentration (18 min) following intra-uterine injection of $PGF_{2\alpha}$ (Fig. 7). Pregnancy at 17 days in cattle does not appear to influence transport of exogenously administered $PGF_{2\alpha}$ from the uterine lumen or its metabolism in the uterus or elsewhere in response to an acute intra-uterine injection.

Basal PGFM concentrations in peripheral plasma were significantly greater in pregnant heifers than in non-pregnant heifers, both prior to and during surgery, on day 17 following breeding or oestrus. Basal concentrations greater than 50 pg/ml of PGFM were observed in 5 of 8 pregnant heifers prior to surgery, but increased concentrations were not detected in any of the non-pregnant heifers during this period. Basal concentrations were amplified slightly in both groups (pregnant > non-pregnant) during the surgery period. This increase in PGFM concentration indicates that early pregnancy modifies endogenous $PGF_{2\alpha}$-PGFM production and/or excretion. Findings of Lewis et al. (1982) and Shemesh et al.

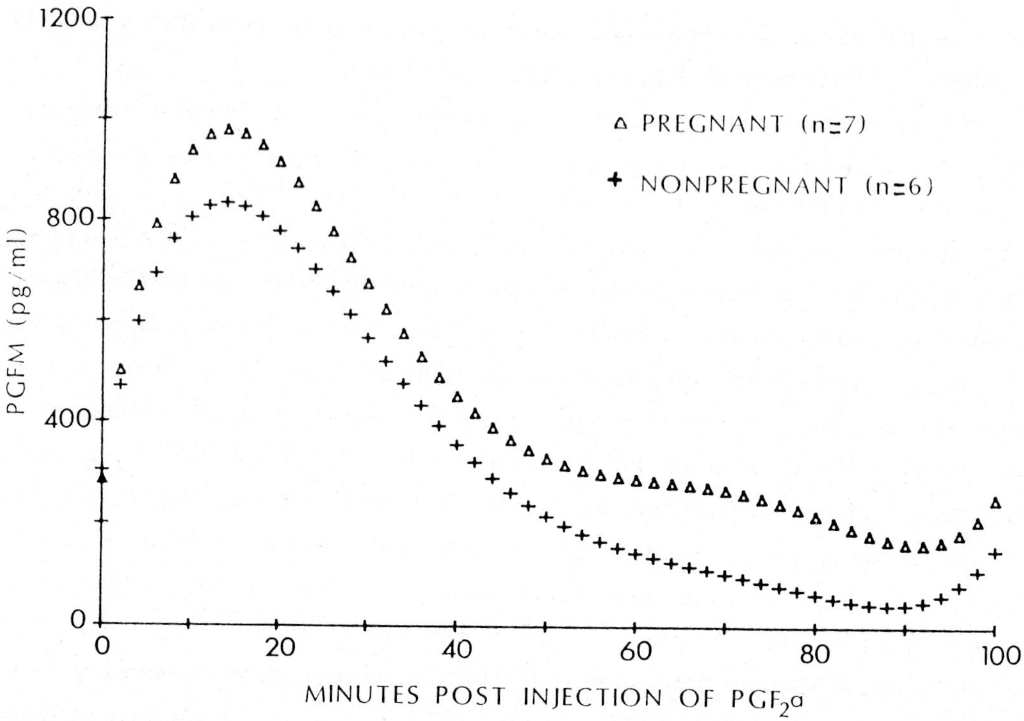

Fig. 7. Regressions of plasma PGFM concentrations in pregnant and non-pregnant heifers following intra-uterine injections of $PGF_{2\alpha}$ (200 µg).

(1979), that bovine blastocysts at this period are capable of significant prostaglandin production, suggest that the blastocyst contributes to this pregnancy increase in basal concentration of PGFM. Increases in concentrations of PGFM of pregnant heifers represented alterations in basal concentrations and were not characteristic of luteolytic peaks (Table I).

A 2 x 2 factorial experiment was designed to determine whether an increase in uterine blood flow induced by E_2 altered $PGF_{2\alpha}$ induced luteolysis in cattle (Rico et al., 1982). On day 13 of the oestrous cycle, 20 Holstein heifers were assigned randomly to four treatments: A, vehicle and saline; B, vehicle and $PGF_{2\alpha}$; C, E_2 and saline; D, E_2 and $PGF_{2\alpha}$. Six ml solutions of E_2 (3 mg) or vehicle (saline:ethanol - 1:1) were administered intravenously (time zero) and $PGF_{2\alpha}$ (5 mg) or saline were injected (5 ml) into the uterine horn ipsilateral to the corpus luteum per cervix 3 h after i.v. injection. A 3 h interval between E_2 injection and $PGF_{2\alpha}$ deposition was chosen, since maximal vasodilatation of uterine vasculature occurs in cattle at about 3 h post-injection of E_2. Jugular blood samples were collected at 8 h intervals beginning at day 12 and continuing until detected oestrus. Heifers were checked for oestrus five times daily from day 13. Mean (±SEM) of hours to luteolysis, determined by progesterone concentrations of less than 1 ng/ml in two consecutive plasma samples, were: 131.3 ± 18.7 h, 52.4 ± 17.1 h, 96.4 ± 20.9 h and 25.5 ± 18.7 h for groups A, B, C and D, respectively. Significant main effects of $PGF_{2\alpha}$ ($P < 0.01$) and E2 ($P < 0.10$) were detected on hours to luteolysis. Time series analysis of day curves for progesterone concentrations indicated that the regression curves for the four groups were not parallel ($P < 0.01$; Fig. 8). Response curves indicated that E_2 + $PGF_{2\alpha}$ treatment (group D) had the most potent luteolytic effect. The vasodilatation effect of E_2 on uterine blood vessels with a subsequent increase in uterine blood flow appeared to enhance $PGF_{2\alpha}$ release from the uterine lumen to the circulation and resulted in an earlier time of luteolysis (e.g. A = 131.3 h versus C=96.4 h; B= 52.4 h versus D=25.5 h). These results confirm that E_2 is luteolytic and accelerates the luteolytic activity of $PGF_{2\alpha}$ that was deposited in the uterus.

Nkuuhe and Manns (1981) reported that addition of PGE_1 to perfusion medium reduced flux of radioactive $PGF_{2\alpha}$ from the uterus of sheep at day 16 of the cycle. Uterine horns were being perfused with radioactive $PGF_{2\alpha}$ in medium 199 for 30 min prior to addition of PGE_1. A reduction of $PGF_{2\alpha}$ flux from the uterus due to PGE_1 may be considered as a potential antiluteolytic effect that would be associated with luteal maintenance. The bovine conceptus synthesized PGE_2 (Shemesh et al., 1979; Lewis et al., 1982), and it was of interest to determine if PGE_2 altered $PGF_{2\alpha}$ movement from the uterine lumen of cattle. To test this concept, the experimental model depicted in Fig. 9 was developed (Wolfenson and Thatcher, 1983b). Surgery was performed on day 14 of the oestrous cycle. The base of the uterine horn, ipsilateral to the CL, was ligated. An affluent catheter was placed into the uterine lumen at the tip of the uterine horn, whereas an effluent catheter was inserted into the uterine lumen at the base of the uterine horn above the main ligation. A catheter was inserted into a uterine branch of the ovarian vein with the catheter tip in the ovarian vein (OV). An arterial catheter was placed into the common carotid (CC) after catheterization of a facial artery. An electromagnetic flow transducer was placed around the uterine artery or a first branch of the uterine artery. A constant infusion rate

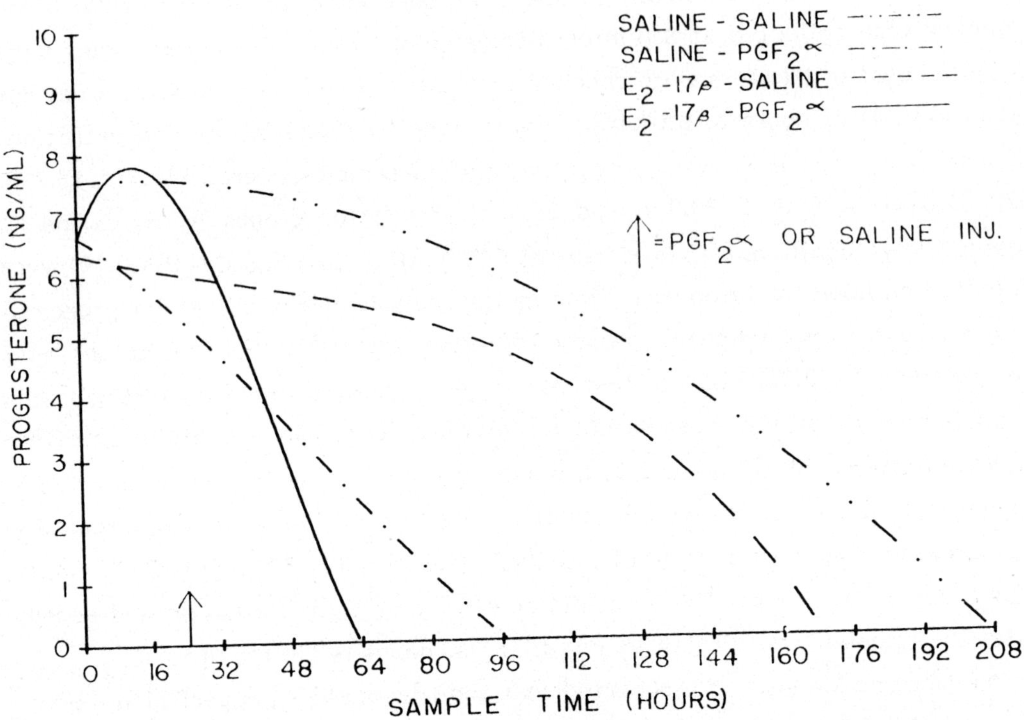

Fig. 8. Regressions of plasma progesterone concentrations in heifers injected systemically with oestradiol (3 mg) or saline followed in 3 h by intra-uterine injections of $PGF_{2\alpha}$ (5 mg) or saline on day 13 of the oestrous cycle.

0.71 ml/min) into the uterine lumen was via an infusion pump. The effluent perfusate was collected by gravity drainage into a fraction collector. Blood flow was monitored continually through a blood flow meter and recording system. Six cows were prepared in this manner. On day 15 of the oestrous cycle, one day following surgery, $PGF_{2\alpha}$ was continually perfused in a phosphate buffered saline solution (pH 7.4) at a rate of 7.54 µg/h (177 ng/ml and 0.71 ml/min) for 9 h (Fig. 10). After 3 h of $PGF_{2\alpha}$ infusion, PGE_2 was added to the perfusate at a concentration of 10 µg/ml (0.71 ml/min or 426 µg/h) and infusion continued for 3 h. At 6 h, PGE_2 was withdrawn, and infusion of $PGF_{2\alpha}$ alone was continued until 9 h (Fig. 10). This system permitted us to monitor uterine blood flow and to calculate uterine prostaglandin production (output due to infusion plus endogenous $PGF_{2\alpha}$ production) from the experimental uterine horn. Simultaneous blood

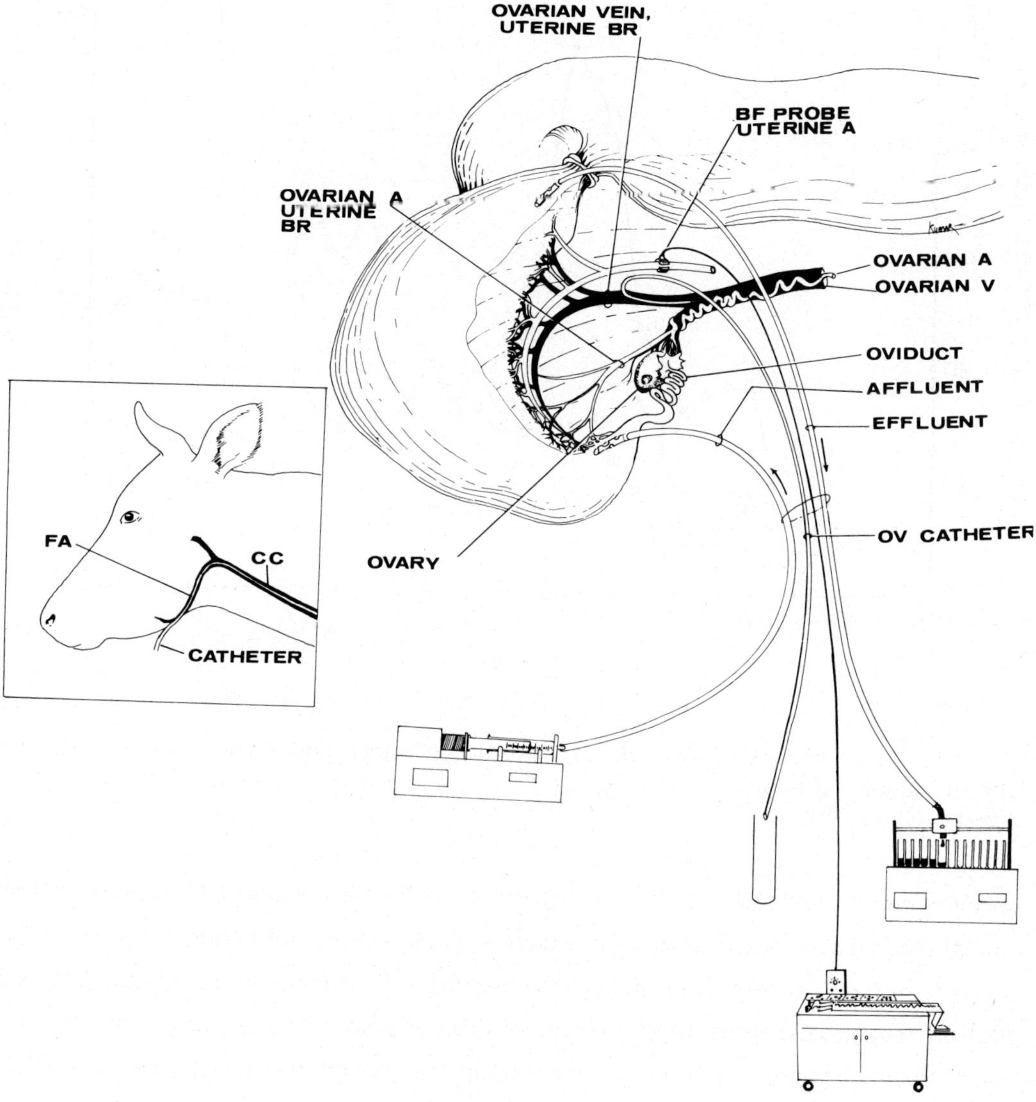

Fig. 9. Diagrammatic presentation of experimental model to test effect of $PGF_{2\alpha}$ and PGE_2 uterine lumenal infusions on uterine production of $PGF_{2\alpha}$ in cyclic cows at day 15. FA = facial artery; CC = common carotid; A = artery; BF Probe = electromagnetic blood flow probe; OV = ovarian vein.

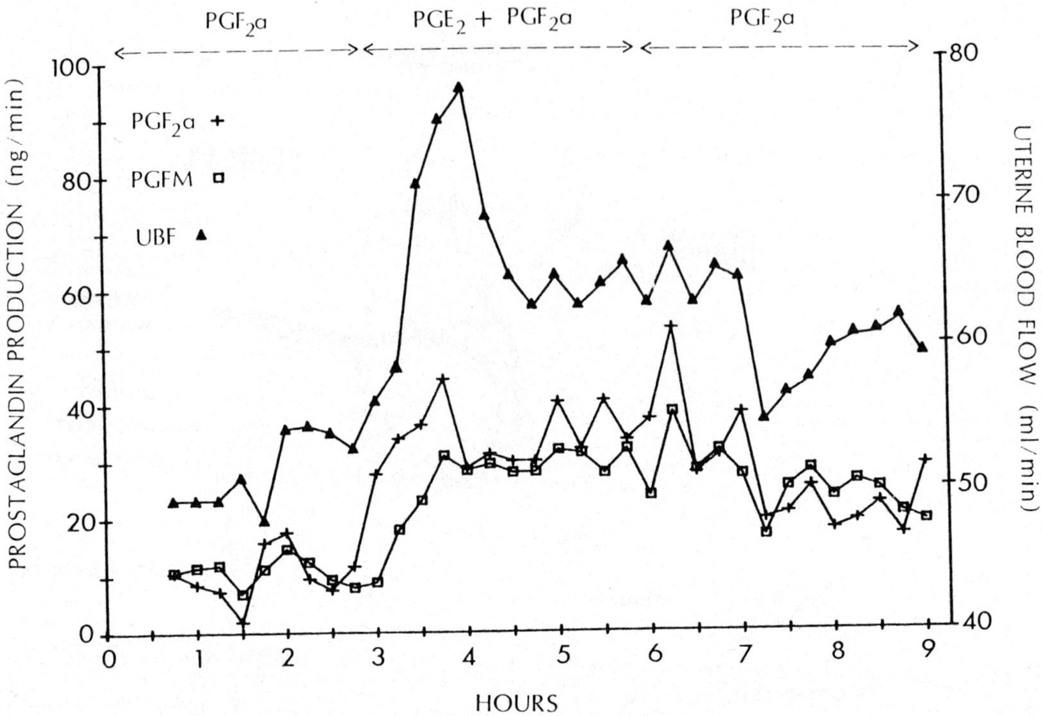

Fig. 10. Time means of uterine prostaglandin and blood flow responses to uterine lumenal infusions of $PGF_{2\alpha}$ and PGE_2 in cyclic cows at day 15. UBF = uterine blood flow.

samples were collected at 15 min intervals from the OV and CC vessels. All six cow preparations contributed to the data set. However, only four cows had functional electromagnetic flow probes to permit calculation of prostaglandin production responses represented in Fig. 10. Data points in Fig. 10, represent the least squares means for each 15 min sampling period for the four cows. PGE_2 infusion increased uterine blood flow ($P < 0.01$) which demonstrated its clear vasodilatory effect on the uterine vasculature in cattle. Associated with increase in uterine blood flow were rises in both $PGF_{2\alpha}$ and PGFM production (or output) ($P < 0.01$) from the uterus. This provided additional evidence for localized metabolism of $PGF_{2\alpha}$ within the uterus as discussed earlier. The increase in $PGF_{2\alpha}$ production (or output) in response to PGE_2 infusion does not support the hypothesis that PGE_2 will inhibit $PGF_{2\alpha}$ movement from the uterus.

In fact, results of these three *in vivo* transport experiments indicate that such agents as PGE_2 and E_2, which are produced by the conceptus, will enhance movement of endometrial or conceptus produced materials from the uterus. It is reasonable that various products of conceptus origin could contribute to local physiological effects. For example, an increase in blood flow to the pregnant uterine horn at day 16 (Ford et al., 1979) may be due to conceptus production of a vasodilatory agent such as PGE_2. A selective increase in uterine blood flow may increase histotrophic nutrients for the developing conceptus. Prostaglandin secretion by conceptus tissues is much more pronounced than is oestrogen secretion. Likewise, a conceptus induced increase in uterine blood flow would enhance delivery of agents from the uterus that may exert a luteotrophic effect on the CL bearing ovary. A luteotrophic factor may be involved, since some factor secreted from the pregnant uterine horn overrides the luteolytic effect of the non-pregnant uterine horn when uterine venous vessels of both horns are merged surgically (Ginther, 1981). Perhaps small molecular weight substances of the prostaglandin series (e.g. PGE_2 produced by the conceptus) may exert a luteotrophic effect. It is interesting that both PGI_2 (prostacyclin; Milvae and Hansel, 1980) and PGE_1 (Kimball and Lauderdale, 1975) will stimulate plasma progesterone concentrations when administered to cattle *in vivo*. In a preliminary report of very limited data, Gimenez and Henricks (1983) reported that intra-uterine infusion with PGE_2 (1.3 mg infused every 4 h from day 9 through 21 post-oestrus) prolonged the luteal phase in two beef cows for a 4 day period beyond that of a control cow. In the pregnant cow, a vasodilatory response to PGE_2 also could enhance transport of proteinase materials that may be luteotrophic. Regardless of these optional luteotrophic postulates, the early pregnant cow appears to have a diminished endometrial secretion of $PGF_{2\alpha}$ that would have been of a luteolytic magnitude. A balance of luteolytic and luteotrophic substances may lead to luteal maintenance, and control of their delivery from the uterus is critical to the process of pregnancy recognition.

ACKNOWLEDGEMENTS

Several areas of research presented were partially supported by the Upjohn Co. Agricultural Division, Kalamazoo, Michigan.

REFERENCES

Bartol, F.F., Thatcher, W.W., Bazer, F.W, Kimball, F.A., Chenault, J.R., Wilcox C.J. and Roberts, R.M., 1981a. Effects of the estrous cycle and early pregnancy on bovine uterine, luteal and follicular responses. Biol. Reprod., 25: 759-776.

Bartol, F.F., Thatcher, W.W., Lewis, G.S., Bliss, E.L., Drost, M. and Bazer, F.W., 1981b. Effect of estradiol-17β on PGF and total protein content in bovine uterine flushings and peripheral plasma concentration of 13,14-dihydro-15-keto-PGF$_{2\alpha}$. Theriogenology, 15: 345-358.

Batra, S.K. and Pandey, R.S., 1983. Relative concentration of 13,14-dihydro-15-keto-prostaglandin F$_{2\alpha}$ in blood and milk of buffaloes during the oestrus cycle and early pregnancy. J. Reprod. Fert., 67: 191-195.

Brunner, M.A., Donaldson, L.E. and Hansel, W., 1969. Exogenous hormones and luteal function in hysterectomized and intact heifers. J. Dairy Sci., 52: 1849-1854.

Chenault, J.R., 1980. Steroid metabolism by the early bovine conceptus I. 5β-reduction of neutral C$_{19}$-steroids. J. Steroid Biochem., 13: 499-506.

Chenault, J.R., Thatcher, W.W., Kalra, P.S., Abrams, R.M. and Wilcox, C.J., 1975. Transitory changes in plasma progestins, estradiol, and luteinizing hormone approaching ovulation in the bovine. J. Dairy Sci., 58: 709-717.

Christian, L.E., Everson, D.O. and Davis, S.L., 1978. A statictical method for detection of hormone secretory spikes. J. Animal Sci., 46: 699-706.

Curl, J.S. and Thatcher, W.W., 1983. In vitro uptake of PGF$_{2\alpha}$ by tissues from the endometrium, ovarian vein and ovarian artery of cyclic and pregnant cattle at day 17. 16th Annual Meeting of Soc. for Study of Reproduction. 8-10 August, 1983, at Case Western Reserve Univ., Cleveland, Ohio. p. 108 (Abstract).

Curl, J.S., Thatcher, W.W. and Bartol, F.F., 1983. In vitro production of PGE$_2$, PGF$_{2\alpha}$ and 13,14-dihydro-15-keto-PGF$_{2\alpha}$ (PGFM) by endometrial tissue from cattle at day 17 of pregnancy or the estrous cycle. 75th Annual Meeting of Amer. Soc. Anim. Sci. 26-29 July, 1983, at Washington State Univ., Pullman, Washington. p. 327 (Abstract).

Eley, R.M., Thatcher, W.W. and Bazer, F.W., 1979. Luteolytic effect of oestrone sulphate on cyclic beef heifers. J. Reprod. Fert., 55: 191-193.

Eley, R.M., Thatcher, W.W., Bazer, F.W. and Fields, M.J., 1983. Steroid metabolism by the bovine uterine endometrium and conceptus. Biol. Reprod., 28: 804-816.

Fairclough, R.J., Smith, J.F. and McGowan, L.T., 1981. Prolongation of the estrous cycle in cows and ewes after passive immunization with PGF antibodies. J. Reprod. Fert., 62: 213-219.

Ford, S.P., Chenault, J.R. and Echternkamp, S.E., 1979. Uterine blood flow of cows during the estrous cycle and early pregnancy: effect of the conceptus on the uterine blood supply. J. Reprod. Fert., 56: 53-62.

Gimenez, T., and Henricks, D.M., 1983. Prolongation of the luteal phase by prostaglandin E$_2$ during the estrous cycle in the cow. A Preliminary Report. Theriogenology, 19: 693-700.

Ginther, O.J., 1981. Local versus systemic uteroovarian relationships in farm animals. Acta vet. scand., Suppl. 77: 103-115.

Greenstein, J.S., Murray, R.W. and Foley, R.C., 1958. Effect of exogenous hormones on the reproductive processes of the cycling dairy heifer. J. Dairy Sci., 41: 1834 (Abstract).

Hafs, H.D., Louis, T.M., Noden, P.A. and Oxender, W.D., 1974. Control of the estrous cycle with prostaglandin $F_{2\alpha}$ in cattle and horses. J. Anim. Sci., Suppl. I. 38: 10.

Hansel, W., Concannon, P.W. and Lukaszewska, J.H., 1973. Corpora lutea of large domestic animals. Biol. Reprod., 8: 222-245.

Hawk, H.W., 1979. Infertility in dairy cattle. In Beltsville Symposia In Agricultural Research. 3. Animal Reproduction, pp. 19-29. Editor H.W. Hawk, Allanheld, Osmum and Co., Montclair, New Jersey.

Hixon, J. and Hansel, W., 1974. Evidence for preferential transfer of prostaglandin $F_{2\alpha}$ to the ovarian artery following intrauterine administration in cattle. Biol. Reprod., 11: 543-552.

Kimball, F.A. and Lauderdale, J.W., 1975. Prostaglandin E_1 and $F_{2\alpha}$ specific binding in bovine corpora lutea: Comparison with luteolytic effects. Prostaglandins, 10: 313-331.

Kindahl, H., Edqvist, L.-E., Bane, A. and Granström, E., 1976. Blood levels of progesterone and 15-keto-13,14-dihydro-prostaglandin $F_{2\alpha}$ during the normal estrous cycle and early pregnancy in heifers. Acta Endocr., 82: 134-139.

Kindahl, H., Lindell, J.-O. and Edqvist, L.-E., 1981. Release of prostaglandin $F_{2\alpha}$ during the oestrous cycle. Acta vet. scand., Suppl. 77: 143-158.

Knickerbocker, J.J., Thatcher, W.W., Foster, D.B., Wolfenson, D., Bartol, F.F. and Caton, D., 1982. Uterine production of $PGF_{2\alpha}$ in response to estradiol-17β in cyclic cattle. Annual Meeting of ASAS and CSAS. 8-11 August, 1982, at University of Guelph, Ontario, Canada. p. 364. (Abstract).

Kubli-Garfias, C.L., Mediano-Conde, L., Beyer, C. and Bondani, A., 1979. In vitro inhibition of rat uterine contractility induced by 5α and 5β progestins. Steroids, 34: 609-617.

Lamothe, P., Bosquet, D. and Guay, P., 1977. Cyclic variation of F prostaglandins in the uterine fluids of the cow. J. Reprod. Fert., 50: 381-382.

Lauderdale, J.W., 1974. Distribution and biological effects of prostaglandins. J. Anim. Sci., Suppl. I. 22:38.

Lewis, G.S., Thatcher, W.W., Bazer, F.W. and Curl, J.S., 1982. Metabolism of arachidonic acid in vitro by bovine blastocysts and endometrium. Biol. Reprod., 27: 431-439.

Malven, P.V. and Hansel, W., 1964. Ovarian function in dairy heifers following hysterectomy. J. Dairy Sci., 47: 1388-1393.

Mapletoft, R.J., Del Campo, M.R. and Ginther, O.J., 1976. Local venoarterial pathway for uterine-induced luteolysis in cows. Proc. Soc. Exp. Biol. Med., 153: 289-294.

Milvae, R.A. and Hansel, W., 1980. The effects of prostacyclin (PGI_2) and 6-keto-$PGF_{1\alpha}$ on bovine plasma progesterone and LH concentrations. Prostaglandins, 20: 641-647.

Milvae, R.A., and Hansel, W., 1983. Luteolytic effect of 13,14-dihydro-$PGF_{2\alpha}$ in heifers. J. Reprod. Fert., 67: 203-207.

Nancarrow, C.D., Buckmaster, J., Chamley, W.A., Cox, R.I., Cumming, I.A., Cummins, L., Drinan, J.P., Findlay, J.K. Goding, J.R., Restall, B.J., Schneider, W. and Thorburn, G.D., 1973. Hormone changes around oestrus in the cow. J. Reprod. Fert., 32: 320-321.

Nkuuhe, J. and Manns, J., 1981. Effect of prostaglandin (PGE_1) on fluid and $PGF_{2\alpha}$ fluxes across the uterine epithelium in the ewe. 14th Annual Meeting of Society for the Study of Reproduction. 10-13 August, 1981 at Oregon State University, Corvallis, Oregon, 24(1) p. 154A (Abstract).

Northey, D.L. and French, L.R., 1980. Effect of embryo removal and intrauterine infusion of embryonic homogenates on the lifespan of the bovine corpus luteum. J. Anim. Sci., 50: 298-302.

Peterson, A.J., Fairclough, R.J., Payne, E. and Smith, J.F., 1975. Hormone changes around bovine luteolysis. Prostaglandins, 10: 675-681.

Rico, L.E., Thatcher, W.W., Drost, M., Wolfenson, D. and Terqui, M., 1981. Plasma PGFM responses to estradiol injection in pregnant and cyclic cows. 73rd Annual Meeting of Amer. Soc. Anim. Sci. 26-29 July, 1981 at North Carolina State University at Raleigh, North Carolina. p. 363 (Abstract).

Rico, L.E., Thatcher, W.W., Drost, M. and Grimmett. J., 1982. Effect of estradiol-17β on $PGF_{2\alpha}$ induced luteolysis. 77th Annual Meeting, Amer. Dairy Sci. Assoc., 27-30 June, 1982, at Pennsylvania State Univ., University Park, Pennsylvania. p. 181 (Abstract).

Riehl, R.M. and Harper, M.J.K., 1981. Preparation of smooth muscle cell suspensions from the rabbit oviduct and prostaglandin binding analysis. Endocrinology, 108: 18-26.

Riehl, R.M., Pathak, R.K. and Harper, M.J.K., 1983. A reliable method for isolating endometrial epithelial cells from rabbits, and preliminary studies of prostaglandin uptake. Biol. Reprod., 28: 363-375.

Shemesh, M. and Hansel, W., 1975. Levels of prostaglandin F (PGF) in the bovine endometrium, uterine venous, ovarian artery and jugular plasma during the estrous cycle. Proc. Soc. Exp. Biol. Med., 148: 123-126.

Shemesh, M., Milaguir, F., Ayalon, N. and Hansel, W., 1979. Steroidogenesis and prostaglandin synthesis by cultured bovine blastocysts. J. Reprod. Fert., 56: 181-185.

Thatcher, W.W. and Chenault, J.R., 1976. Reproductive physiological responses of cattle to exogenous prostaglandin $F_{2\alpha}$. J. Dairy Sci., 59: 1366-1375.

Thatcher, W.W., Terqui, M., Thimonier, J., Ravault, J.P. and Mauleon, P., 1979. 15 keto PGF and hormonal responses to injection of estradiol in heifers. Proc. S. Div. Amer. Soc. Anim. Sci. p 43. New Orleans, Louisiana (Abstract).

Villa-Godoy, A., Ireland, J.J., Wortman, J.A., Ames, N.K. and Fogwell, R.L., 1981. Luteal function in heifers following destruction of ovarian follicles at three stages of diestrus. 73rd Annu. Meeting of Amer. Soc. Anim. Sci., 26-29 July, 1981 at North Carolina State University, Raleigh, North Carolina p. 372 (Abstract).

Williams, W.F., Lewis, G.S., Thatcher, W.W. and Underwood, C.S., 1980. Plasma PGFM in pregnant and open heifers prior to and during laparotomy and following intrauterine injection of $PGF_{2\alpha}$. 72nd Annual Meeting of Amer. Soc. of Anim. Sci., Cornell Univ., Ithaca, New York 50(1): 337 (Abstract).

Wiltbank, J.N. and Casida, L.E., 1956. Alteration of ovarian activity by hysterectomy. J. Anim. Sci., 15: 134-140.

Wiltbank, J.N., Ingalls, J.E. and Rowden, W.W., 1961. Effects of various forms and levels of estrogen alone or in combinations with gonadotropins on the estrous cycle of beef heifers. J. Anim. Sci., 20: 341-346.

Wolfenson, D. and Thatcher, W.W., 1983a. Concentrations of $PGF_{2\alpha}$ in the ovarian artery of early pregnant and cyclic cattle (unpublished results).

Wolfenson, D. and Thatcher, W.W., 1983b. Effect of uterine lumenal perfusion of $PGF_{2\alpha}$ and PGE_2 on uterine output of $PGF_{2\alpha}$ in cyclic cattle (unpublished results).

Animal Reproduction Science, 7 (1984) 177—194
Elsevier Science Publishers B.V., Amsterdam — Printed in The Netherlands

CONTROL OF PROSTANOID SYNTHESIS IN BOVINE TROPHOBLAST AND PLACENTOME

M. Shemesh+, W. Hansel*, J. Strauss, III++, A. Rafaeli+, S. Lavi+ and F. Mileguir+

\+ Department of Hormone Research, Kimron Veterinary Institute, Beit Dagan, P.O. Box 12, (Israel).

* Department of Physiology, New York State College of Veterinary Medicine, Cornell University, Ithaca, New York 14853, (USA).

++ Departments of Pathology and Obstetrics and Gynecology, University of Pennsylvania School of Medicine, Philadelphia, PA 19104, (USA).

ABSTRACT

Shemesh, M., Hansel, W., Strauss, J., Rafaeli, A., Lavi, S. and Mileguir, F., 1984. Control of prostanoid synthesis in bovine trophoblast and placentome. Anim. Reprod. Sci., 7: 177-194.

Dispersed bovine trophoblast from pre-implantation embryos produced PGE_2 and $PGF_{2\alpha}$ during the initial 8 days of culture. Concomitant with the appearance of syncytial masses, prostanoid production was reduced by day 12 of culture. Addition of 8-bromo-cyclic AMP hastened the appearance of syncytial masses and reduced prostanoid production. Dispersed fetal cotyledons and maternal caruncles produced less prostanoid on a per cell basis than pre-implantation trophoblast. Appreciable net synthesis of PGI_2 and $PGF_{2\alpha}$ by maternal caruncle cells only occurred with tissue removed at term. The reduced prostanoid formation by placental tissue may be due, in part, to an endogenous inhibitor of prostanoid synthesis. A heat-labile factor from cytosolic fractions of placentomes was found to inhibit prostaglandin formation by dispersed luteal cells, bovine granulosa cells and dispersed fetal cotyledon. The factor also inhibited arachidonate metabolism by seminal vesicle microsomes. We conclude that temporal changes in prostaglandin production are related to differentiation of the tissues, and are modulated by an endogenous inhibitor.

INTRODUCTION

During the past decade an impressive body of literature has been generated indicating that prostanoids play a critical role in a variety of reproductive processes including ovulation, regression of the corpus luteum, implantation, regulation of utero-placental blood flow, ripening of the cervix, and myometrial contractile activity at term. The ovaries, uterus and placental tissues each synthesize a number of products from arachidonic acid, including prostaglandin E2 (PGE_2), prostaglandin $F_{2\alpha}$ ($PGF_{2\alpha}$) and prostaglandin I_2 or prostacyclin (PGI_2). These prostanoids have been shown to influence the function of the reproductive tract in in vivo and in vitro experiments.

The role of prostanoids in the control of corpus luteum function has received particular attention. In the cyclic animal, this gland has a lifespan characteristic for each species; it secretes progesterone for a limited period and then regresses. Following fertilization, the functional and structural demise of the corpus luteum (luteolysis) is delayed, and pregnancy ensues. Maintenance of the function of the corpus luteum during pregnancy requires pituitary gonadotropins, in some species ovarian oestradiol, as well as tropic hormones derived from the placenta. In certain species a suppression of the formation of luteolytic factors must also occur.

In laboratory and some domestic animals, the factor responsible for the demise of the corpus luteum of the non-fertile cycle is $PGF_{2\alpha}$ which emanates from the uterus (McCracken et al., 1972). In the cycle of conception, uterine $PGF_{2\alpha}$ synthesis is reduced, and it has been suggested that the presence of conceptuses in the uterine lumen causes local inhibition of $PGF_{2\alpha}$ production (Rowson and Moor, 1967; Thorburn et al., 1973; Martal et al., 1979). Indeed, infusion of homogenates of trophoblast into the uterus of cycling ewes leads to maintenance of corpus luteum function. This observation is especially intriguing since release of prostaglandins by the trophoblast may be important for implantation (Kennedy, 1977; Rankin et al., 1979). Although the mechanisms by which suppression of uterine luteolytic activity are achieved remain to be elucidated, one possibility that seemed particularly attractive was that the conceptuses and placentae produce an inhibitor of prostaglandin synthesis. Luteolysis must also occur prior to parturition in species in which progesterone, that maintains pregnancy, is derived mainly from corpora lutea. Uterine $PGF_{2\alpha}$ initiates corpus luteum regression in this situation (Currie et al., 1973; Umo et al., 1976) and hence, whatever suppressive effect the conceptuses and placentae have on uterine luteolytic activity during gestation must be relieved by term.

Because of the apparently complex interactions between the products of conception and the uterus with respect to synthesis of prostanoids, we have begun to examine some factors controlling prostaglandin synthesis by bovine trophoblast and placentomes. As an initial step we studied prostanoid synthesis by dispersed trophoblast and placentome cells maintained in culture. In addition, evidence for a placental inhibitor of prostaglandin synthesis was sought.

MATERIALS AND METHODS

Animals

Embryos and placental tissue were collected from multiparous Holstein-Friesan cows at an abbatoir. Gestational age was estimated by the size of the embryo or crown-rump length of the fetus.

Culture of dispersed trophoblast and placentomes

Trophoblast from embryos at 16-18 days of gestation, or maternal caruncles or fetal cotyledons collected at 120-275 (term) days of gestation were enzymatically dispersed using collagenase by a modification of the method of Simmons et al. (1976). Dispersed cells (2-5 x 10^5 cells/vessel) were cultured in plastic 25 cm^2 flasks (Falcon) in medium 199 (Gibco) supplemented with 5% calf serum. The cells were maintained at 37 C in an atmosphere of 5% CO_2:95% air. Culture media were changed every 4 days. In each experiment aliquots of cells were taken on the day of plating to determine cellular prostanoid content.

Preparation of placentome extracts

Extracts were routinely prepared from placentomes of 100-200 days gestational age and, in some instances, tissue obtained at term by a previously reported method (Ailenberg and Shemesh, 1983). The placentomes were collected within 30 min of slaughter and placed on ice. Fetal cotyledons were sometimes separated from the maternal caruncles.

Pools of tissue (300-500 g) were homogenized in 3 vol of chilled 0.9% saline. Homogenates were centrifuged at 10,000 x g for 30 min and the supernatants from this centrifugation were treated with ammonium sulfate (20% w/v). The precipitated material was collected by centrifugation, suspended in distilled water and dialyzed for 96 h against distilled water at 4 C. The dialyzed solution was adjusted to pH 7 and lyophilized. The lyophilized material was tested for inhibitory activity on prostanoid synthesis.

Assay of the inhibitor of prostanoid synthesis

The inhibitory activity of placentome extracts on prostanoid synthesis was assessed using three different incubation systems: 1) suspensions of bovine granulosa cells; 2) dispersed luteal cells from mid-cycle (day 12) bovine corpora lutea; and 3) dispersed fetal cotyledon cells. The incubation of bovine granulosa cells has been described previously (Shemesh, 1979). Briefly, 2-5 x 10^5 cells were routinely used, with prostanoid secretion being monitored over 24 h periods. The procedure of Simmons et al. (1976) was used to prepare dispersed bovine luteal cells, which were used in short term incubations (2 h). Similar methods were used to prepare dispersed cells from fetal cotyledons. Medium 199 was used as the basic incubation medium in all studies. Each treatment group consisted of five to eight replicate incubations in all experiments. Prostanoid levels were determined in the cells at zero time and then in the incubation medium after the indicated time of incubation.

Measurement of prostaglandin synthetase activity

Bovine seminal vesicles were obtained fresh from a slaughter house. The glands were cleaned of surrounding fat and homogenized in 0.1 M potassium phosphate buffer, pH 8.1, (5 ml buffer to 1 g of tissue) with a Sorvall Omnimixer. The crude homogenate was centrifuged at 800 xg and the supernatant centrifuged at 108,000 xg for 1 h. The resulting precipitate, designated as microsomes, was resuspended to the original volume in 0.1 M potassium phosphate buffer, pH 8.1. Prostaglandin synthetase activity was determined by incubating 1 ml aliquots of the suspension with 0.2 ml of buffer containing 10 µg arachidonic acid (Sigma, St. Louis) and 200 nCi of $[1-^{14}C]$ arachidonic acid (Amersham, 56.4 mCi/mmol) at 37C for 30 min. Incubations were terminated by the addition of 2 M citric acid (0.25 ml) as described previously by Shemesh et al. (1981). The products formed were extracted twice with 10 vol of ether and separated by thin-layer chromatography using a system of chloroform: methanol: acetic acid:

water (90:9:1:0.65, by vol.). To determine the effects of tissue extracts on enzymatic activity, 1 to 6 mg of boiled or non-boiled placentome extract were added to the incubation tubes.

Analytical methods

PGE_2, $PGF_{2\alpha}$ (Shemesh, 1979) and 6-keto prostaglandin $F_{1\alpha}$, the stable product of PGI_2 (Salmon, 1978) were quantitated by specific radioimmunoassays.

Statistical methods

Student's t-test or Dunnett's t-test for comparison of multiple treatment groups against a single control group were used to determine significant differences between treatment groups.

RESULTS

Prostanoid production by dispersed trophoblast, maternal caruncules and fetal cotyledons

Dispersed trophoblast from 16-18 day embryos consisted of four main cell types in culture: small round cells, mono or multi-nucleated giant cells, syncytial cells and fibroblast-like cells. The predominant cell type in each of the cultures changed with time. In the case of dispersed trophoblast, during the first week of culture the majority of cells were of the small round type (Fig. 1a). After 2-3 weeks of culture there had been a dramatic change in the morphology of the cells with the disappearance of small round cells and the appearance of syncytial masses with multiple nuclei, one or more nucleoli and granular cytoplasm. The number of fibroblast-like cells also increased during this time (Fig. 1b).

Dispersed trophoblast released PGE_2 and $PGF_{2\alpha}$ into the culture medium (Fig. 2). Net synthesis of both of these prostanoids was observed as the amount of PGE_2 and $PGF_{2\alpha}$ released exceeded the "zero-time" cellular contents. Lower amounts of PGE_2 were released during the first 4 day period and greatest release took place over the subsequent 4 days, with secretion declining to low levels by the twelfth day of culture. Secretion of $PGF_{2\alpha}$ was less than PGE_2, but it was also greatest during the first 8 days of culture and then declined to low levels. The marked decline in the secretion of prostanoids coincided with the appearance of syncytial masses in the cultures.

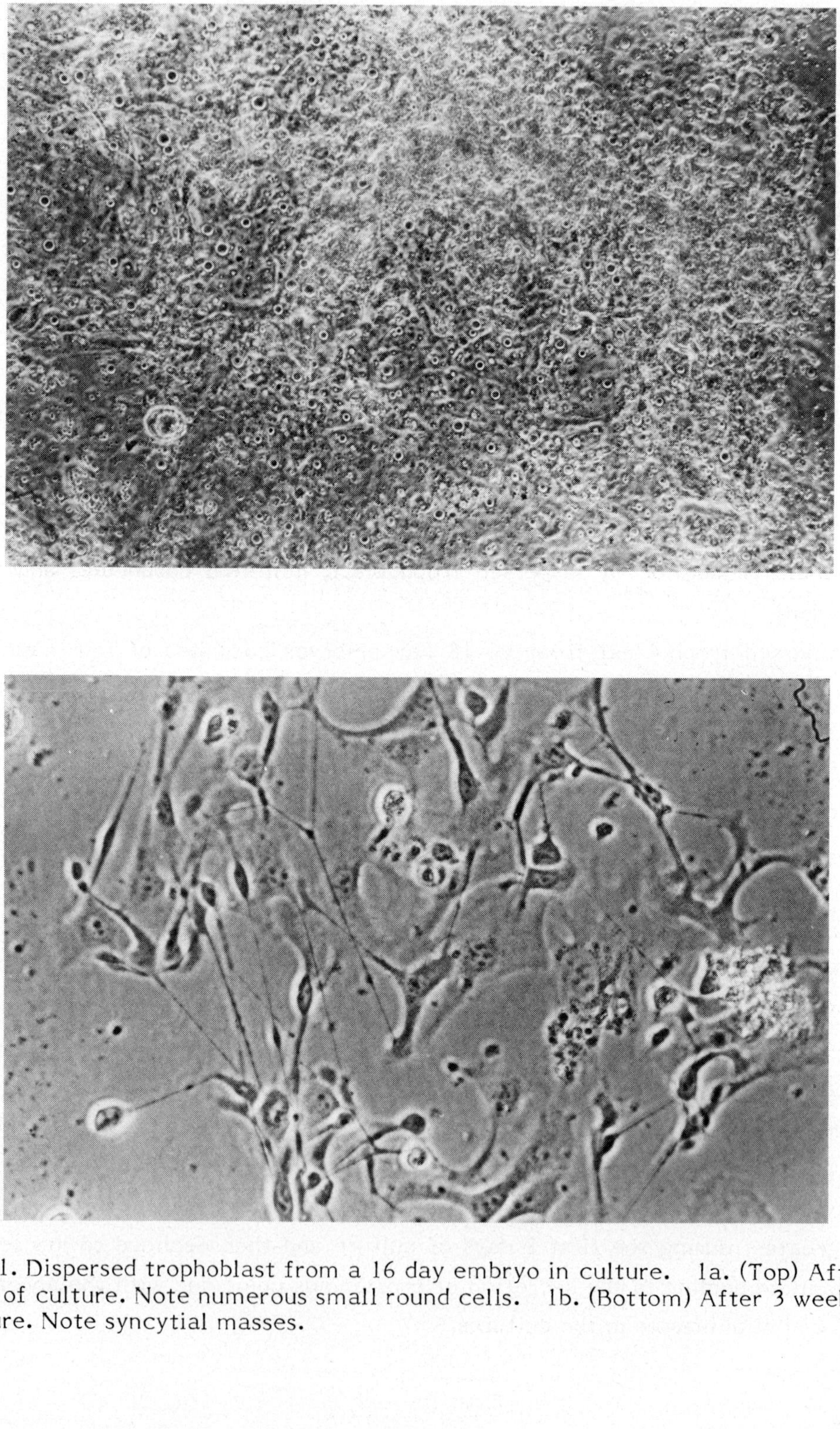

Fig. 1. Dispersed trophoblast from a 16 day embryo in culture. 1a. (Top) After 7 days of culture. Note numerous small round cells. 1b. (Bottom) After 3 weeks of culture. Note syncytial masses.

Cultures of dispersed caruncules and cotyledons also contained small round cells, giant cells, syncytial cells, and fibroblast-like cells. The syncytial masses were more abundant in fetal cotyledon cultures. The ultrastructure of the fetal cotyledon cells was distinctive. The cell cytoplasm usually contained numerous vacuoles, a prominent Golgi system, and distended rough endoplasmic reticulum formations. A fibrillar extracellular matrix was seen between the cells.

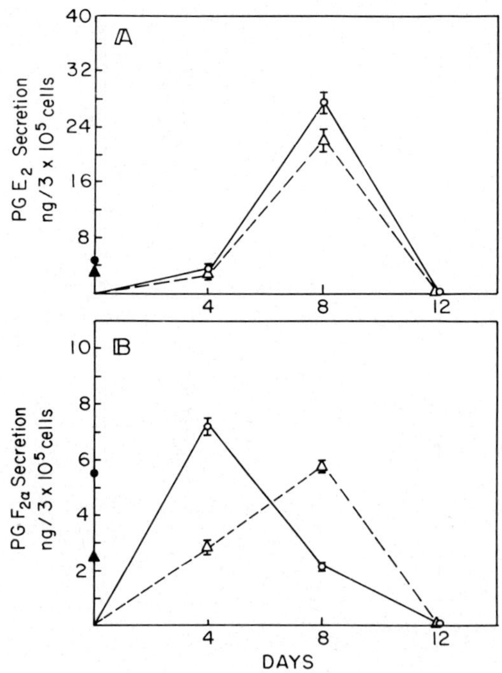

Fig. 2. PGE_2 (A) and $PGF_{2\alpha}$ (B) production by dispersed trophoblast from 16-18 days embryos. Dispersed trophoblast was maintained in culture as described in the text. Prostanoid levels in cells at plating (●,▲) and in the medium (○,△) at the indicated times were determined. Values presented are means ± SEM of 5 replicates from two separate experiments.

The production of prostanoids by dispersed maternal caruncules and fetal cotyledons was considerably less, on a per cell basis, than for dispersed trophoblast. With cells prepared from tissue collected on day 260 of gestation, there appeared to be minimal net synthesis of PGI_2 by maternal caruncle cells while net production of PGI_2 by the fetal cotyledon cells was demonstrable. A representative study is shown in Fig. 3. The same was true for $PGF_{2\alpha}$, with minimal net formation being found with maternal caruncle cells and demonstrable net synthesis observed with fetal cotyledon. However, with tissue harvested at term (day 275 of gestation), there was evidence for net synthesis of PGI_2 and $PGF_{2\alpha}$ by maternal caruncles. A representative study is shown in Fig. 4.

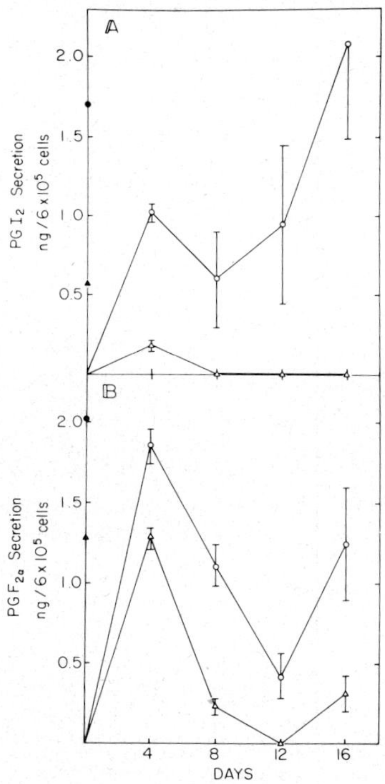

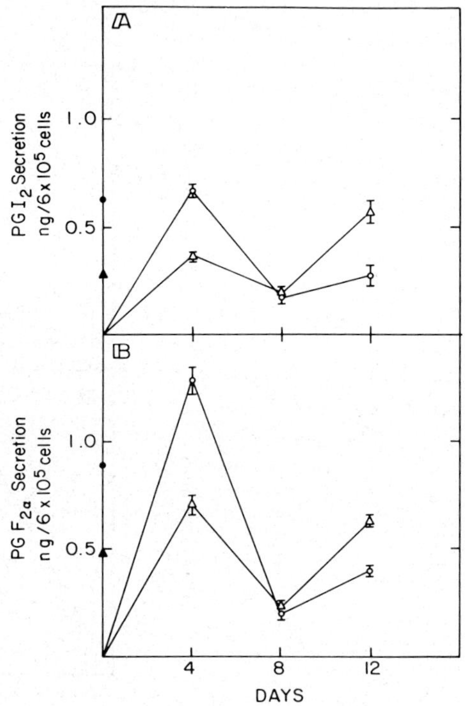

Fig. 3. (Left) PGI_2 (A) and $PGF_{2\alpha}$ (B) production by dispersed maternal caruncle and fetal cotyledon from 260 days placentomes. Dispersed maternal caruncle (△) and fetal cotyledon (○) were maintained in culture as described in the text. Prostanoid levels were determined in the cells and in the medium at the indicated times. Prostanoid levels in cells at plating (●, ▲). Values are means ± SEM of 5 replicate cultures.

Fig. 4. (Right) PGI_2 (A) and $PGF_{2\alpha}$ (B) production by dispersed maternal caruncle and fetal cotyledon from term placentomes. Dispersed maternal caruncle (△) and fetal cotyledon (○) were maintained in culture as described in the text. Prostanoid levels were determined in the cells and in the medium at the indicated times. Prostanoid levels in cells at plating (●, ▲). Values are means ± SEM of 5 replicate cultures.

We were interested in determining if the function of the trophoblast could be modulated in vitro by tropic agents. We chose for our initial studies to examine the effects of cyclic nucleotides. Addition of the analogue of cAMP, 8-bromo-cyclic AMP, to cultured trophoblast cells resulted in an accelerated appearance of syncytial masses such that they were present by day 8 in treated cultures but not in controls. These masses became more pronounced in appearance with time (Fig. 5). The morphological change in the cultures was associated with a significant reduction in net synthesis of prostanoids (Table I). These observations suggest that cyclic AMP may alter prostaglandin formation by affecting the differentiation of the cells in culture.

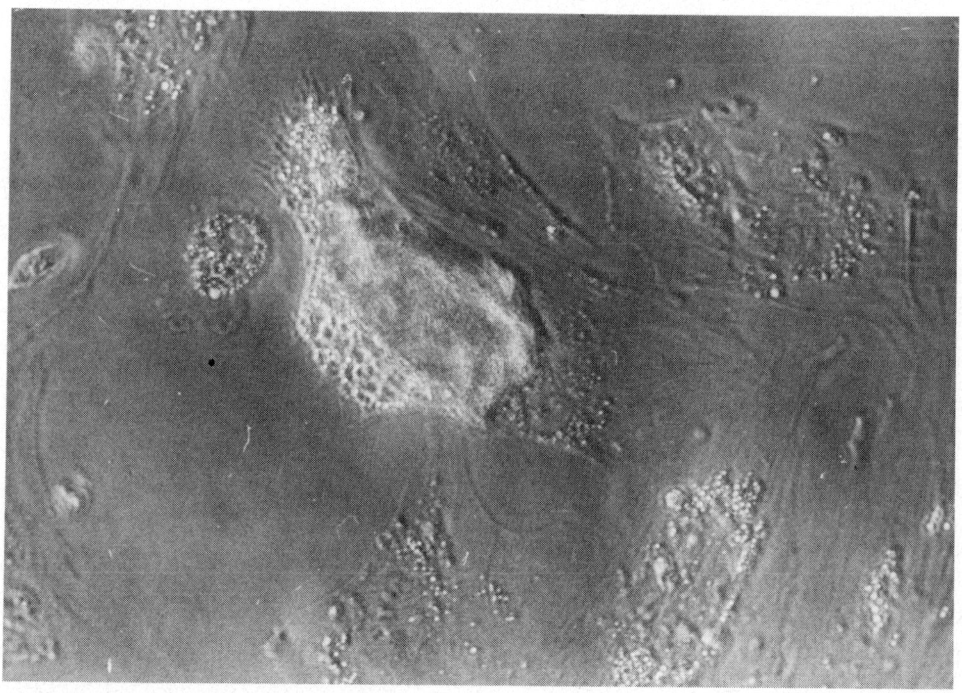

Fig. 5. Effect of 8-bromo-cAMP on dispersed trophoblast in culture. Cells were exposed to 1 mM 8-bromo-cAMP. Syncytial masses were marked in cultures exposed to the cycle nucleotide. This photomicrograph shows one such mass in cultures treated for 3 weeks.

Table I. The effect of 8-bromo-cAMP on formation of $PGF_{2\alpha}$ and PGE_2 by cultured trophoblast cells

Prostaglandin production ng/3 x 10^5 cells/4 days			
$PGF_{2\alpha}$		PGE_2	
Control	8-bromo-cAMP	Control	8-bromo-cAMP
5 ± 0.2	0.5	25 ± 3	5 ± 0.4

Trophoblast from 16-18 day embryos was dispersed and placed in culture as described in the text. Medium was changed after 4 days of culture and 1 mM 8-bromo-cAMP was added to the medium. Prostaglandin secretion during the subsequent 4 days was then determined. Values presented are means ± SEM of net synthesis from a single experiment in which there were 17 replicate cultures in each group. 8-Bromo-cyclic AMP reduced ($p < 0.001$) secretion of both prostanoids.

Presence of an inhibitor of prostanoid synthesis in bovine placentomes

To determine if placentomes contain an inhibitor of prostanoid synthesis which might account for the suppression of uterine luteolytic activity during pregnancy, we examined the effects of extracts from 100-200 day placentae on prostaglandin synthesis by a variety of cells. Results from a representative experiment in which the effects of the extracts on prostanoid production by bovine luteal cells were examined are shown in Table II. In the presence of 2 mg/ml of caruncle extract there was complete inhibition of PGI_2 and $PGF_{2\alpha}$ secretion during 2 h of incubation. If treated luteal cells were collected and washed by centrifugation and then reincubated for an additional 20 h, the inhibitory action of placental extract was only partially removed (Table II). Thus, the inhibitory effect was relatively long-lived. Dispersed fetal cotyledon or maternal caruncle cells from 243 day placentomes produced significant reductions (69 to 87%) in $PGF_{2\alpha}$ and PGI_2 in the incubation medium when co-incubated with dispersed luteal cells (Table III). This suggests that the placental cells actually secrete the inhibitory material.

The effects of the 20% ammonium sulfate precipitates from cytosolic extracts of whole placentomes collected at 100-150 days of gestation or at term on

Table II. Effect of placental extracts on PGI_2 and $PGF_{2\alpha}$ formation by bovine luteal cells

Treatments	Prostanoid production pg/2 x 10^5 cells			
	2 h incubation with extract		20 h incubation after washing cells	
	6-keto-$PGF_{1\alpha}$	$PGF_{2\alpha}$	6-keto-$PGF_{1\alpha}$	$PGF_{2\alpha}$
Control	176 ± 10	89 ± 17	4037 ± 94	1983 ± 89
Placentome extract	N.D.	N.D.	618 ± 37	340 ± 23

Dispersed luteal cells were incubated initially with 2 mg/ml of whole placentome extract from 150 day placentae or buffer for 2 h. The cells were then washed by centrifugation and resuspended in incubation medium and reincubated for 20 h. Prostanoid secretion was determined during each of these incubations. Values are means ± SEM for net prostanoid synthesis of 5 replicates in each treatment group. The placentome extract significantly ($p < 0.01$) reduced prostanoid production in the short term and long term incubations.

N.D. = not detectable

$PGF_{2\alpha}$ production by bovine granulosa cells are shown in Table IV. These extracts produced a dose-dependent inhibition of prostanoid formation. Addition of arachidonic acid (2 µg/ml) failed to overcome the inhibitory effects of the placental extracts, suggesting that the inhibitory factor was not acting by reducing the supply of prostanoid precursor. The inhibitory activity in the placental extracts was destroyed by boiling (10 min) as well as by acidification to pH 3.

Dispersed fetal cotyledon from a 243 day placenta was found to produce small quantities of $PGF_{2\alpha}$ and PGI_2 during short-term incubation (Table V). The addition of 2 mg/ml of maternal caruncle or fetal cotyledon extract from a 150 day placenta produced a $\geq$ 50% suppression of $PGF_{2\alpha}$ and PGI_2 release by the cells.

We next examined the direct effect of the placental extracts on the conversion of arachidonic acid into prostanoids by bovine seminal vesicle microsomes. Three distinct radioactive peaks were found upon thin-layer chromatography of reaction products. One peak had an R_f value similar to that of an authentic

Table III. Effect of dispersed caruncle and cotyledon cells from a 243 day placentomes on PGI_2 and $PGF_{2\alpha}$ formation by dispersed bovine luteal cells

Treatments	Prostanoid production pg/5 x 10^5 cells/3 h					
	6-keto-$PGF_{1\alpha}$	% Inhibition	P	$PGF_{2\alpha}$	% Inhibition	P
Luteal cells	358 ± 35			693 ± 146		
Luteal cells + maternal caruncle cells	80 ± 57	78	< 0.001	175 ± 12	75	< 0.001
Luteal cells + fetal cotyledon cells	111 ± 48	69	< 0.01	91 ± 43	87	< 0.001

Dispersed luteal cells from mid-cycle corpora lutea (5 x 10^5/cells tube) were incubated in the absence or presence of 1.5 x 10^6 dispersed cells from either caruncles or cotyledons. PGI secretion, assessed as 6-keto-$PGF_{1\alpha}$, and $PGF_{2\alpha}$ were determined after a 3 h incubation. Values presented are means ± SEM of 5 replicates from a single experiment.

Table IV. Effect of placentome extracts on $PGF_{2\alpha}$ formation by bovine granulosa cells

$PGF_{2\alpha}$ production pg/4 x 10^5 cells/24 h				
Gestational age of placentome		Amount of placental extract added		
	Control	0.5 mg/ml	1 mg/ml	2 mg/ml
100-150 days	560 ± 39	-	385 ± 18 (31%)[a]	280 ± 15 (50%)[a]
Term	560 ± 39	316 ± 17 (43%)[a]	-	50 (90%)[a]

Bovine granulosa cells were incubated with various doses of the placental extracts from mid gestation (100-150 days) or term placentae for a 24 h period. $PGF_{2\alpha}$ released into the incubation was determined. Values represent means ± SEM for net synthesis by 5 replicates in each treatment group.

[a]Percent inhibition

Table V. Effect of extracts from caruncles and fetal cotyledons from a 150 day placenta on $PGF_{2\alpha}$ and PGI_2 formation by dispersed fetal cotyledon cells from 243 day placentomes

	Prostanoid production pg/5 x 10^5 cells/3 h	
Treatment	$PGF_{2\alpha}$	6-keto-$PGF_{1\alpha}$
Control	149 ± 88	111 ± 31
Maternal caruncle extracts (2 mg/ml)	None detected	None detected
Fetal cotyledon extracts (2 mg/ml)	None detected	None detected

Dispersed fetal cotyledon cells from a 243 day placenta were incubated in the absence or presence of extracts of caruncle or fetal cotyledons from a 150 day placenta. $PGF_{2\alpha}$ and PGI_2, measured as 6-keto-$PGF_{1\alpha}$, secretion during a 3 h incubation were determined. Values presented are means ± SEM for net synthesis by 5 replicates from a single experiment.

$PGF_{2\alpha}$ standard (R_f = 0.30). Another peak had an R_f similar to that of thromboxane B_2 (R_f = 0.42) and the third peak had an R_f equivalent to that of PGE_2 (R_f = 0.56). These products were not formed when the microsomal preparations were boiled prior to incubation. By determining the sum of radioactivity accumulating in the three peaks we obtained an estimate of metabolism of arachidonate to prostanoids. Using this measure of enzymic activity, the addition of placental extract was found to cause a dose-dependent inhibition of arachidonate metabolism (Table VI). Inhibitory activity was present in both mid-gestation and term placentomes.

Table VI. Effect of placental extract on arachidonate metabolism by seminal vesicle microsomes

Treatment	% Inhibition of ^{14}C-arachidonate metabolism dose of extract (mg/ml)		
	1	3	6
Mid gestation	52%	62%	84%
Term	60%	88%	97%

Metabolism of ^{14}C-arachidonate was determined in the presence of various concentrations of extract of mid-gestation or term placentomes as described in the text.

DISCUSSION

Our results demonstrate that bovine trophoblast and placentomes are capable of producing prostanoids following enzymatic dispersion and incubation in culture. The ability of trophoblast from 16-18 day embryos to generate prostanoids during the initial days in culture, on a per cell basis, exceeded that of dispersed fetal cotyledon or cells obtained from maternal caruncles. As prostaglandins have been implicated in the implantation process (Kennedy, 1977; Rankin et al., 1979), which occurs around 30 days of gestation in the cow, it is of interest that pre-implantation trophoblast cells reveal a high rate of prostanoid formation. However, the high rate of prostaglandin synthesis is not maintained and declines markedly concomitant with the formation of syncytial masses in the cultures.

These findings suggest that the prostaglandin-synthesizing activity changes in parallel with differentiation of trophoblast cells. 8-Bromo cyclic AMP appears to hasten the formation of syncytial masses in culture and also leads to a reduction in prostaglandin production by the trophoblast cells.

Our observations of reduced formation of prostaglandins by trophoblast cells as their differentiation occurs in vitro are consistent with our finding of a relatively low rate of prostanoid formation by dispersed fetal cotyledon. While fetal cotyledon cells demonstrated some net synthesis of prostanoids when they were derived from day 260 or term placentomes, maternal caruncle cells only displayed an ability to generate prostanoids when they were prepared from placentomes collected at term. This increase in the synthesizing capacity of maternal caruncle cells at term may have significance in the control of parturition. Even though the amount of net prostanoid synthesis per cell was relatively small at this time, the amount of caruncular tissue in the uterus at term (kilograms) could be responsible for the formation of a substantial quantity of prostanoid. Evans et al. (1982) have recently reported that dispersed placentomes from sheep increase production of prostanoids as pregnancy approaches term. The observations of these investigators are consistent with our findings.

The relatively low prostanoid production by placental tissue from mid gestation and late pregnancy compared with trophoblast may be in part accounted for by the presence of an endogenous inhibitor of prostaglandin synthesis. A heat-labile protein, present in the cytosolic fraction of caruncles and cotyledons, suppressed prostanoid synthesis when added to bovine luteal and granulosa cells as well as dispersed fetal cotyledon. The material may be secreted by the placental cells as the addition of dispersed caruncle or cotyledon to dispersed bovine lutein cells resulted in suppression of prostanoid levels in the culture fluid. The placental factor also reduced metabolism of labeled arachidonic acid by bovine seminal vesicle microsomes.

The exact nature of this inhibitory material is not known nor is its mechanism of action certain. Since exogenous arachidonic acid did not overcome its action, it is unlikely that it affects the availability of prostanoid precursor. Since the placental extracts directly inhibited the metabolism of arachidonate by microsomes, it is possible that the inhibitor acts on cyclo-oxygenase or one of the essential cofactors for arachidonate metabolism to prostanoids. If this is its primary mode of action, then it must be capable of entering cells. While our preliminary data suggest the factor is a protein as it is precipitable by ammonium sulfate and is heat-labile and acid-labile, this remains to be proven. Indeed, the factor could be a small molecule which associates with proteins.

The concept of an endogenous inhibitor of prostaglandin synthesis has recently been proposed by several investigators working with other systems. Of considerable relevance to the present studies is the report of Harrowing and Williams (1979) of a factor in homogenates of rat and human placenta which inhibits arachidonate metabolism. Wlodawer et al. (1976) have also shown that the bovine endometrium contains an inhibitor of prostaglandin biosynthesis. An inhibitor of prostaglandin synthesis has been found in bovine follicular fluid (Shemesh, 1979). It is not known whether the various prostaglandin synthesis inhibitors which have been described are similar in structure or activity.

A number of interesting questions remain unexplored with respect to the function of the endogenous prostaglandin inhibitor found in the placenta. Does 8-bromo-cyclic AMP act to suppress prostanoid production by pre-implantation trophoblast by inducing inhibitor in the tissue? Does this inhibitor only function locally in the placenta or uterus, or is it a secretory product capable of affecting other organs such as the ovary? What is the role of this material near term? At this time uterine prostaglandin production increases and we have found enhanced prostanoid formation by maternal caruncle cells immediately prior to parturition. Is the action of the inhibitor overcome at term and if so, by what mechanism?

Clearly, the interactions of conceptus and uterus on prostanoid syntheses are complex. Our initial investigations point out several important areas for future exploration. These include: 1) the mechanism by which the conceptus suppresses uterine prostaglandin synthesis while requiring prostanoid production for implantation; 2) the factors controlling prostaglandin synthesis by differentiating trophoblast; 3) the potential role of cyclic nucleotides in modulating trophoblast and placental prostanoid production; and 4) the role of endogenous inhibitors of prostaglandin production in the regulation of placental and uterine prostaglandin synthesis.

CONCLUSIONS

The interactions of the conceptus and the uterus with respect to prostaglandin synthesis appear to change with the state of differentiation of the tissues. One factor which may play a significant role in modulating prostanoid synthesis during this time is a heat-labile endogenous inhibitor of prostanoid synthesis found in placentomes.

ACKNOWLEDGEMENTS

This research was supported by USPHS grant HD 16103 and a grant from the United States-Israel (Bi-National) Agricultural Research and Development Fund-BARD (I-1-80). The technical assistance of Margaret Hammond and Ray Satman is gratefully acknowledged. We thank Ms. Dana Kelly for her assistance in the preparation of this manuscript.

REFERENCES

Ailenberg, M. and Shemesh, M., 1983. Partial purification of a chorionic gonadotropin-like protein from bovine cotyledons. Biol. Reprod., 28: 517-522.

Currie, W.B., Wong, M.S.F., Cox, R.I. and Thorburn, G.D., 1973. Spontaneous or dexamethasone induced parturition in the sheep and goat: Changes in plasma concentrations of maternal prostaglandin F and foetal oestrogen sulphates. Mem. Soc. Endocrinol., 20: 95-118.

Evans, C.A., Kennedy, T.G. and Challis, J.R.G., 1982. Gestational changes in prostanoid concentrations in intrauterine tissues and fetal fluids from pregnant sheep, and the relation to prostanoid output in vitro. Biol. Reprod., 27: 1-11.

Harrowing, P.D. and Williams, K.I., 1979. Homogenates of rat placenta contain a factor(s) which inhibits arachidonic acid metabolism. Brit. J. Pharm., 67: 482P.

Kennedy, T.G., 1977. Evidence for a role for prostaglandin in the initiation of blastocyst implantation in the rat. Biol. Reprod., 16: 286-291.

Martal, J., Lacroix, M.C, Loudes, C., Saunier, M. and Winterberger-Torres, S., 1979. Trophoblastin an antiluteolytic protein present in early pregnancy in sheep. J. Reprod. Fert., 56: 63-73.

McCracken, J.A., Carlson, J.C., Glew, M.E., Goding, J.R., Baird, D.T., Green, K. and Samuelsson, B., 1972. Prostaglandin $F_{2\alpha}$ identified as a luteolytic hormone in sheep. Nature, London 238: 129-134.

Rankin, J.C., Ledford, B.E., Jonsson, H.J. and Baggett, B., 1979. Prostaglandins, indomethacin and the decidual cell reaction in the mouse uterus. Biol. Reprod., 20: 399-404.

Rowson, L.E.M. and Moor, R.M., 1967. The influence of embryonic tissue homogenate infused into the uterus on life-span of the corpus luteum in the sheep. J. Reprod. Fert., 13: 511-516.

Salmon, J.A., 1978. A radioimmunoassay for 6-keto-prostaglandin $F_{1\alpha}$. Prostaglandins, 15: 383-397.
Shemesh, M., 1979. Inhibiting action of follicular fluid on progesterone and prostaglandin synthesis in bovine follicles. J. Endocr., 82: 27-31.
Shemesh, M., Ailenberg, M., Lavi, S. and Mileguir, F., 1981. Regulation of prostaglandin biosynthesis by an endogenous inhibitor from bovine placenta. In: N.B. Schwartz and M. Hunzicker-Dunn (Editors), Dynamics of Ovarian Function, p. 161-166, Raven Press, New York.
Simmons, K.R., Caffrey, J.L., Phillips, J.L., Abel, J.H. and Niswender, G.D., 1976. A simple method for preparing suspensions of luteal cells. Proc. Soc. Exp. Biol. Med., 152: 366-371.
Thorburn, G.D., Cox, R.I., Currie, W.B., Restall, B.J. and Schneider, W., 1973. Prostaglandins F and progesterone concentrations in the utero-ovarian venous plasma of the ewe during the oestrous cycle and early pregnancy. J. Reprod. Fert., Suppl. 18: 151-158.
Umo, I., Fitzpatrick, R.J. and Ward, W.R., 1976. Parturition in the goat: Plasma concentrations of prostaglandin F and steroid hormones and uterine activity during late pregnancy and parturition. J. Endocrinol., 68: 383-389.
Wlodawer, P., Kindahl, H. and Hamberg, M., 1976. Biosynthesis of prostaglandin $F_{2\alpha}$ from arachidonic acid and prostaglandin endoperoxides in the uterus. Biochem. Biophys. Acta., 431: 603-614.

Animal Reproduction Science, 7 (1984) 195—216
Elsevier Science Publishers B.V., Amsterdam — Printed in The Netherlands

THE INFLUENCE OF PREGNANCY ON $PGF_{2\alpha}$ SECRETION IN CATTLE. I. Concentrations of 15-keto-13,14-dihydroprostaglandin $F_{2\alpha}$ and progesterone in peripheral blood of recipients of transferred embryos

K.J. Betteridge*, G.C.B. Randall, M.D. Eaglesome and E.A. Sugden

Agriculture Canada, Animal Diseases Research Institute, P.O. Box 11300, Station H, Nepean, Ontario, K2H 8P9, (Canada).

ABSTRACT

Betteridge, K.J., Randall, G.C.B., Eaglesome, M.D. and Sugden, E.A., 1984. The influence of pregnancy on $PGF_{2\alpha}$ secretion in cattle. 1. Concentrations of 15-keto-13,14-dihydro-prostaglandin $F_{2\alpha}$ and progesterone in peripheral blood of recipients of transferred embryos. Anim. Reprod. Sci., 7: 195-216.

Three experiments were performed to study the influence of viable or formalin-killed embryos, transferred late in the oestrous cycle, on the maintenance of the corpus luteum and prostaglandin production by the uterus. In Experiment 1, day-12 to day-16 embryos were transferred to synchronous recipients (± 24h) on day 12 to day 17. All five recipients of formalin-killed embryos and nine recipients of viable embryos exhibited a decline in progesterone to baseline levels (0.8 ng/ml) which was followed by oestrus in some of them. In a further nine animals, subsequently shown to be pregnant, luteal phase levels of progesterone were maintained. In six animals (not pregnant at day 42) progesterone levels were maintained for varying periods following transfers between days 13 and 17. In Experiment 2, progesterone and 15-keto-13,14-dihydroprostaglandin $F_{2\alpha}$ (PGFM) concentrations were determined in plasma samples collected hourly for 15 days from seven animals receiving fresh or formalin-killed embryos on day 16 or 17. Marked hour-to-hour changes in progesterone concentrations were seen in all animals initially and persisted in those in which luteal phase levels were maintained. In three cows progesterone concentrations declined to baseline levels during the period of hourly sampling. Progesterone levels were maintained in the others although they subsequently declined in two cases. In all seven cows a large PGFM peak (189-828 pg/ml) was associated with surgical transfer. The luteolysis in three recipients was associated with a series of PGFM peaks (165-396 pg/ml)

---

* Present address: Centre de Recherche en Reproduction Animale, Faculté de Médecine Vétérinaire, Université de Montréal, C.P. 5000, St-Hyacinthe, Québec, J2S 7C6, (Canada).

whereas only occasional single peaks were observed in those cows in which luteal phase concentrations of progesterone were maintained. In Experiment 3 the cause of the PGFM peak associated with surgery was investigated. While extensive manipulation of the genital tract, at different times in the cycle, or oxytocin injections on day 16 had little effect on PGFM concentrations, a peak (310-1000 pg/ml) was observed immediately after the intravenous injection of sodium thiopentone.

It was concluded that the corpus luteum can be maintained by the introduction of a viable embryo as late as day 17 of the cycle and that the decline in progesterone concentration at luteolysis is associated with a series of PGFM peaks which were absent or reduced in animals in which progesterone levels were maintained.

INTRODUCTION

The transfer of hatched "embryos" (blastodermic vesicles) up to 17 days after oestrus in cattle leads to either normal pregnancy or a return to oestrus after an extended (> 24-day) or normal (~21-day) oestrous cycle (Betteridge et al., 1980). Pregnancy failures, representing embryonic death, are of great economic significance, especially to the embryo transfer industry after various kinds of embryo manipulation or storage (Betteridge, 1982).

The pathogenesis of embryonic death is poorly understood: for example, it is not known whether the death of a transferred embryo causes, or is caused by, the associated luteolysis and return to oestrus. Two hormones in the maternal circulation which could help define the real sequence of events during early pregnancy failure are progesterone and prostaglandin $F_{2\alpha}$: progesterone as a reflection of corpus luteum (CL) function; $PGF_{2\alpha}$ as a generally accepted important factor in luteolysis in cattle. In view of the extremely short half-life of $PGF_{2\alpha}$, the metabolite, 15-keto-13,14-dihydroprostaglandin $F_{2\alpha}$ (PGFM), is a more practical indicator of $PGF_{2\alpha}$ production and has been used in investigations of how the embryo prevents luteolysis during early pregnancy in both sheep and cattle. There are two schools of thought as to how pregnancy affects PGFM levels in the maternal blood in the sheep and cow. In sheep, long-term frequent sampling of peripheral blood has shown that the PGFM peaks associated with luteolysis in cyclic animals are very much reduced at equivalent times after oestrus in pregnant animals (Peterson et al., 1976; Webb et al., 1981; Moore et al., 1982). Shorter-term frequent sampling of uterine venous blood, on the other hand, revealed no difference in PGFM concentrations between pregnant and non-pregnant ewes 11-16 days after oestrus (Lewis et al., 1977). In cattle, Kindahl et al. (1976, 1981) reported that pregnancy abolished the PGFM peaks of the oestrous cycle whereas Williams et al. (1980) described peripheral concentrations of PGFM as being greater in pregnant that in non-pregnant heifers. Inskeep and Murdoch (1980) have urged caution in the use of metabolites as an indicator of uterine secretion of PGF2α but there is evidence in cattle that fluctuations in peripheral plasma concentrations of PGFM reflect, at least in part, similar fluctuations in uterine synthesis and release of PGF2α (Bartol et al., 1981). In buffaloes, Batra and Pandey (1983), sampling at 6 h intervals, have shown a pronounced PGFM peak in peripheral plasma about 44 h before the onset of oestrus. The peak seems to be much smoother and longer sustained than in cattle. The same authors, with less frequent sampling, found no such peak and no rise in PGFM levels in plasma for the first 40 days of pregnancy in buffaloes.

This paper reports a study of progesterone and PGFM concentrations in cattle used as recipients of embryos transferred 12-17 days after oestrus and among which, therefore, a considerable amount of early embryonic death was to be expected. Much of the work has been described in preliminary form in two abstracts (Betteridge et al., 1978; Randall et al., 1980).

MATERIALS AND METHODS

The study comprised three experiments: in Experiment 1, 29 recipients were sampled daily to follow progesterone levels before and after surgical transfer; in Experiment 2, seven recipients were sampled hourly for 15 days to follow progesterone and PGFM levels before and after transfer; in Experiment 3, the cause of an apparently "unphysiological" PGFM peak found to be associated with anaesthesia in Experiment 2 was investigated in 19 animals and also in vitro.

Animals and surgical procedures

The beef and dairy heifers used, as well as their management and surgical methods of embryo collection and transfer, have been described elsewhere (Betteridge et al., 1980). Preoperatively, food was withheld for 48 h and water for 24 h. Blood samples were taken from the jugular vein using evacuated tubes ("Vacutainers", Becton Dickinson) containing EDTA as an anticoagulant. They were kept in iced water until centrifuged at 4°C (within about 1 hour of collection) after which plasma was aspirated and stored in two aliquots at -20°C in screw-topped glass vials until thawed for assay.

Assays

Progesterone was measured by a radioimmunoassay previously described (Betteridge et al., 1977). Within-assay and between-assay coefficients of variation were 6.6% and 14.5%, respectively. The lower limit of detection in the assay as used was 0.8 ng/ml.

15-Keto-13,14-dihydro-$PGF_{2\alpha}$ (PGFM) was assayed in unextracted 0.5 ml volumes of plasma using the technique described by Granström and Kindahl (1976). The radioimmunoassay for PGFM used a goat antibody (Reference 11560-JCC-140D, goat #23) kindly supplied by Dr. K.T. Kirton of the Upjohn Company. Its crossreactivity has been determined by the supplier to be 100% with 15-keto-13,14-dihydroprostaglandin $F_{2\alpha}$; 20% with 15-ketoprostaglandin $F_{2\alpha}$; $\simeq$0.5% with 13,14-dihydroprostaglandin $F_{2\alpha}$; <0.1% with C_{16} urinary prostaglandin metabolites; and < 0.1% with prostaglandin $F_{2\alpha}$. Standard PGFM was also kindly supplied by the Upjohn Company and radio-labelled tracer PGFM (^3H, specific activity 100 Ci/mmol) was purchased from the New England Nuclear Company. The sensitivity of the assay was 30 pg PGFM/ml plasma, and, at plasma levels of 230 pg/ml, the within-assay coefficient of variation was 10% while that between-assays was 11.1%.

Assays of both progesterone and PGFM were performed in duplicate and results were excluded if values for paired aliquots differed from their mean by more than 20%.

Within assays, peaks were defined as those values exceeding mean basal levels by more than two standard deviations (Brinkley, 1981).

Experiment 1

Embryos were obtained from 14 superovulated donors (one day-12, three day-13, five day-15 and five day-16; day 0 = first day of oestrus) as previously described (Betteridge et al., 1980). Day-15 embryos from two of the donors were fixed in formolsaline to provide control, killed embryos which were then washed by prolonged immersion in five changes of excess culture medium (TCM-199, Gibco) over 3 days at 5° C before transfer in a minimal volume of the same medium to five control recipients (one day-14, three day-15 and one day-16). Each recipient received one embryo in each uterine horn under general anaesthesia (Betteridge et al., 1980). The remaining donors provided embryos which were similarly transferred but in the normal, living state to 24 recipients at either day 12, 13, 14, 15, 16 or 17 (2, 3, 11, 3, 2 and 3 animals, respectively). All recipients were synchronous (± 24 h) with their respective donors. Eighteen of those 24 recipients received pairs of embryos which had been biopsied for sex determination (Hare et al., 1976) in connection with other work; three of the 24 received pairs of embryos already described in another context (Betteridge et al., 1980); one recipient received four day-16 embryos. Blood samples for progesterone analysis were taken daily (with occasional exceptions) from the day of transfer (or 1-2 days before) until day 27-29 (pretransfer oestrus = day 0). Animals were observed for signs of oestrus as far as their post-operative condition would allow.

Experiment 2

Seven recipients were bled in tie-stalls at hourly intervals for 15 days from the day preceding bilateral surgical transfer of freshly recovered day-16 embryos (three day-16 and two day-17 recipients) or of control day-15 embryos fixed in formol-saline and washed as in Experiment 1 (one day-16 and one day-17 recipient). Less frequent samples (once or twice daily) were taken before and after the hourly sampling session. In each animal, the haematocrits of the first one hundred samples were calculated from measurements taken directly from the vacutainer sampling tubes in order to gauge the possible effects of haemoconcentration on hormone concentrations.

Experiment 3

The cause of apparently artefactual PGFM peaks in Experiment 2 was investigated in five parts, a-e.

In Experiment 3(a), blood samples were taken from three day-7 superovulated donor heifers and three day-7 recipients at 30 min intervals over a 10-12 h period which included routine transcervical collection or transfer procedures under epidural anaesthesia (5 ml lidocaine hydrochloride). Additional samples were taken during the procedures which themselves lasted 13 min for an attempted collection, 60 and 75 min for collections and 5 min for transfers.

In Experiment 3(b), blood samples were taken from two day-12, two day-16, one day-17 and one day-19 cyclic heifers at 30 min intervals over 8 h periods which included extensive manipulation of the genital tract under epidural anaesthesia. Additional samples were taken during the manipulations themselves which lasted 15-20 min and included introduction of a vaginal speculum, dilatation of the cervix with a steel rod and thorough transrectal palpation of the uterine horns and ovaries.

In Experiment 3(c), the same blood sampling schedule as in Experiment 3(b) was used in two day-16 cyclic heifers which were injected i.v. with 10 and 100 i.u. oxytocin in two separate injections 3.5 h apart. In one heifer the lower dose was given first; in the other heifer, the higher dose. Additional samples were taken immediately before, and within 2-3 min after, the oxytocin injections.

In Experiment 3(d), similar blood sampling schedules to those of Experiments 3(b) and 3(c) were used in five heifers anaesthetized in the same way as recipients in Experiment 2 with sodium thiopentone (approximately 11 mg/kg body weight) given by rapid i.v. injection as a 25% solution. Two of the heifers were at day 7 of their oestrous cycle, two were at day 16 and the fifth had been ovariohysterectomized 3 months previously. Anaesthesia was maintained (in the cyclic animals only) by halothane in oxygen in closed-circuit following endotracheal intubation. The midline was clipped, shaved and scrubbed as for surgery in these four animals. Additional samples were taken during the induction and intubation procedures.

In Experiment 3(e), sodium thiopentone solution was added directly to plasma samples in vitro at various concentrations (0.125, 0.25, 0.5, 1.0, 2.0, 4.0 and 8.0 mg/ml) before radioimmunoassay.

RESULTS

Experiment 1

All five control recipients of formalin-killed embryos exhibited a decline in progesterone concentrations to baseline levels by days 19-21 (Fig. 1, Panels A and B). Three animals also showed oestrous behaviour and post-oestral blood was seen in a fourth. Progesterone concentration in four of the five rose again during the ensuing luteal phase, but in a fifth animal (B172K, a day-15 recipient), progesterone concentrations remained basal in the daily samples between days 19 and 28 as well as in intermittent samples taken on days 33 and 37. This animal was in oestrus on day 36.

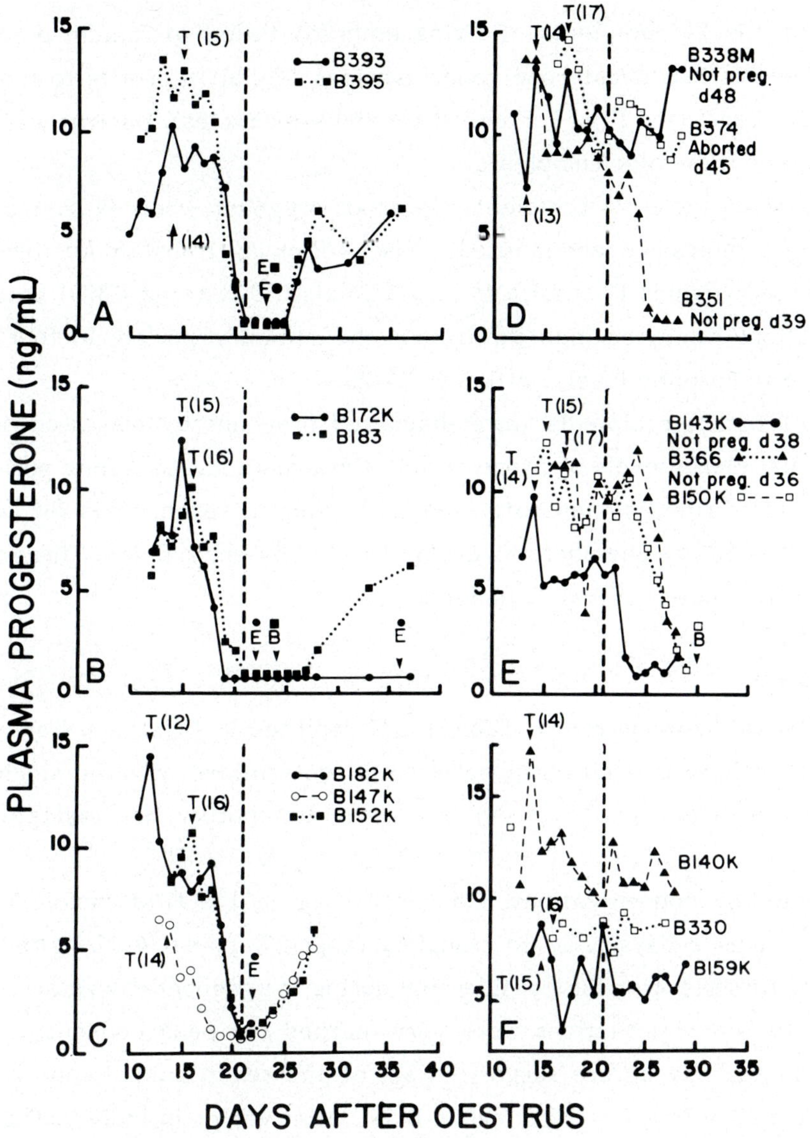

Fig. 1. Progesterone profiles, Experiment 1. The day of transfer (T) is indicated in brackets and signs of oestrus (E) or post-oestral blood (B) are indicated with the same shape of symbol as used for the graph to which they apply. For purposes of comparison, day 21 is indicated with a broken vertical line in each panel. Panels A and B: profiles for 4 of the 5 control transfers, each showing a return to oestrus at the normal time. Panel C: three profiles exemplifying the nine recipients returning to oestrus at the normal time. Panels D and E: profiles for the six recipients showing an extension of their luteal phases. Panel F: three profiles exemplifying animals which remained pregnant. (Note the changed vertical scale in this panel).

Nine of the 24 recipients of living embryos failed to remain pregnant and showed declines in progesterone concentrations similar in profile to those of the control recipients (see Fig. 1, Panel C for three examples). Oestrus was observed in only two of these nine recipients.

Six more of these 24 recipients were not pregnant when examined at about day 42, but exhibited extended luteal phases following transfers between days 13 and 17 (Fig. 1, Panels D and E). In two animals (B338M and B374) progesterone concentrations remained high throughout the sampling period; in four they did not decline to baseline levels until days 24-29.

Nine of the 24 recipients were diagnosed pregnant by palpation and maintained luteal phase levels of progesterone throughout the sampling period. Pregnancy was confirmed by laparotomy or slaughter in seven cases (in connection with other work), by one abortion at day 44, and by one calving. Three representative profiles appear in Fig. 1, Panel F.

Experiment 2

Results are summarized in Table I and depicted in Figs. 2-5. For each cow, the plotted values of each hormone were all determined within a single, or two successive, assays. Changes from one assay to another are indicated on the graphs.

Progesterone concentrations: Concentrations in the first sample from each recipient, taken 2 days prior to transfer, ranged between 4-12 ng/ml. Without exception, these levels rose considerably during the preoperative starvation period. Hour-to-hour fluctuations were very marked but peak concentrations were reached on the day before surgery in six of the seven animals and ranged between 9 and 20 ng/ml, 39-152% higher than initial values in individual cases. The haematocrits of the seven animals also increased over the same period (Table II) and, in the five animals actually sampled during surgery, individual values were 34.5%, 52.2%, 51.1%, 52.3% and 52.3%. Hour-to-hour changes continued to be a feature of the progesterone profiles post-operatively, with variations ranging from just under three-fold (B205K) to as much as five-fold (B207K).

Both control recipients of formalin-killed embryos showed declines in plasma progesterone concentrations during the hourly sampling period. However, whereas in animal B470 this occurred on days 17 and 18 as expected, animal B205K had

Table I. Summary of the results of Experiment 2 which are depicted graphically in Figs. 2-5

RECIPIENT NO.	PGFM PEAKS*								DAYS OF LUTEOLYSIS
	MINIMAL PEAK IN ASSAY** (pg/ml)		BEFORE TRANSFER		AT TRANSFER		AFTER TRANSFER		
			Days	Levels (pg/ml)	Minutes after induction	Level (pg/ml)	Days	Levels (pg/ml)	
B470 (control)	1st:	225	16	226, 227	53	431	17-18	298, 305, 396, 312, 229	17,18
	2nd:	147							
B205K (control)	1st:	220	15	256	40	828	23 25-26 28	195 357, 257, 356, 197 196	25, 26
	2nd:	192							
B444	1st:	140	16	142	42	189	20-21	192, 165, 169, 203	20, 21
	2nd:	120							
B226K	1st:	187	15-16	189, 230, 203	~60	321	20 25 38	191, 220 192 189	37, 38
	2nd:	180							
B484	1st:	227	-	-	69	320	23 29-30	229 186, 284, 191	42-46
	2nd:	167							
B426M	1st:	225	15	280, 232	75	336	24-25 27 28-29	176, 186 188 190, 180, 207, 210, 195	Aborted day 50
	2nd:	175							
B207K	1st:	217	-	-	~30	741	28	210, 202	Calved
	2nd:	193							

* Concentrations exceeding basal mean of respective assay by > 2 standard deviations (Brinkley, 1981).
** For samples in each assay, cf. Figs. 2-5.

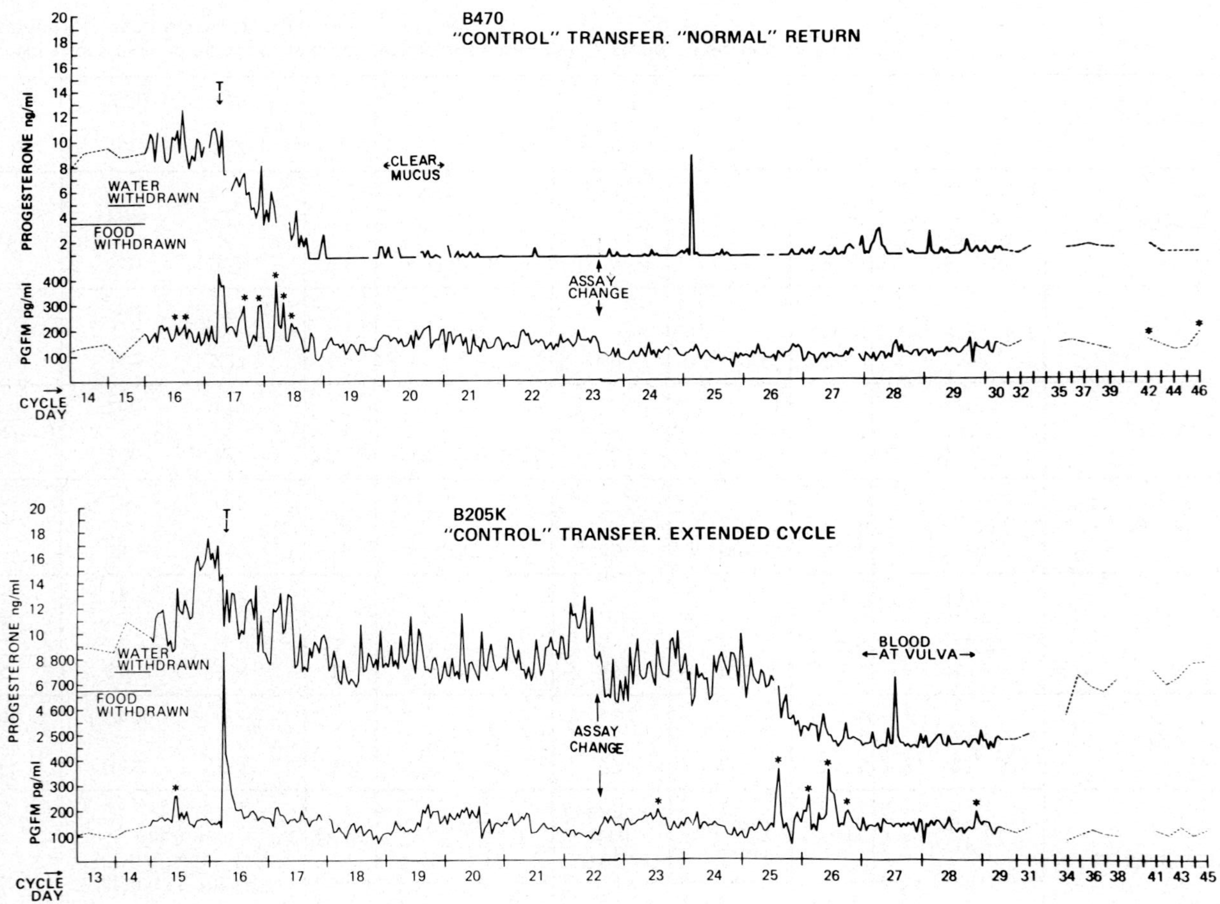

Fig. 2. Plasma progesterone and PGFM profiles of two control animals (B470, upper panel; B205K, lower panel) which received formalin-killed embryos (see text). T indicates the time of transfer. Asterisks mark peaks of PGFM other than the one at surgery (cf. Table I).

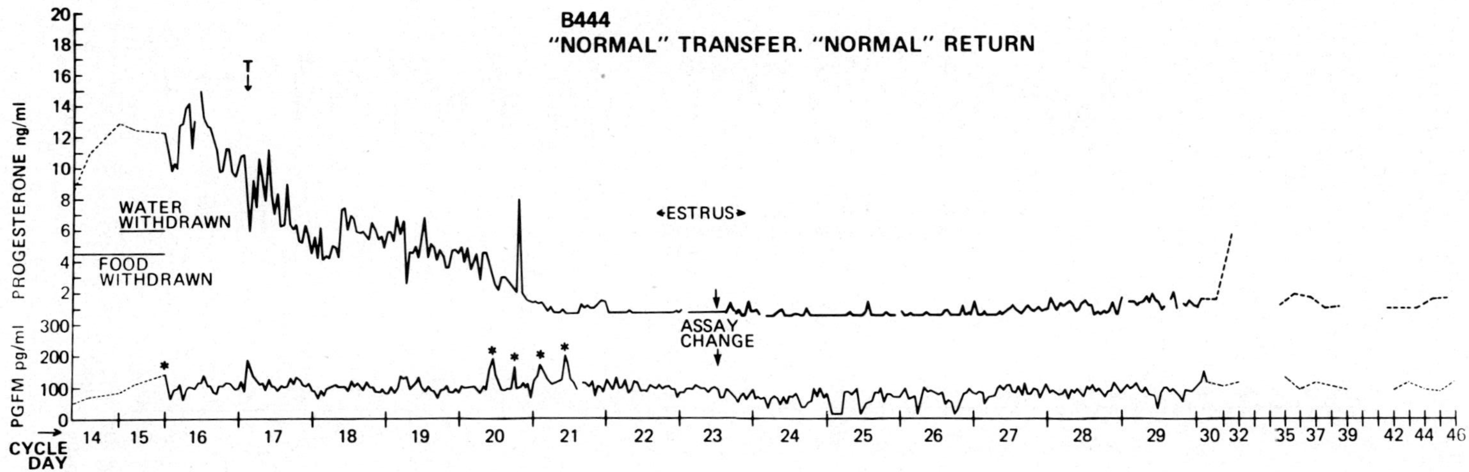

Fig. 3. Plasma progesterone and PGFM profiles in recipient B444 which returned to oestrus on day 23. T indicates the time of transfer. Asterisks mark peaks of PGFM other than the one at surgery (cf. Table I).

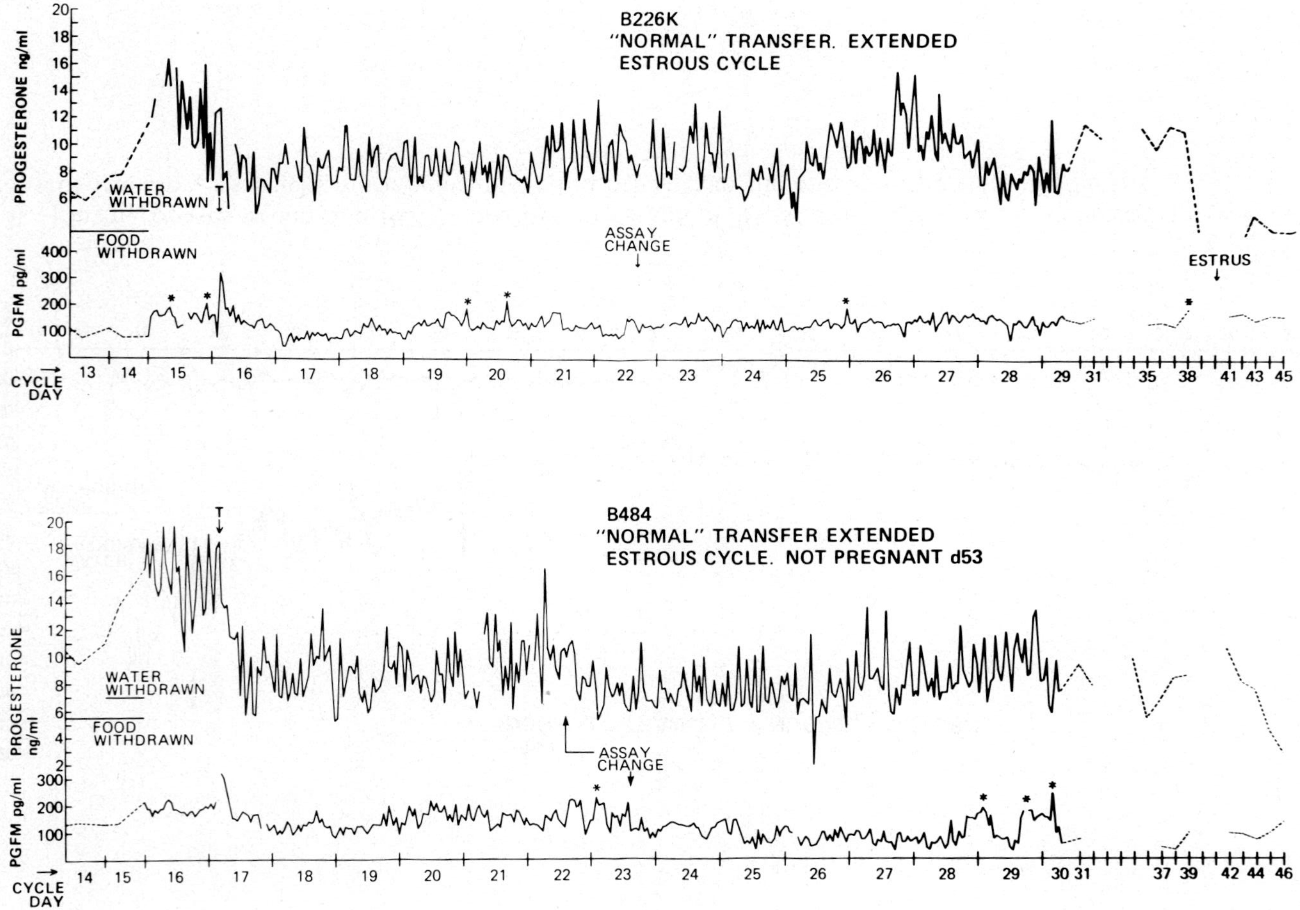

Fig. 4. Plasma progesterone and PGFM profiles in two recipients (B226K, upper panel; B484, lower panel) with prolonged luteal phases. T indicates the time of transfer. Asterisks mark peaks of PGFM other than the one at surgery (cf. Table I).

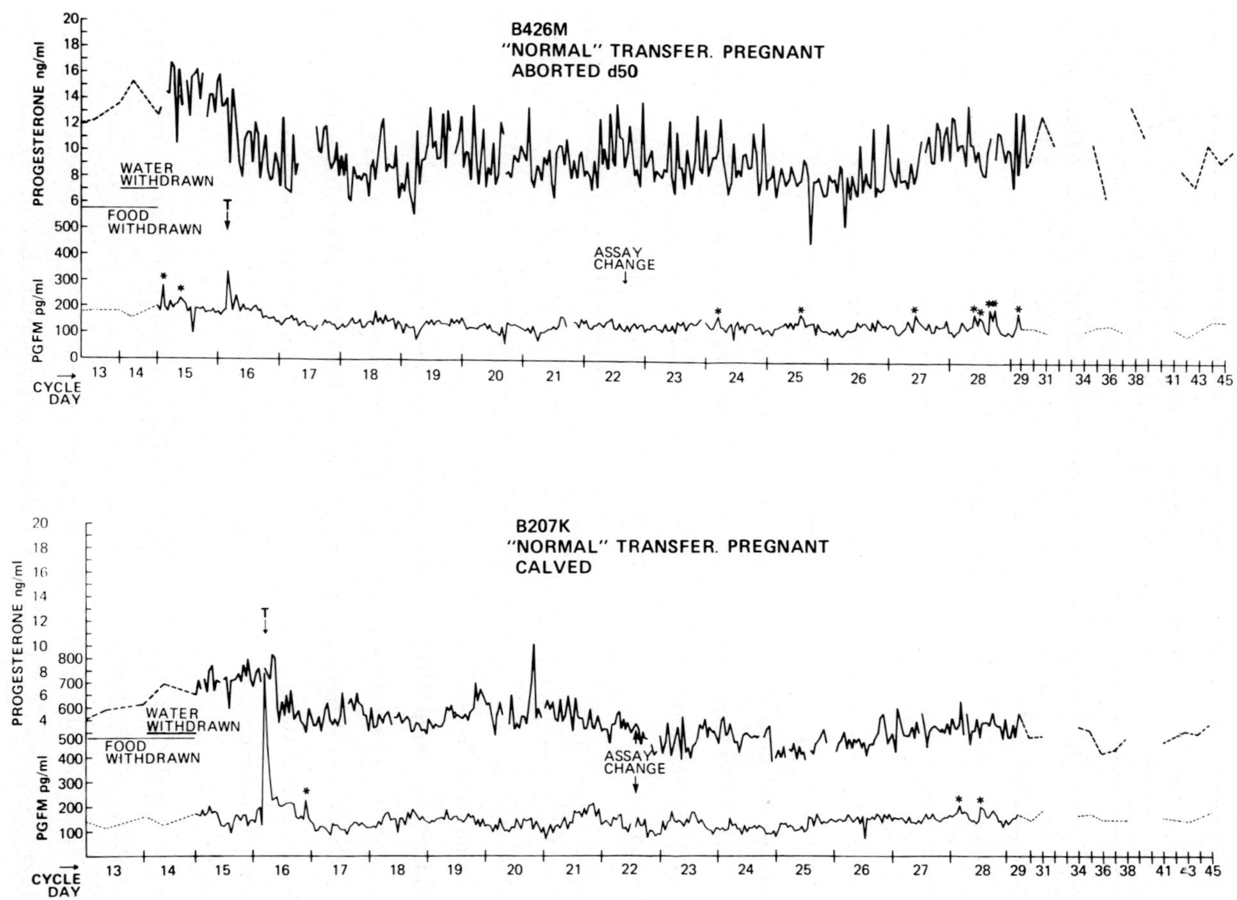

Fig. 5. Plasma progesterone and PGFM profiles in two recipients (B426M, upper panel; B207K, lower panel) proven to have become pregnant. B426M aborted on day 50 and B207K calved one calf normally at term. T indicates the time of transfer. Asterisks mark peaks of PGFM other than the one at surgery (cf. Table I).

Table II. Mean haematocrits (%) of the seven recipients in Experiment 2 for the periods indicated between three days before and two days after surgical transfer

DAY -3		DAY -2		DAY -1		DAY 0		DAY +1		DAY +2	
Normal feeding		No food		No food or water		Sur-gery	Re-covery	Normal feeding		Normal feeding	
0800	1600	0800	1600	0800-1900	2000-0700	0800-1900	2000-0700	0800-1900	2000-0700	0800-1900	2000-0700
35.3	32.7	35.5	36.2	38.6	39.4	43.5	40.7	38.5	36.6	34.7	35.8

a prolonged luteal phase with luteolysis not being evident until days 25 and 26. Oestrous behaviour was not observed in either recipient but B470 shed clear mucus on day 20 and post-oestral blood was seen on days 27 and 28 in B205K. The luteal phase following the return to oestrus was normal in B205K, with progesterone levels ranging between 3.5 and 7.6 ng/ml between days 34 and 45 (approximately days 8 and 19 of her new cycle). However, progesterone levels in B470 remained between the lower limit of the assay (0.8 ng/ml) and 1.5 ng/ml up until day 46 except for one isolated sample at 8.8 ng/ml and 9 other samples between 1.5 and 3.0 ng/ml, suggesting a deficient luteal phase (Fig. 2). A similar deficiency was seen in B444 (Fig. 3).

PGFM concentrations: Mean basal concentrations from which peaks were calculated ranged between 80 and 163 pg/ml (Table I). Five of the seven animals showed evidence of small peaks preoperatively on days 15 and/or 16. All animals showed peaks in association with surgery but the time relationship of a given hourly sample to induction of anaesthesia varied, making it impossible to compare these peaks.

All three recipients which underwent luteolysis during the hourly blood sampling session (B470 and B205K, Fig. 2; and B444, Fig. 3; Table I) showed 4-5 concomitant PGFM peaks. These varied in prominence, the recipient with the lowest baseline levels of PGFM (B444) showing correspondingly lower peak concentrations. Only occasional, and generally small, peaks were seen in the two animals in which progesterone declines occurred during the later period of daily sampling (B226K and B484, Fig. 4; Table I). In the two animals which remained pregnant throughout the experimental period, (B426M and B207K) neither showed (by regression analysis) any significant rise in baseline levels of PGFM up until days 28 or 29 of pregnancy but did show (especially B426M) some relatively low peaks between days 24 and 29 (Fig. 5; Table I).

Experiment 3

Extensive manipulation of the genital tract per rectum under epidural anaesthesia at various stages of the oestrous cycle (Experiments 3a and 3b) failed to provoke any consistent increase in PGFM in the peripheral blood, concentrations falling within the range 85-225 pg/ml (results not shown).

Oxytocin injection at day 16 (Experiment 3c) produced a small rise in PGFM concentrations (from 110 to 225 pg/ml) in one animal at the higher dose (Fig. 6, lower panel).

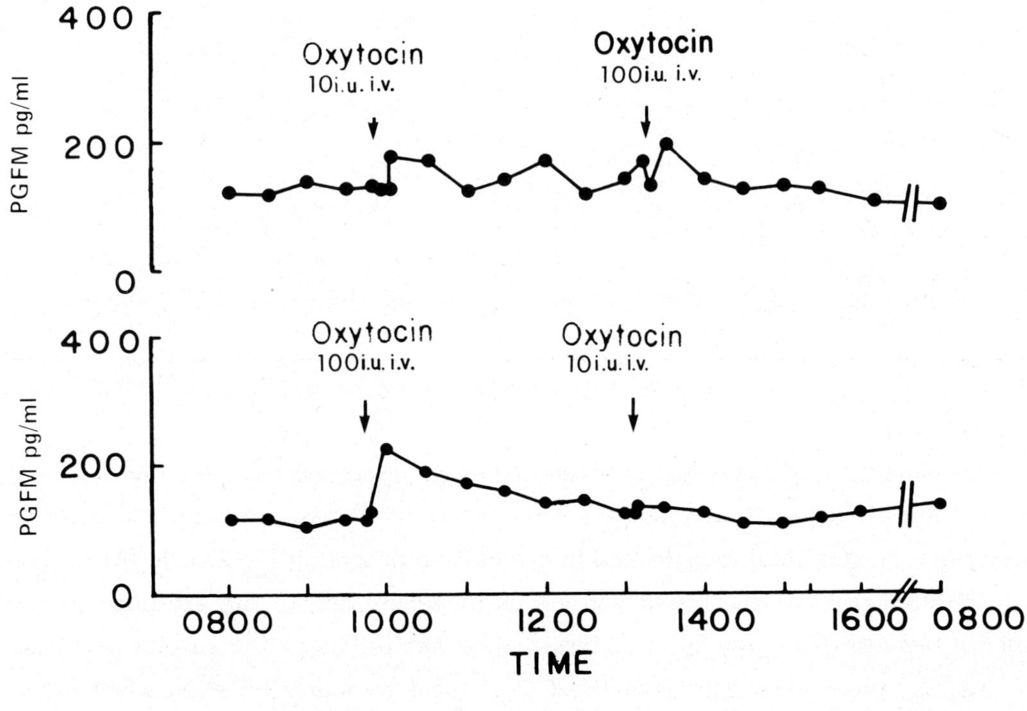

Fig. 6. PGFM concentrations in two heifers on day 16 of their oestrous cycles before and after i.v. injections of 10 i.u. or 100 i.u. oxytocin at 3.5 h intervals.

Peripheral blood, taken from all four cyclic animals (two day-7 and two day-16) as well as the ovariohysterectomized heifer immediately after injection of sodium thiopentone, showed peaks in the PGFM assay (Experiment 3d, Fig. 7). The size of the peak varied between animals (310-1000 pg/ml) but with no apparent relationship to the stage of the oestrous cycle nor to other manipulations such as endotracheal intubation or scrubbing in preparation for surgery.

Adding sodium thiopentone directly to plasma samples in vitro led to an apparent rise in the amount of PGFM detected in the plasma (Table III), but the rise was not proportional to the amount of thiopentone added.

Table III. Effect of adding sodium thiopentone to bovine plasma on "apparent" PGFM concentration assayed

Sodium thiopentone added (mg/ml)	PGFM concentration assayed (pg/ml)	
	Plasma 233K	Plasma 238K
0	108.7	144.5
0.125	105.7	121.5
0.25	122.5	157.6
0.50	150.8	135.5
1.00	208.9	190.4
2.00	294.6	267.5
4.00	367.3	305.8
8.00	235.8	301.6

DISCUSSION

Experiment 1 indicated that the spectrum of responses to late embryo transfer (normal pregnancy, extended oestrous cycles or immediate return to oestrus; Betteridge et al., 1980) is reflected in the plasma progesterone profiles of recipients. Prolongation of the luteal phase was evidently due to the presence of the embryo because five control transfers had no such effect. The results also confirmed that the embryo can save the CL at least temporarily, even when introduced into the uterus as late as day 17, very close to the time at which luteolysis would normally begin. However, the results gave no indication of the causes of

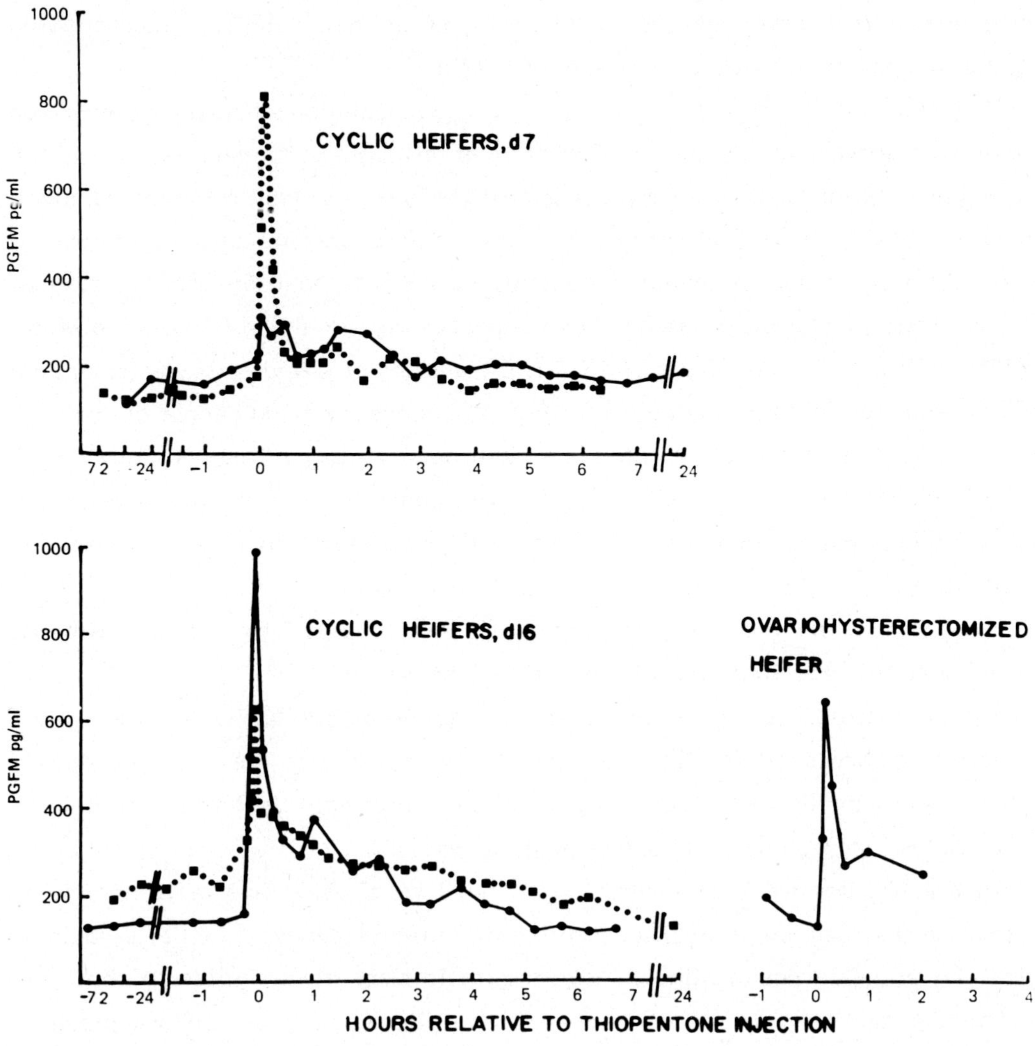

Fig. 7. PGFM-reactive material in the peripheral blood following i.v. thiopentone injections in: two cyclic heifers on day 7 of the oestrous cycle; two on day 16; and one heifer which had been ovariohysterectomized for 3 months.

subsequent luteolysis in the six recipients with extended cycles. Had luteolysis already been initiated and therefore only incompletely suppressed by the late introduction of embryos?

Experiment 2 aimed at answering this question by simultaneous analysis of more frequent plasma samples from similar recipients for PGFM as well as progesterone. It was remarkable how well the seven animals tolerated the repeated jugular venepunctures with a minimum of restraint.

The wide hour-to-hour fluctuations in progesterone concentrations confirmed previous observations in cattle (Peterson et al., 1975; Kindahl et al., 1976), underlining the hazards of interpreting profiles based on less frequent sampling (Peterson et al., 1975). The preoperative rise in progesterone concentrations was evidently not just due to haemoconcentration, but may have been brought about by the stress of starvation as has been reported also in sheep (Green and Moor, 1977). The fact that one of the two control transfers of killed embryos (to animal B205K) resulted in an extended cycle (Fig. 2) is confusing and cannot be explained. Her four previous cycle lengths had each been 19 days. Taken together with Experiment 1, she was the only one of seven controls to show such a response, but whether infection or unusual trauma to the uterus was involved must remain conjectural.

Another unexpected and unexplained finding revealed by the progesterone determinations was that of deficient luteal phases in two recipients that returned to oestrus after receiving either killed embryos (B470, Fig. 2, upper panel) or normal embryos (B444, Fig. 3) on day 17 in Experiment 2 and in one day-15 control recipient (B172K, Fig. 1, Panel B) in Experiment 1. The latter animal, nevertheless, showed normal oestrus again at day 36.

PGFM profiles were somewhat confounded by a peak invariably associated with surgery. Excluding that peak (see below), the 4-5 peaks of PGFM associated with the progesterone decline in three animals were very similar to those described during luteolysis in non-pregnant cattle (Peterson et al., 1975; Kindahl et al., 1976). The very variable progesterone concentrations made it impossible to say whether falls in progesterone preceded elevations in PGFM or whether the reverse was true. The absence or reduction of PGFM peaks in plasma up to day 24 in all four animals in which the CL persisted through the period of frequent sampling extends the findings of Kindahl et al. (1976, 1981) indicating that early pregnancy suppresses that episodic type of PGFM formation or release during the period when luteolysis would occur in cyclic animals. A number of relatively low PGFM peaks occurred after day 24, but it is perhaps surprising that no significant rise in baseline levels could be detected within the first 28 or 29 days of pregnancy because a rise in urinary levels of 11-ketotetranor PGF metabolites has been detected from about day 20 (Harvey et al., 1984).

Data for clarifying whether transferred embryos can suppress PGFM peaks that have already been initiated were incomplete. There was a suggestion of such an effect in B226K and B426M (Figs. 4 and 5; Table I), but against this must be set similar findings in the control animals (Fig. 2; Table I) after transfer of killed embryos. Similarly, information for defining the sequence of endocrine events associated with embryonic death was not obtained in this study because the luteolysis which presumably accompanied the loss of transferred embryos in B226K and B484 (Fig. 4), and the abortion in B426M (Fig. 5, upper panel), occurred after the hourly blood sampling had ended.

The PGFM peak associated with surgery was at first presumed to be associated with manipulation of the genital tract and consequent oxytocin release because Roberts et al. (1975) have shown that either mechanical stimulation of the ovine uterus in situ or oxytocin infusion produces marked $PGF_{2\alpha}$ release late in the luteal phase though not at earlier stages of the oestrous cycle. However, Experiments 3(a) and 3(b) showed no such effect of tract manipulation in cattle. Further, PGFM peaks produced by single i.v. injections of oxytocin (Experiment 3c, Fig. 6) were considerably lower than those seen during surgical transfers. Closer inspection of the sampling schedule then revealed that elevated PGFM levels were found in samples taken from anaesthetized animals before laparotomy. The results of Experiment 3(e) showed that it was the sodium thiopentone injection rather than any subsequent procedure that was involved and that the uterus is not necessary for the effect (Fig. 7). Although it should be remembered that haematocrits in anaesthetized animals often exceeded 50%, simple haemoconcentration obviously would not account for the 4- to 9-fold elevations in concentrations seen in some animals. However, it is still not clear whether the observed peaks are entirely of PGFM or whether the injection of a bolus of about 5 g sodium thiopentone as a 25% solution could also sufficiently "contaminate" plasma samples to produce the artefactual assay "concentrations" produced by direct addition of the barbiturate to plasma in Experiment 3(e). In this connection, it can be calculated that 5 g of sodium thiopentone fully diluted in the blood plasma of a 500 kg animal would produce a concentration of about 0.2 mg/ml, about the second lowest concentration tested in vitro (Table III). In either case, the findings indicate that PGFM determinations made on unextracted plasma from cattle under barbiturate anaesthesia need to be interpreted with caution.

In conclusion, although studies of hormonal changes in blood plasma in early pregnancy are obviously essential to our understanding of the mechanisms involved in the establishment of pregnancy and of factors leading to embryonic mortality, they are beset by practical difficulties. Infrequent (e.g. daily) sampling is relatively easy to apply over prolonged periods, but is of limited use where hour-to-hour fluctuations are marked (as they are for progesterone) or secretions are episodic (as they are for PGFM). Frequent sampling (e.g. hourly) gives much more complete information, but at the cost of great inconvenience and markedly increased assay expense which, in practice, limit the length of time over which a given animal can be studied. It was for this reason that alternative methods have been sought for investigation of prostaglandin metabolism in relation to embryonic mortality (Harvey et al., 1984).

ACKNOWLEDGEMENTS

We are grateful for the exceptional technical assistance of Mrs. J. Hierlihy, Mrs. S. Miller, and Messrs. Y. Barbeau, R. Bériault, G. Raby, J. Scholten and J. Shackleton in maintaining the prolonged sampling schedules and for skilled analytical help; to Miss E. Cathcart and Messrs. S. Shearer and D. Mulvihill for other help with the animals; to Mr. G. Goyette for assay assistance; to Dr. D. Harvey for help in Experiment 3(d); to Dr. A.K. Goff for assay and statistical aid; to Mme D. Theberge for typing the manuscript.

REFERENCES

Bartol, F.F., Thatcher, W.W., Lewis, G.S., Bliss, E.L., Drost, M. and Bazer, F.W., 1981. Effects of estradiol-17β on PGF and total protein content in bovine uterine flushings and peripheral plasma concentrations of 13,14-dihydro-15-keto-$PGF_{2\alpha}$. Theriogenology, 15: 345-358.

Batra, S.K. and Pandey, R.S., 1983. Relative concentrations of 13,14-dihydro-15-keto-prostaglandin $F_{2\alpha}$ in blood and milk of buffaloes during the oestrous cycle and early pregnancy. J. Reprod. Fert., 67: 191-195.

Betteridge, K.J., 1982. Embryonic mortality in relation to embryo transfer in farm animals: a position paper. Proc. of the Second European Conference on Embryo Transfer in Mammals, Annecy. (In press).

Betteridge, K.J., Eaglesome, M.D., Sugden, E.A. and Mitchell, D., 1977. Synchronization of estrus and ovulation in cattle with prostaglandin analogue AY 24655. Can. J. Anim. Sci., 57: 23-32.

Betteridge, K.J., Eaglesome, M.D., Randall, G.C.B., Mitchell, D. and Sugden, E.A., 1978. Maternal progesterone levels as evidence of luteotrophic or antiluteolytic effects of embryos transferred to heifers 12-17 days after estrus. Theriogenology, 9: 86.

Betteridge, K.J., Eaglesome, M.D., Randall, G.C.B. and Mitchell, D., 1980. Collection, description and transfer of embryos from cattle 10-16 days after oestrus. J. Reprod. Fert., 59: 205-216.

Brinkley, H.J., 1981. Endocrine signaling and female reproduction. Biol. Reprod., 24: 22-43.

Granström, E. and Kindahl, H., 1976. Radioimmunoassay for prostaglandin metabolites. In "Advances in Prostaglandin and Thromboxane Research", Vol. 1. (Editors B. Samuelsson and R. Paoletti), Raven Press, New York, pp. 81-92.

Green, D. and Moor, R.M., 1977. The influence of anaesthesia on the concentrations of progesterone and cortisol in peripheral blood of sheep. Res. Vet. Sci., 22: 122-123.

Hare, W.C.D., Mitchell, D., Betteridge, K.J., Eaglesome, M.D. and Randall, G.C.B., 1976. Sexing two week-old bovine embryos by chromosomal analysis prior to surgical transfer: Preliminary methods and results. Theriogenology, 5: 243-253.

Harvey, D., Basu, S., Betteridge, K.J., Goff, A.K. and Kindahl, H., 1984. The influence of pregnancy on $PGF_{2\alpha}$ secretion in cattle. II. Urinary levels of 11-ketotetranor PGF metabolites and plasma progesterone concentrations during the oestrous cycle and early pregnancy. Anim. Reprod. Sci., 7: 217-234. (This symposium).

Inskeep, E.K. and Murdoch, W.J., 1980. Relation of ovarian functions to uterine and ovarian secretion of prostaglandins during the estrous cycle and early pregnancy in the ewe and cow. Int. Rev. Phys., 22: 325-355.

Kindahl, H., Edqvist, L.-E., Bane, A. and Granström, E., 1976. Blood levels of progesterone and 15-keto-13,14-dihydro-prostaglandin $F_{2\alpha}$ during the normal oestrous cycle and early pregnancy in heifers. Acta endocrinol., 82: 134-149.

Kindahl, H., Lindell, J.-O. and Edqvist, L.-E., 1981. Release of prostaglandin $F_{2\alpha}$ during the oestrous cycle. Acta vet. scand., Suppl. 77: 143-158.

Lewis, G.S., Wilson, L., Jr., Wilks, J.W., Pexton, J.E., Fogwell, R.L., Ford. S.P., Butcher, R.L., Thayne, W.V. and Inskeep, E.K., 1977. $PGF_{2\alpha}$ and its metabolites in uterine and jugular venous plasma and endometrium of ewes during early pregnancy. J. Anim. Sci., 45: 320-327.

Moore, L.G., Watkins, W.B., Peterson, A.J., Tervit, H.R., Fairclough, R.J., Havik, P.G. and Smith, J.F., 1982. Embryonic suppression of oxytocin-associated neurophysin release in early pregnant sheep. Prostaglandins, 24: 79-88.

Peterson, A.J., Fairclough, R.J., Payne, E. and Smith, J.F., 1975. Hormonal changes around bovine luteolysis. Prostaglandins, 10: 675-684.

Peterson, A.J., Tervit, H.R., Fairclough, R.J., Havik, P.G. and Smith, J.F., 1976. Jugular levels of 13,14-dihydro-15-keto-prostaglandin F and progesterone around luteolysis and early pregnancy in the ewe. Prostaglandins, 12: 551-558.

Randall, G.C.B., Betteridge, K.J., Eaglesome, M.D. and Sugden, E.A., 1980. Progesterone and prostaglandin-F-metabolite levels in the peripheral blood of cattle used as recipients 16 or 17 days after estrus. Theriogenology, 13: 108.

Roberts, J.S., Barcikowski, B., Wilson, L. (Jr.), Skarnes, R.C. and McCracken, J.A., 1975. Hormonal and related factors affecting the release of $PGF_{2\alpha}$ from the uterus. J. Ster. Biochem., 6: 1091-1097.

Webb, R., Mitchell, M.D., Falconer, J. and Robinson, J.S., 1981. Temporal relationships between peripheral plasma concentrations of oxytocin, progesterone and 13,14-dihydro-15-keto-prostaglandin $F_{2\alpha}$ during the oestrous cycle and early pregnancy in the ewe. Prostaglandins, 22: 443-451.

Williams, W.F., Lewis, G.S., Thatcher, W.W. and Underwood, C.S., 1980. Plasma PGFM in pregnant and open heifers prior to and during laparotomy and following intrauterine injection of $PGF_{2\alpha}$. J. Anim. Sci., 51, Suppl. 1: 337-338.

Animal Reproduction Science, 7 (1984) 217—234
Elsevier Science Publishers B.V., Amsterdam — Printed in The Netherlands

THE INFLUENCE OF PREGNANCY ON $PGF_{2\alpha}$ SECRETION IN CATTLE. II. Urinary levels of 11-ketotetranor PGF metabolites and plasma progesterone concentrations during the oestrous cycle and early pregnancy

D. Harvey, S. Basu*, K.J. Betteridge, A.K. Goff and H. Kindahl*

Centre de Recherche en Reproduction Animale, Faculté de Médecine Vétérinaire, Université de Montréal, C.P. 5000, St-Hyacinthe, Québec, J2S 7C6, (Canada).

* Department of Obstetrics and Gynaecology, College of Veterinary Medicine, Swedish University of Agricultural Sciences, 750 07 Uppsala, (Sweden).

ABSTRACT

Harvey, D., Basu, S., Betteridge, K.J., Goff, A.K. and Kindahl, H. 1984. The influence of pregnancy on $PGF_{2\alpha}$ secretion in cattle. II. Urinary levels of 11-ketotetranor PGF metabolites and plasma progesterone concentrations during the oestrous cycle and early pregnancy. Anim. Reprod. Sci., 7: 217-234.

Radioimmunoassay of the urinary 11-ketotetranor PGF metabolites after i.v. infusion of prostaglandin $F_{2\alpha}$ ($PGF_{2\alpha}$) showed that these metabolites could be detected in the urine in concentrations proportional to the dose of $PGF_{2\alpha}$ infused. Daily output of the 11-ketotetranor PGF metabolites in the urine was measured in four catheterized heifers during the oestrous cycle and peaks in their excretion were detected at the time of luteolysis, indicated by the decline in plasma progesterone. In five additional catheterized heifers, levels of the metabolites in daily urine samples were determined throughout the oestrous cycle, during early pregnancy following artificial insemination, during cloprostenol-induced abortion, and after embryo transfer. During early pregnancy, there was an absence of the clear peaks of the urinary metabolites seen during luteolysis. However, there was a gradual increase in their concentration starting at about day 20. The results indicate that the measurement of urinary 11-ketotetranor PGF metabolites once or twice daily may serve as a good method for following $PGF_{2\alpha}$ secretion in long-term studies, especially in regard to changes that take place during pregnancy and embryonic mortality.

0378-4320/84/$03.00

INTRODUCTION

It has been well established that peaks of prostaglandin $F_{2\alpha}$ ($PGF_{2\alpha}$) occur towards the end of the oestrous cycle of the cow. These are thought to be responsible for luteolysis and the subsequent sharp decline of plasma progesterone (Nancarrow et al., 1973; Shemesh and Hansel, 1975). If conception occurs, the embryo must signal its presence to the mother not later than about the 16th day following oestrus in order to block luteolysis (Northey and French, 1980; Betteridge et al., 1980). However, the mechanism whereby the corpus luteum (CL) of pregnancy is "protected" is still not clearly understood. Furthermore, whether or not the release of $PGF_{2\alpha}$ is abolished during early pregnancy is still under debate (Betteridge et al., 1984).

Both $PGF_{2\alpha}$ and its initial main plasma metabolite (15-keto-13,14-dihydro-$PGF_{2\alpha}$ or 15-ketodihydro-$PGF_{2\alpha}$) have short half-lives in the circulation. Therefore, very frequent blood sampling is necessary if either compound is to be used to study the cyclic fluctuations in uterine secretion of $PGF_{2\alpha}$. The practical difficulties that this imposes on long-term studies in cattle have been discussed in the preceding paper (Betteridge et al., 1984). Although 15-keto-13,14-dihydro-$PGF_{2\alpha}$ can also be found in saliva, long-term collection of this secretion in cattle is difficult (Harvey, 1983). It was therefore of interest to determine whether metabolites in other body fluids could be useful for such investigations. The way in which prostaglandins are excreted differs widely between species. Humans, for example, eliminate 90% of $PGF_{2\alpha}$ as metabolites in the urine, whereas only 45% takes this route in guinea pigs (Granström and Samuelsson, 1971). Neff et al. (1981) have shown that after i.m. administration of tritiated $PGF_{2\alpha}$ in the cow, more than 60% was excreted in urine, 30% in the faeces and less than 1% in milk. Presumably the elimination of endogenous prostaglandins follows the same pattern.

Granström and Kindahl (1976) have developed a radioimmunoassay that measures some $PGF_{2\alpha}$ urinary metabolites (the 11-ketotetranor PGF metabolites) and have suggested that it should be useful for long-term studies, one or two urine specimens per day substituting for the alternative hourly blood samples. The RIA of urinary $PGF_{2\alpha}$ metabolites has already proven useful for evaluating total $PGF_{2\alpha}$ production rates in man (Ohki et al., 1975; Granström and Kindahl, 1976; Kitamura and Ishihara, 1980).

The goals of the present study were: a) to verify that prostaglandin metabolites could be detected in the urine after i.v. infusion of $PGF2\alpha$ in cows; b) to correlate urinary prostaglandin metabolite levels during the oestrous cycle with luteolysis, as measured by the fall in plasma progesterone concentration; and c) to compare changes in urinary prostaglandin metabolite levels during normal early pregnancy with those associated with embryonic mortality.

MATERIALS AND METHODS

The study comprised three sequential experiments. In Experiment 1, prostaglandin metabolites were measured in blood and urine samples taken frequently following i.v. infusion of $PGF2\alpha$. In Experiment 2, continuous urine collection and daily blood sampling were undertaken throughout the oestrous cycle so that urinary 11-ketotetranor PGF metabolite levels could be correlated with plasma progesterone levels. In Experiment 3, an improved method of urine collection was used to monitor the same compounds during four consecutive experimental periods: firstly, a complete oestrous cycle without insemination; secondly, the initial 30 days of pregnancy resulting from artificial insemination; thirdly, the return to oestrus following cloprostenol-induced abortion; and, finally, the cycle or the pregnancy ensuing after embryo transfer on the 13th-15th day following that oestrus.

Animals

Holstein heifers, 15 to 24 months of age and weighing between 350 and 400 kg were used throughout. They were housed and fed a maintenance ration of hay and grain. Water was available ad libitum. The heifers were observed twice daily for signs of oestrus and the day of oestrus was designated as day 0 of the cycle.

Blood samples

Daily samples for progesterone determinations were drawn between 0800 and 0900 from the coccygeal vein (or occasionally the artery) into 7 ml evacuated tubes containing EDTA ("Vacutainers", Becton Dickinson). The frequent serial samples in Experiment 1 were taken by syringe through an indwelling catheter in the jugular vein and then transferred into similar EDTA tubes. Plasma separation and storage was as described in the preceding paper (Betteridge et al., 1984).

Urine samples

Three distinct methods were used. For Experiment 1, a latex Foley catheter (30 French guage, 1 cm outside diameter, o.d.) was introduced into the urinary bladder through the urinary meatus, held in place by inflation of the cuff with 30 ml air, and closed at the external end with a pair of haemostats. Every 20 min the catheter was opened and gently agitated to induce the micturition reflex. Two 3 ml aliquots of the urine collected were immediately frozen. For Experiment 2, the same method of catheterization was used with two modifications for long-term collection: first, leakage was prevented by ligating the meatus around the catheter with a single nylon suture; and second, the catheter was linked to a 20 l polypropylene container by a 3 m x 6 mm rubber tube suspended from pulleys behind the heifer. This container was emptied twice daily after recording the volume of urine collected and three 10 ml aliquots were frozen. For Experiment 3, an improved chronic collection system was used (Fig. 1). The bladder was catheterized with a soft plastic tube of only 2.5 mm o.d. ("Bardic" stomach tube, Levin type no. 12, Murray Hill, New Jersey) which was anchored to the vestibular floor by a single suture. To prevent faecal contamination, the catheter was passed out of the vagina through the right ischiorectal fossa by means of a trochar and cannula under local anaesthesia. There it was coupled via a unidirectional plastic valve ("Security Valve", Simport Plastics, Beloeil, Québec) to a rubber tube running along the back to a 1 l polyethylene bottle suspended from a neck collar. The system collected only a few ml each time the animal urinated, the rest being eliminated normally. The bottle was emptied each morning and two 3-4 ml aliquots were frozen. Urine samples from all experiments were stored at -70° before being transported from Canada to Sweden in dry ice for assay of 11-ketotetranor PGF metabolites.

Experimental protocols

For Experiment 1, one heifer (No. 5), ovariohysterectomized two months previously, was used. An indwelling catheter was inserted into each jugular vein; one for i.v. infusions of $PGF_{2\alpha}$, the other for blood sampling. An infusion pump (Harvard Apparatus, Millis, Massachussets) was used for the administration of $PGF_{2\alpha}$ (dinoprost tromethamine, "Lutalyse", Upjohn). Blood samples were taken every 10 min, and urine samples every 20 min, throughout the 7 h of the experiment. A

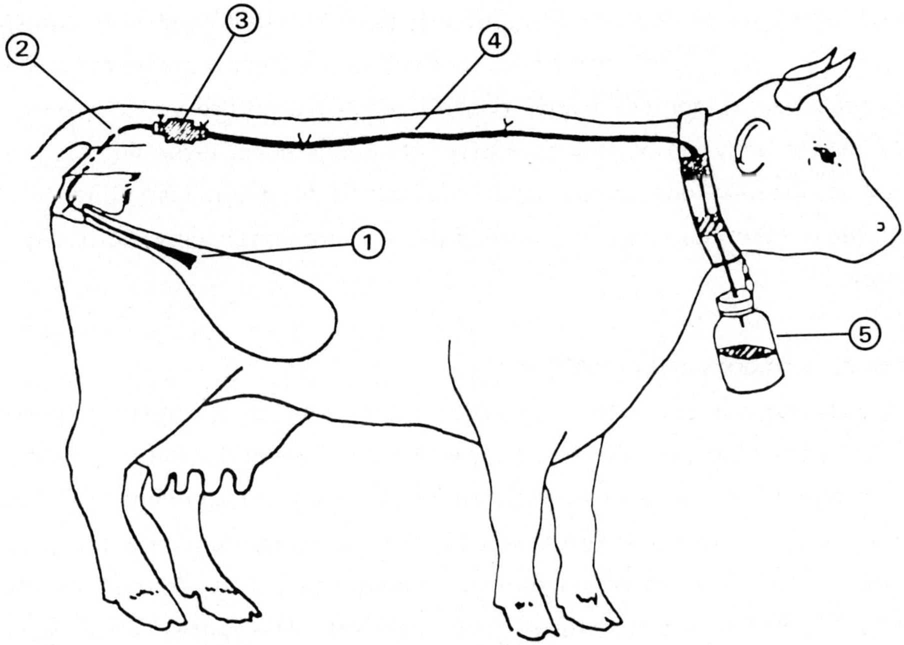

Fig. 1. Urine collection system used in Experiment 3. 1: soft catheter in the urinary bladder; 2: point of exit in the ischiorectal fossa; 3: one way valve; 4: rubber tube anchored along the back; 5: collection bottle. For details, see text.

first infusion, of 125 µg $PGF_{2\alpha}$ over 1 h, was initiated at the beginning of the second hour and followed, 3 h after its completion, by a second infusion of 500 µg $PGF_{2\alpha}$ over 1 h.

In Experiment 2, four intact heifers (Nos. 55, 171, 206 and 291) were sampled daily (blood) and twice daily (urine) over a 3-week period beginning at mid-cycle, thus spanning the periods of luteolysis and oestrus.

In Experiment 3, five intact heifers were used (Nos. 307, 311, 322, 323 and 335). Daily sampling of blood and urine was initiated 2 days after catheterization, corresponding approximately to the middle of the oestrous cycle. The five heifers were inseminated during their first subsequent oestrus and, 30 days later, they received 500 µg of cloprostenol ("Estrumate", I.C.I.) i.m. to induce abortion and a return to oestrus. Thirteen to fifteen days after this induced oestrus, four of the heifers served as recipients in a surgical embryo transfer. The day-14 embryos used were collected from a superovulated donor by mid-line surgery under general anaesthesia as described by Betteridge et al. (1980) and were transferred singly, to the uterine horn ipsilateral to the CL, through a flank incision under local anaesthesia. After the transfer, the heifers were observed twice daily for signs of oestrus and, one or two days after its occurrence, urine and blood sampling was stopped. In one animal which remained pregnant, sampling continued until 3 weeks after calving, but here data are presented only until day 60 of pregnancy.

Analyses of hormones and creatinine

Progesterone was extracted from 200 µl of plasma using redistilled petroleum ether. Recovery rates were between 75 and 85%. Concentration of progesterone was determined using a specific antiserum obtained from Immunotech, Montreal. Crossreactivity for this antiserum was 100% for progesterone; 1% for pregnenolone; and < 0.1% for androstenedione, testosterone, 17-hydroxy-progesterone, oestrone, oestradiol, corticosterone and cortisol. The intra- and inter-assay coefficients of variation were both < 16%.

15-Keto-dihydro-$PGF_{2\alpha}$ was measured directly in 200 µl aliquots of plasma using a specific radioimmunoassay. The antiserum used was generously supplied by Dr. K.T. Kirton, Upjohn, Kalamazoo, U.S.A. and its given cross-reactivities were: 100% for 15-keto-13,14-dihydro-$PGF_{2\alpha}$, 20% for 15-keto-$PGF_{2\alpha}$, and 0.5% for 13,14-dihydro-$PGF_{2\alpha}$. This antibody also cross-reacted 10% with the 11-ketotetranor PGF metabolites and < 0.2% with $PGF_{2\alpha}$ (A.K. Goff and H. Kindahl, unpublished observations). The intra- and inter-assay coefficients of variation were both < 12%.

11-Ketotetranor PGF metabolites were analyzed in urine according to a method described earlier (Granström and Kindahl, 1976, 1982). The antibody was raised against 5α,7α-dihydroxy-11-ketotetranorprostane-1,16-dioic acid. The corresponding monocarboxylic acid, 5α,7α-dihydroxy-11-ketotetranorprostanoic acid, was used as standard and labelled ligand. Cross-reactivity for this antiserum was: 100% for 5α,7α-dihydroxy-11-ketotetranorprostane-1,16-dioic acid, between 51 and 98% for various tetranor PGF metabolites, between 1.0 and 4.6% for dinor PGF metabolites, and < 0.1% for C_{20} prostaglandins. Urine was diluted 100 and 500 fold, and 50 and 100 µl were taken for radioimmunoassay. The intra-assay coefficient of variation varied between 6.3 and 10.1% for different ranges of the standard curve. The inter-assay coefficient of variation was 12%.

Creatinine was measured colorimetrically by the technique of Fabiny and Ertingshausen (1971) using a kit obtained from Worthington Diagnostics, Freehold, N.J.

Presentation of Results of 11-ketotetranor PGF metabolite determinations in urine varied according to the different collection methods used in the three experiments. Thus concentrations were considered appropriate to Experiment 1 and daily production to Experiment 2. In Experiment 3, however, concentrations were related to creatinine levels because only an indeterminate fraction of the urine was collected. The concentrations were corrected as follows: a) the mean urinary creatinine concentration was divided by the mean plasma concentration to obtain the mean ratio for each animal over the experimental period; b) the daily urinary creatinine concentration was divided by the daily plasma concentration to obtain the daily ratio; c) the daily correction factor was obtained by dividing the mean ratio by the daily ratio; d) each concentration of the 11-ketotetranor PGF metabolites was multiplied by its corresponding daily correction factor.

The presence of peaks of urinary 11-ketotetranor PGF metabolites was determined by the method of Brinkley (1981), and peaks are defined as those values exceeding the mean basal levels by more than two standard deviations.

In Experiment 2, daily output is expressed as the sum of the pairs of 12 h outputs. In the graphs from Experiments 2 and 3, a level depicted for a given day is that found in the morning collection of urine which had accumulated during the preceding 24 h.

RESULTS

Experiment 1: Appearance of PGF metabolites in the blood and urine after i.v. infusion of $PGF_{2\alpha}$

The plasma level of 15-ketodihydro-$PGF_{2\alpha}$ varied from 100 to 140 pg/ml (mean value of approximately 125 pg/ml) except during the two $PGF_{2\alpha}$ infusions. These infusions resulted in peaks in concentrations of the metabolite, each proportional to the dose of $PGF_{2\alpha}$ infused (Fig. 2). A similar relationship was observed for urinary 11-ketotetranor PGF metabolites with levels rising to over

100 ng/ml during the first PGF2α infusion and over 500 ng/ml during the second one. These urinary peaks appeared 30 to 40 min after the plasma peaks of 15-ketodihydro-PGF2α metabolite (Fig. 2).

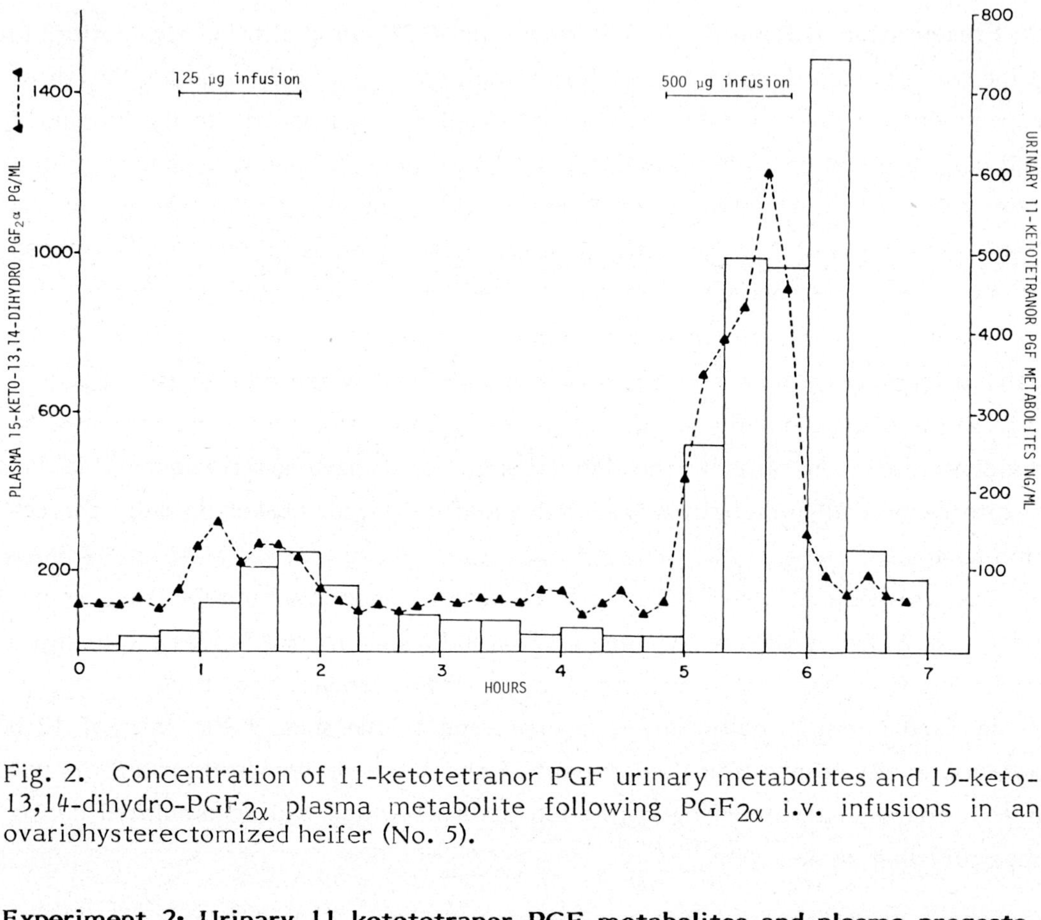

Fig. 2. Concentration of 11-ketotetranor PGF urinary metabolites and 15-keto-13,14-dihydro-PGF$_{2\alpha}$ plasma metabolite following PGF$_{2\alpha}$ i.v. infusions in an ovariohysterectomized heifer (No. 5).

Experiment 2: Urinary 11-ketotetranor PGF metabolites and plasma progesterone concentrations during the oestrous cycle

The four animals had normal plasma progesterone profiles during their oestrous cycles. Their mean basal outputs of the 11-ketotetranor PGF metabolites ranged between 80 and 240 µg/day. All four animals exhibited peaks of urinary 11-ketotetranor PGF metabolites around the time of luteolysis, ranging from 200 to 650 µg/day. A representative profile is shown in Fig. 3. Levels of 11-ketotetranor PGF metabolites were elevated for 2 days in 3 animals and for 1 day in the fourth. Peaks coincided with, but did not precede, the fall in plasma

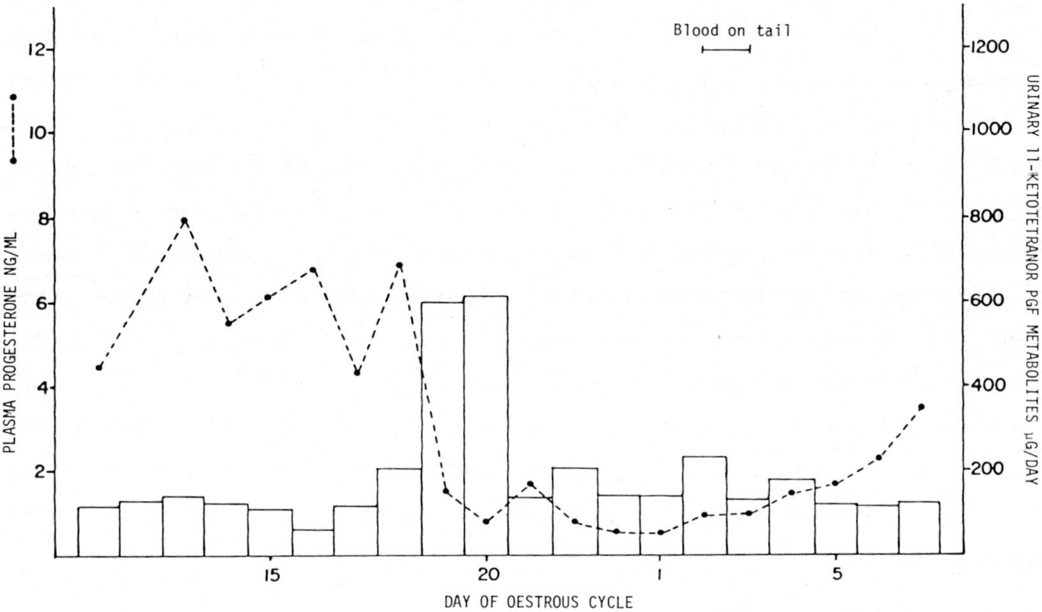

Fig. 3. Daily production of the urinary 11-ketotetranor PGF metabolites and peripheral plasma levels of progesterone in a heifer (No. 55) during the oestrous cycle.

progesterone. In two of the animals, a subsequent peak was observed, coincident with post-oestrus bleeding. Apart from these, no other peaks were observed during the oestrous cycle.

The urine collection technique used in this experiment proved to be simple but uncomfortable for the animals, causing vaginal contamination and irritation. Moreover, the total volume of urine collected per day was highly variable (between 1.8 and 10 l), and far below the mean 14 l per day reported by Dukes (1955).

Experiment 3: Urinary 11-ketotetranor PGF metabolites during the oestrous cycle and early pregnancy and following induced abortion and embryo transfer

The urine collection system used in this experiment was easy to maintain and was well-tolerated by the animals. The urine of heifer No. 322 was collected for 350 consecutive days (with 2 minor interruptions of up to 8 days due to cystitis or mechanical problems) during which period she calved. Heifer No. 335 suffered from a chronic cystitis and her catheter had to be removed on day 42. The volume of urine collected from all the animals generally exceeded 1 l although smaller volumes, indicating a problem in the collection system occurred intermittently (less than 15% of samples).

All animals displayed normal 21-day oestrous cycles at the beginning of the experiment. Following artificial insemination all five heifers became pregnant as indicated by high (> 5 ng/ml) plasma progesterone levels beyond day 24 (Figs. 4-7). Following embryo transfer, however, only one heifer (Fig. 7) remained pregnant and calved normally. Of the other three, two had normal 21-day cycles (Figs. 4 and 5) and one had a 24-day cycle (Fig. 6).

The eight periods of spontaneous luteolysis in the five heifers (five initial cycles and three returns to oestrus after transfer) were all accompanied by peaks of urinary 11-ketotetranor PGF metabolites ranging between 70 and 150 ng/ml (Figs. 4-7).

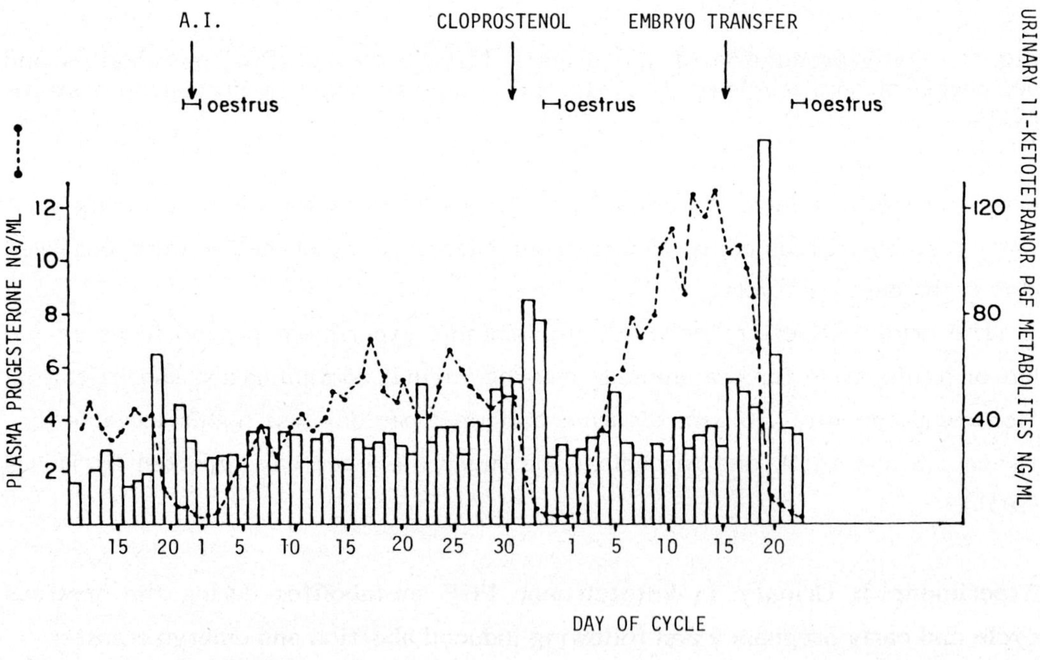

Fig. 4. Concentrations of urinary 11-ketotetranor PGF metabolites and peripheral plasma levels of progesterone in heifer No. 307 during the oestrous cycle and early pregnancy, following induced abortion, and after embryo transfer at day 15. Urinary metabolite values corrected for variations in glomerular filtration rate.

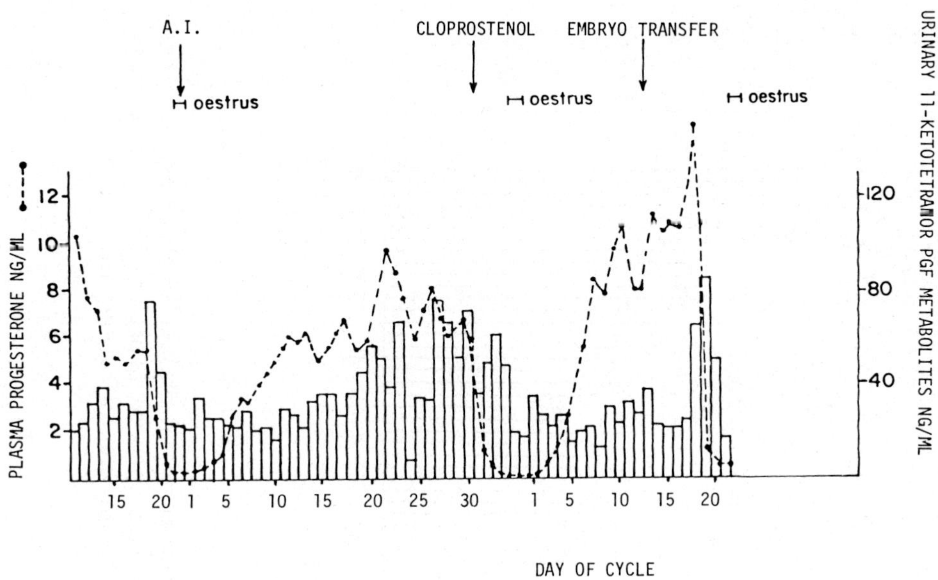

Fig. 5. Concentrations of urinary 11-ketotetranor PGF metabolites and peripheral plasma levels of progesterone in heifer No. 323 during the oestrous cycle and early pregnancy, following induced abortion, and after embryo transfer at day 13. Urinary metabolite values corrected for variations in glomerular filtration rate.

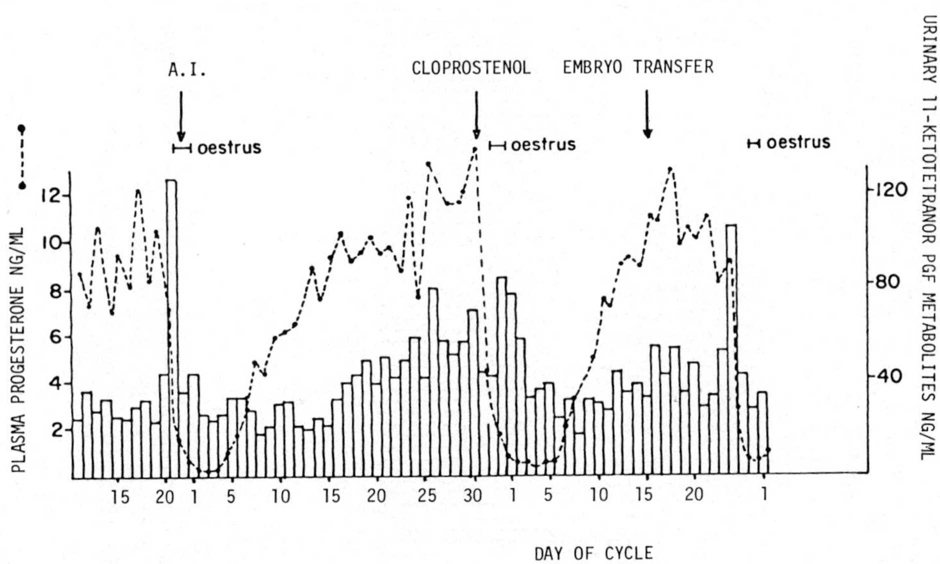

Fig. 6. Concentrations of urinary 11-ketotetranor PGF metabolites and peripheral plasma levels of progesterone in heifer No. 311 during the oestrous cycle and early pregnancy, following induced abortion, and after embryo transfer at day 15. Urinary metabolite values corrected for variations in glomerular filtration rate.

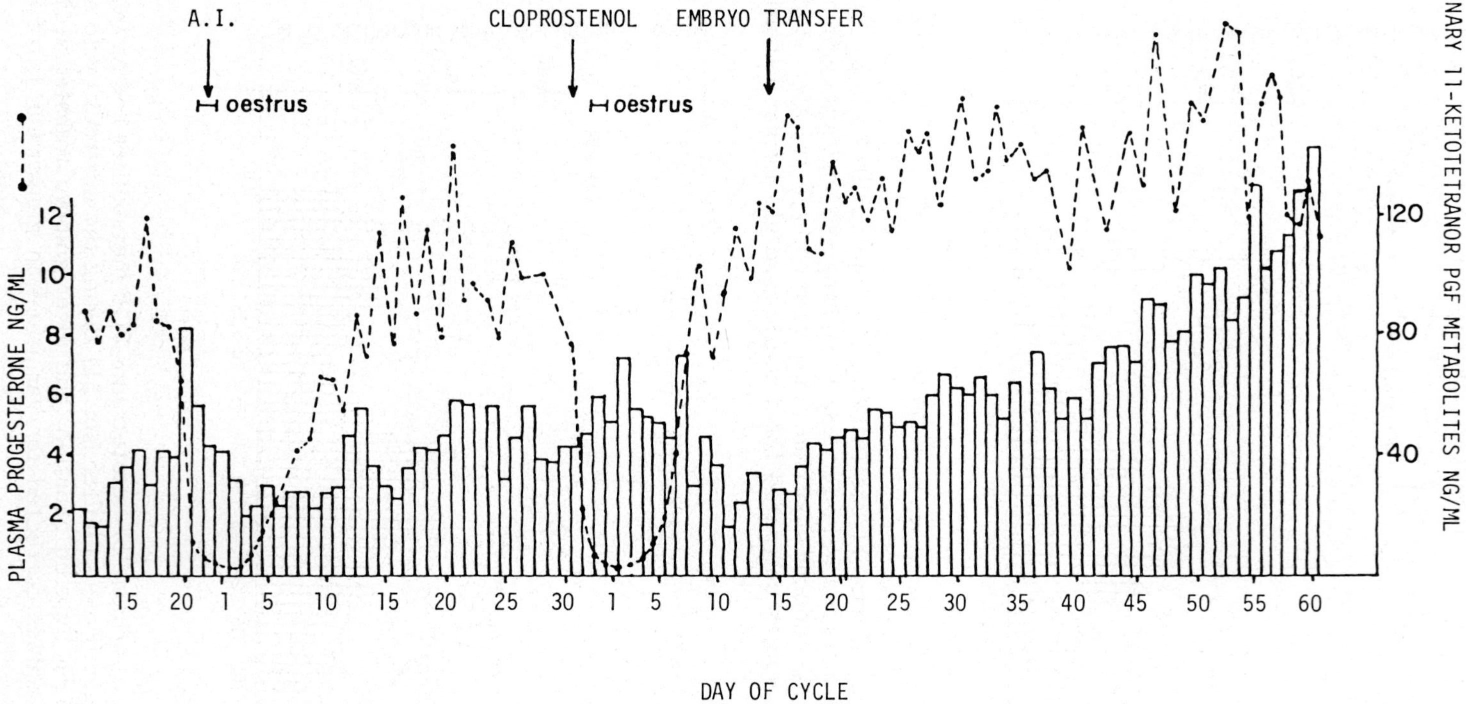

Fig. 7. Concentrations of urinary 11-ketotetranor PGF metabolites and peripheral plasma levels of progesterone in heifer No. 322 during the oestrous cycle and early pregnancy, following induced abortion, and after embryo transfer at day 14. Urinary metabolite values corrected for variations in glomerular filtration rate.

After all five successful inseminations and the one successful embryo transfer, concentrations of the urinary 11-ketotetranor PGF metabolites increased significantly above basal values beginning around day 20, but with considerable variation between animals. This increase continued steadily until the time of induced abortion and, in the pregnant recipient (Fig. 7), until day 60. By day 30-35 of pregnancy, concentrations of the metabolites approached those seen during luteolysis.

Cloprostenol injection at day 30 induced an abrupt decline in plasma progesterone concentrations within 24 h and oestrus within 48 to 96 h in all four animals. A degenerating conceptus was recovered from the vulva 96 h after injection in one animal in which oestrus began at the same time. Peaks of urinary 11-ketotetranor PGF metabolites were also observed after injection of cloprostenol. These peaks occurred 2 to 3 days after injection, after the decline in progesterone and, in two cases (Figs. 6 and 7) during or after the induced oestrus. The overall higher levels of the metabolites seen during early pregnancy tended to obscure these peaks.

DISCUSSION

The results from Experiment 1 met the first goal of this study, for $PGF_{2\alpha}$ infusions in heifer No. 5 (Fig. 2) clearly showed that the magnitudes of plasma 15-ketodihydro-$PGF_{2\alpha}$ and urinary 11-ketotetranor PGF metabolite peaks are proportional to the dose of $PGF_{2\alpha}$ infused. In the case of the 15-ketodihydro-$PGF_{2\alpha}$, this agreed with the findings of Kindahl et al. (1976). Finding that the urinary metabolite level increased several-fold during infusion of $PGF_{2\alpha}$ is in accordance with the observations of Neff et al. (1981) and Kindahl (1980), who showed that a large proportion of $PGF_{2\alpha}$ is excreted in the urine. As was to be expected, an interval (30-40 min) separates the peaks of $PGF_{2\alpha}$ metabolites found in the blood and urine.

The plasma concentrations of 15-ketodihydro-$PGF_{2\alpha}$ reached during the first infusion were similar to the peak concentrations which occur during luteolysis (Kindahl et al., 1976). Since the concentration of the 11-ketotetranor PGF metabolites in the urine rose well above basal levels during this infusion, it was predicted that the urinary level of these metabolites would also be greater at the time of luteolysis than during the rest of the oestrous cycle. The second experiment, which addressed this question as the second objective of the study, confirmed the prediction in all four animals studied. In blood, the appearance of 15-ketodihydro-PGF$_{2\alpha}$ has been shown to begin the day before the decline in progesterone levels and to terminate 2 to 3 days later. The larger peaks are usually found during and immediately after the progesterone decline (Peterson et al., 1975; Kindahl et al., 1976; Thatcher, 1981, 1984). The peak in the urinary 11-ketotetranor PGF metabolites also occurred during and after, rather than before, the decline of progesterone, but this was probably largely because urine was only analyzed every 24 h. Authors generally agree that, during luteolysis in the cow, there is an increase in PGF$_{2\alpha}$ concentration in the uterine fluid (Lamothe et al., 1977; Bartol et al., 1981), as well as in endometrial tissue and uterine vein blood (Shemesh and Hansel, 1975). Analogous increases have also been well-described for plasma 15-ketodihydro-PGF$_{2\alpha}$ (Peterson et al., 1975; Kindahl et al., 1976). Since 11-ketotetranor PGF metabolites, concentrations in urine rise similarly at the time of luteolysis, it appears that these urinary metabolites are a valuable indicator of PGF$_{2\alpha}$ secretion by the cow's uterus. The same conclusions were made for the guinea pig (Granström and Kindahl, 1976). However, the uterus is probably not the only source of 11-ketotetranor PGF metabolites because low, variable, but clearly detectable levels have also been found in the urine of an ovariohysterectomized heifer studied one month after surgery (unpublished observations) and in the hysterectomized guinea pig (Granström and Kindahl, 1976).

The urine collection system used during Experiment 2 was not very satisfactory, since it caused some discomfort to the animals. Also, the urine volume collected was variable and below the expected volume, suggesting some loss. The collection system used in Experiment 3 proved easier to maintain and was well-tolerated by the animals. Although no daily measurement of urine volume was necessary, variable urine output had to be taken into consideration by correcting for creatinine excretion. Using this system, interesting patterns of urinary 11-ketotetranor PGF metabolites were obtained.

During early pregnancy, there was a significant gradual increase in the urinary concentration of the 11-ketotetranor PGF metabolite after about day 20. By day 30-35, levels often approached those observed at luteolysis. Despite this increase, the profile in pregnant animals was clearly different from that found at the time of luteolysis, when large, sharp peaks lasting only 1 or 2 days were observed. The fact that urinary 11-ketotetranor PGF metabolites increase during early pregnancy suggests that, as in the ewe (Ellinwood et al., 1979), total $PGF_{2\alpha}$ output in the cow is higher in pregnant than in non-pregnant animals. This is especially interesting in view of conflicting evidence on the question of whether pregnancy increases or decreases plasma levels of $PGF_{2\alpha}$ or 15-ketodihydro-$PGF_{2\alpha}$. Thus, long-term studies with frequent sampling in cows (Kindahl et al., 1976, 1981; Betteridge et al., 1984) or less frequent sampling in buffaloes (Batra and Pandey, 1983) found lower plasma levels of the metabolite during the third week after oestrus in pregnant than in non-pregnant animals undergoing luteolysis. Acute studies, however, have shown basal peripheral plasma levels of the same metabolite (Williams et al., 1980; Thatcher et al., 1984) or concentrations of $PGF_{2\alpha}$ in uterine vein blood (Lukaszewska and Hansel, 1980) to be higher in pregnant than in non-pregnant heifers at equivalent stages after oestrus. The reasons for these discrepancies are not clear. The biological significance of an overall increased $PGF_{2\alpha}$ output early in pregnancy must also be interpreted with caution. Its effect on the CL, for example, is evidently mitigated by many other factors so that the net concentration of $PGF_{2\alpha}$ in ovarian arterial plasma is lower during the third week of pregnancy than during luteolysis when pulsatile releases of $PGF_{2\alpha}$ are of special significance (Thatcher et al., 1984). The source of $PGF_{2\alpha}$ during early pregnancy probably includes the conceptus as well as the endometrium since both are capable of producing it as well as 15-ketodihydro-$PGF_{2\alpha}$ and, in the case of the conceptus, production increases with age (Shemesh et al., 1979; Lewis et al., 1982; Thatcher et al., 1984). Continuous urine collection and assay of the 11-ketotetranor PGF metabolites, perhaps at more frequent intervals than in the present study, could be useful in resolving some of these questions.

Cloprostenol was incorporated into this study merely to terminate pregnancies so that the same animals could be used again as recipients. However, continuous urine collection during the treatment yielded useful data. The patterns of progesterone decline and return to oestrus closely resembled those reported by Lindell et al. (1980/81) in five heifers treated with the same dose of cloprostenol between days 39 and 71. The recovery of a conceptus from the vulva in one animal in the present study, at the time of the induced oestrus (96 h after injection), also conforms to the observations of Lindell et al. and suggests that final expulsion of the conceptus took place at corresponding times, unobserved, in the other heifers, too. Lindell et al. showed that cloprostenol induced intermittent releases of endogenous PGF2α (measured as 15-keto-13,14-dihydro-PGF2α) into the blood and that the release terminated at the time of abortion. The pre-abortion 15-ketodihydro-PGF2α levels were higher than the levels observed after the abortion. The interpretation of post-injection concentrations of urinary 11-ketotetranor PGF metabolites in the present study was complicated by the fact that levels were already high at the time of treatment (day 30). However, the observed post-injection peaks in the urinary metabolite support the findings of Lindell et al. (1980/81). In discussing the source of the endogenous PGF2α evidently released by cloprostenol, Lindell et al. favour the notion that it is produced from the uterus, and that this is a result of removal of a conceptus-induced block to PGF2α synthesis and release rather than just progesterone withdrawal. However, as discussed above, the conceptus itself could also be a source of the $PGF_{2\alpha}$ and its metabolites.

The use of late embryo transfer in this study was designed to meet the third objective, determination of the sequence of events around the time of embryonic mortality. As no clear-cut case of embryonic mortality was observed, more work on this point will be necessary. Continuous urine collection and assay of urinary 11-ketotetranor PGF metabolites as an indicator of $PGF_{2\alpha}$ secretion should be useful in these, and similar, long-term studies.

ACKNOWLEDGEMENTS

This study was supported by grants from NSERC and Agriculture Canada and by the Swedish Council for Forestry and Agricultural Research. We thank Mlle M. Lafrance for skilled analytical assistance, the farm staff of the CRRA for help with the animals, and Mme D. Théberge for typing the manuscript.

REFERENCES

Bartol, F.F., Thatcher, W.W., Bazer, F.W., Kimball, F.A., Chenault, J.R., Wilcox, C.J. and Roberts, R.M., 1981. Effects of the estrous cycle and early pregnancy on bovine uterine, luteal and follicular responses. Biol. Reprod., 25: 759-776.

Batra, S.K. and Pandey, R.S., 1983. Relative concentrations of 13,14-dihydro-15 keto-prostaglandin $F_{2\alpha}$ in blood and milk of buffaloes during the oestrous cycle and early pregnancy. J. Reprod. Fert., 67: 191-195.

Betteridge, K.J., Eaglesome, M.D., Randall, G.C.B. and Mitchell, D., 1980. Collection, description and transfer of embryos from cattle 10-16 days after oestrus. J. Reprod. Fert., 59: 205-216.

Betteridge, K.J., Randall, G.C.B., Eaglesome, M.D. and Sugden, E.A., 1984. The influence of pregnancy on PGF$_{2\alpha}$ secretion in cattle. I. Concentrations of 15-keto-13,14-dihydroprostaglandin F$_{2\alpha}$ and progesterone in peripheral blood of recipients of transferred embryos. Anim. Reprod. Sci., 7: 195-216. (This symposium).

Brinkley, H.J., 1981. Endocrine signaling and female reproduction. Biol. Reprod., 24: 22-43.

Dukes, H.H., 1955. In: The Physiology of Domestic Animals, 7th ed. Cornell University Press, Ithaca, pp. 504 and 563.

Ellinwood, W.E., Nett, T.M. and Niswender, G.D., 1979. Maintenance of the corpus luteum of early pregnancy in the ewe. Prostaglandin secretion by the endometrium in vitro and in vivo. Biol. Reprod., 21: 845-856.

Fabiny, D.L. and Ertingshausen, O., 1971. Automated reaction - rôle method for determination of serum creatinine with the centrichem. Clin. Chem., 10: 391-397.

Granström, E. and Samuelsson, B., 1971. On the metabolism of prostaglandin F$_{2\alpha}$ in female subjects. J. Biol. Chem., 246: 5254-5263.

Granström, E. and Kindahl, H., 1976. Radioimmunoassay for urinary metabolites of prostaglandin F$_{2\alpha}$. Prostaglandins, 12: 759-783.

Granström, E. and Kindahl, H., 1982. Radioimmunoassay of 5α,7α-dihydroxy-11-ketotetranorprostane-1,16-dioic acid, a major prostaglandin F metabolite in blood and urine. Methods Enzymol., 86: 339-353.

Harvey, D., 1983. Etude des changements de niveau des métabolites de la prostaglandine F$_{2\alpha}$ et de la progestérone pendant le cycle oestral et au début de la gestation chez la vache. Mémoire de Maîtrise, Université de Montréal.

Kindahl, H., Edqvist, L.-E., Bane, A. and Granström, E., 1976. Blood levels of progesterone and 15-keto-13,14-dihydro-prostaglandin F$_{2\alpha}$ during the normal oestrous cycle and early pregnancy in heifers. Acta endocrin., 82: 134-149.

Kindahl, H., 1980. Prostaglandin biosynthesis and metabolism. J. Amer. Vet. Med. Ass., 176: 1173-1177.

Kindahl, H., Lindell, J.-O. and Edqvist, L.E., 1981. Release of prostaglandin F$_{2\alpha}$ during the oestrous cycle. Acta vet. scand., Suppl. 77: 143-158.

Kitamura, S. and Ishihara, Y., 1980. Radioimmunoassay of the main urinary metabolite of prostaglandin F$_{2\alpha}$ in normal subjects after oral administration of prostaglandins E$_2$ and F$_{2\alpha}$. In: B. Samuelsson and P.W. Ramwell (Editors), Advances in Prostaglandin and Thromboxane Research. Raven Press, New York, Vol. 6, pp. 213-217.

Lamothe, P., Bousquet, D. and Guay, P., 1977. Cyclic variation of F prostaglandins in the uterine fluids of the cow. J. Reprod. Fert., 50: 381-382.

Lewis, G.S., Thatcher, W.W., Bazer, F.W. and Curl, J.S., 1982. Metabolism of arachidonic acid in vitro by bovine blastocysts and endometrium. Biol. Reprod., 27: 431-439.

Lindell, J.-O., Kindahl, H. and Edqvist, L.-E., 1980/81. Prostaglandin induced early abortions in the bovine. Clinical outcome and endogenous release of prostaglandin F$_{2\alpha}$ and progesterone. Anim. Reprod. Sci., 3: 289-299.

Lukaszewska, J. and Hansel, W., 1980. Corpus luteum maintenance during early pregnancy in the cow. J. Reprod. Fert., 59: 485-493.

Nancarrow, C.D., Buckmaster, J., Chamley, W., Cox, R.I., Cumming, I.A., Cummins, L., Drinan, J.P., Findlay, J.K. Goding, J.R., Restall, B.J., Schneider, W. and Thorburn, G.D., 1973. Hormonal changes around oestrus in the cow. J. Reprod. Fert., 32: 320-321.

Neff, A.W., Ruppel, C.C., Gosline, R.E., Jaglan, P.S. and McGrath, J.P., 1981. $PGF_{2\alpha}$ residue studies in beef and dairy cattle. Acta vet. scand., Suppl. 77: 11-27.

Northey, D.L. and French, L.R., 1980. Effect of embryo removal and intrauterine infusion of embryonic homogenates on the lifespan of the bovine corpus luteum. J. Anim. Sci., 50: 298-302.

Ohki, S., Imaki, K., Hirata, F., Hanya, T. and Natazawa, N., 1975. Radioimmunoassay of main urinary metabolite of prostaglandin-$F_{2\alpha}$. Prostaglandins, 10: 549-555.

Peterson, A.J., Fairclough, R.J., Payne, E. and Smith, J.F., 1975. Hormonal changes around bovine luteolysis. Prostaglandins, 10: 675-684.

Shemesh, M. and Hansel, W., 1975. Levels of prostaglandin F (PGF) in bovine endometrium, uterine venous, ovarian arterial and jugular plasma during oestrous cycle. Proceeding of the Society for Experimental Biology and Medicine, 148: 123-126.

Shemesh, M., Milaguir, F., Ayalon, N. and Hansel, W., 1979. Steroidogenesis and prostaglandin synthesis by cultured bovine blastocysts. J. Reprod. Fert., 56: 181-185.

Thatcher, W.W., 1981. Control of oestrous cycle and pregnancy recognition in cattle. In: Graduate Student Symposium on the Peri-implantation Period in Domestic Animals. University of Florida School of Medicine, pp. 11-18.

Thatcher, W.W., Wolfenson, D., Curl, J.S., Rico, L.E., Knickerbocker, J.J., Bazer, F.W. and Drost, M., 1984. Prostaglandin dynamics associated with development of the bovine conceptus. Anim. Reprod. Sci., 7: 149-176. (This Symposium).

Williams, W.F., Lewis, G.S., Thatcher, W.W. and Underwood, C.S., 1980. Plasma PGFM in pregnant and open heifers prior to and during laparotomy and following intra-uterine injection of $PGF_{2\alpha}$. J. Anim. Sci., Suppl. 1, 51: 337-338.

Animal Reproduction Science, 7 (1984) 235—244
Elsevier Science Publishers B.V., Amsterdam — Printed in The Netherlands

SUSCEPTIBILITY OF THE BOVINE CORPUS LUTEUM TO EXOGENOUS PROSTAGLANDINS DURING PREGNANCY - PRACTICAL CONSIDERATIONS

R.H. Schultz, D.D. Copeland and T.S. Wollen

From BAYVET Division, Miles Laboratories, Inc. P.O. Box 390, Shawnee Mission, Kansas 66201, (USA).

ABSTRACT

Schultz, R.H., Copeland, D.D. and Wollen, T.S., 1984. Susceptibility of the bovine corpus luteum to exogenous prostaglandins during pregnancy - practical considerations. Anim. Reprod. Sci., 7: 235-244.

Although several indications exist calling for terminating of pregnancy in cattle, by far the greatest demand is for pregnant heifers entering feedlots in the United States and Canada. Management problems associated with calving in the feedlot, along with economic losses incurred by price penalties for heifers pregnant at slaughter are the main factors responsible for the need to induce abortion in these animals. Three prostaglandins have received regulatory approval in North America allowing field experience to be generated for this indication. Incidence of pregnant heifers entering the lots, economic considerations, post-treatment complications and convenience are main factors utilized by veterinarians in their determination as to whether or not they induce abortion with prostaglandins in these animals. Cattle can be safely and effectively aborted using various prostaglandins up to 100 to 150 days of pregnancy. Also, in the past two years, trials have been reported which indicate that the combination of dexamethasone and cloprostenol provides a means for successful abortion induction during advanced pregnancy.

Indications and economic considerations

Dairy practitioners occasionally find heifers (and less frequently cows) pregnant at routine fertility examinations or when called to examine animals because they are not being detected in oestrus. The problem usually results from a bull gaining access to the animals or mistakes being made in reproductive record keeping. Often these animals are too young to calve safely or they may be bred to the wrong, or an unknown, bull. Thus, the need to induce abortion is present.

0378-4320/84/$03.00

In the rapidly growing field of embryo transfer there is occasional need to terminate pregnancy. Animals purchased to serve as recipient animals sometimes arrive pregnant. Surprisingly, even some cows presented as donor animals are also found to be pregnant. In either case, the animals must be safely aborted and returned to a cycling state. Following embryo collection, from the superovulated donor animals, multiple corpora lutea exist and it is also possible that embryos may have been missed in the collection procedure and remain in the uterus. It is frequently desired to use these donors for repeated collections and therefore they need to be safely aborted and returned to a cycling state.

By far the greatest indication for termination of pregnancy in the United States and Canada is in feedlot heifers. In these countries steers and heifers are typically pastured following weaning and then sold (most often through auction sales) to feedlot operators for fattening and finishing prior to slaughter. At what age these animals enter the feedlots often depends on cattle prices, prices and availability of feedstuffs and weather conditions. Management of heifers prior to entering the feedlot varies tremendously. Many times they run on open range or pastures with ready access to bulls. In many situations, under marginal management, bulls will be allowed to run continuously with the entire herd throughout the year. Thus, many of the heifers entering the feedlots are pregnant.

That these pregnant heifers are an economic and management liability has long been recognized. In recent years there are increased efforts to define the levels of economic significance of these animals for the feedlot operator and to develop reasonable schemes for managing pregnant heifers as they enter the feedlot. Many factors must be taken into account in these efforts. They include: the pregnancy rate of heifers entering the feedlot, the ages of these pregnancies (i.e., will they calve in the feedlot or be advanced pregnant at the time of slaughter), the effect pregnancy has on daily gains and feed conversion, the methods used to market the finished animals (prices set on the hoof or prices established after slaughter), price premiums for guaranteed open heifers (or, more realistically, penalties for pregnant heifers), efficacy of the abortifacient used, veterinary costs and drug costs.

The incidence of pregnant animals entering feedlots varies by season, origin of the heifers and the economic conditions of the cattle industry. Wilson (1983) found between 23 and 46% pregnant in four groups of heifers examined for entry into one trial. Bennett (1983) and Bennett and Rupp (1983) surveyed feedlot operators in Colorado and found the incidence to range from zero to 25% (mean 16.5%). In a multi-location field trial Syntex Agribusiness (1982) examined 1441 heifers for pregnancy and found 644 (44.6%) pregnant. Kansas State workers (Simms and Kuhl, 1982; Keay et al., 1983) found approximately 20% pregnancies in animals tested. McAllister and Lauderdale (1979) found incidence rates ranging from zero up to 50% in 11 different feedlot locations. At the 11 locations, 547 of 2299 (24%) were pregnant with 70 to 100% of the pregnancies less than 150 days.

Discrimination against pregnant heifers by packer buyers is based on the reduced dressing percentage in pregnant heifers. In their survey, Bennett and Rupp (1983) found that the loss in dressing percentage of slaughter heifers due to pregnancy ranged from one to six percent with an average of 3.38%. Similar values were found by Simms and Kuhl (1982). Using these differences in dressing percentages as they affect final carcass yield along with differing abortion rates and costs involved with inducing abortions, these authors demonstrated positive economic results from aborting groups of heifers at 20, 30 and 40% pregnancy rates. Included in the costs were veterinary palpation fees and drug costs.

Pregnant heifers that calve in the feedlot present serious management problems for the manager of the feedlot as well as considerable economic losses for the owner. Severe dystocia is a very common problem in these animals due to reduced birth canal size as a result of fat deposition. Also, the unsanitary conditions in the lots contribute to post-calving problems such as complications from retained placenta and acute metritis. Many of these animals become very sick and die. Those that do survive grow very poorly and realize no profit for the feedlot operation. These potential problems and losses must be taken into account when deciding whether or not to induce abortion in heifers in advanced stages of pregnancy. With the recent development of successful means for inducing abortion in heifers in the last half of pregnancy it will be possible and necessary to study the management and economics associated with this practice.

Methods available for abortion induction in cattle

Enucleation of the corpus luteum, although effective for inducing abortion in pregnancies up to about 150 days (McDonald et al., 1953; Tanabe, 1966; Estergreen et al., 1967), cannot be recommended. It is difficult to express the corpus luteum of pregnancy and therefore damage to the ovary with resultant haemorrhage is very often the sequellae. It is particularly important not to use this method if the animals are to be rebred following the induced abortion. Bursal and ovarian adhesions often occur as a result of the associated trauma followed by reduced or lost fertility.

Ovariectomy (spaying) will also result in abortion in pregnancies up to approximately 150 to 200 days (Estergreen et al., 1967; Lindell et al., 1981). Spaying heifers destined for the feedlot has been employed to varying degrees. It is used primarily to prevent oestrus and the increased activity associated with oestrus. However, since many of these animals are pregnant, abortion will occur when the corpus luteum of pregnancy is removed. Although there are reports of successful methods for spaying heifers, the trouble, expense and tissue damage resulting in reduced carcass quality have prevented this practice from becoming widely used (Rupp and Kimberling, 1982; Bennett, 1983).

Physical damage to the conceptus and its membranes usually results in abortion (Ball and Carroll, 1963; Millar, 1974). Examples of this include crushing of the amnionic vesicle and decapitation. Rupture of the fetal membranes by passing a rod or pipette through the cervix has also been reported (Parmigiani et al., 1978). This procedure can only be done in early pregnancy while the uterus is still retractable. One can speculate that the mechanism of action for abortion caused by physical damage is that the destruction of the foetus and membranes allows the production of prostaglandin causing luteolysis and emptying of the uterus. The variability in percent responding and in response time places these techniques in a category of questionable value.

Diethylstilboestrol (DES) was used extensively by feedlot managers to induce abortion until the United States Food and Drug Administration enforced its ban on the use of DES. Other oestrogens are still available for use although they are not as effective as DES was reported to be. Little in the way of actual figures have been published on the efficacy of oestrogens as abortifacients in cattle. Hill and Pearson (1958) reported on the use of DES. Brand et al. (1975) used oestradiol valerate, Refsal et al. (1976) and Seguin et al. (1978) reported on the use of oestradiol cypionate. None reported exceptionally good results and one must be cognizant of the undesirable side effects following oestrogen usage.

Corticosteroids, notably dexamethasone, will terminate pregnancy during the last month of pregnancy, but was shown by Adams and Wagner (1970) to give inconsistent results prior to day 255. Thus, although these products are used for parturition induction, they are not adequate abortifacients when used alone.

Prostaglandin $F_{2\alpha}$ and its analogues have been proven to be effective and safe abortifacients for pregnancies in cattle up to about 150 days of gestation (Lauderdale, 1972; Lamond et al., 1973; Brand et al., 1975; Jackson and Cooper, 1976; Copeland, 1978; Schultz and Copeland, 1981). They have rapidly displaced use of methods listed above due to their safety, efficacy and ease of administration even though they may be more expensive than some of the methods listed. With this expanded use more information is being gained concerning factors affecting their efficacy.

Factors affecting efficacy of prostaglandins

Dose titration studies have been conducted to establish effective levels of the various prostaglandins being used for induction of abortion in cattle. Tables I, II and III illustrate dose titration data generated to establish recommended dose levels for field use of prostaglandin $F_{2\alpha}$, for which 25 mg/head was chosen; cloprostenol, for which 375 μg/head was chosen; and fenprostalene, for which 1.0 mg/head was chosen. Obviously, there is no magic cut off point in terms of an optimum effective dose.

Table I. Dose titration study using prostaglandin $F_{2\alpha}$ (1). (Stage of pregnancy up to 100 days). (McAllister, J.F. and Lauderdale, J.W., 1979)

Dose (mg)	No. treated	No. aborting (2)	%
0	47	8	17
5	44	26	59
10	50	37	74
20	50	45	90
40	47	40	85
80	50	49	98

Dose selected: 25 mg/head.
(1) Data from trial conducted at 7 locations.
(2) By 35 days post treatment.

Table II. Combined summary data from 3 trials using cloprostenol to induce abortion in feedlot heifers. (Stage of pregnancy up to 150 days). (Combined data from Copeland et al., 1978; Fields et al., 1977 and Youngquist et al., 1977)

	Dose of cloprostenol (μg)				
	62.5	125	250	375	500
No. Heifers treated	27	49	109	88	83
No. Heifers aborted	16	31	89	85	78
% Abortion rate	59.3	63.3	81.7	96.6	94.0

Dose selected: 375 μg/head.

Table III. Dose titrations studies using fenprostalene. (Stage of pregnancy up to 150 days). (Syntex Agribusiness, 1982)

	% Aborted		
Dose (mg)	Trial I Herefords n=65	Trial II Sev. breeds n=52	Trial III Her/Ang X n=65
0.0625	31		
0.125	31		
0.250	62	46	60
0.500	92	69	91
1.000		92	82
2.000			100

Dose selected: 1.0 mg/head.

Using the prostaglandins available it is difficult to establish a good dose-weight relationship in terms of efficacy for induction of abortion in feedlot heifers. Using alfaprostol for example, Keay et al. (1983) suggested that using 0.7 mg/100 pounds (~50 kg) of body weight resulted in heavier animals receiving a larger total dose which, in turn resulted in higher efficacy than in lighter weight animals. This occurred even though the lighter weight animals were at 83 days of pregnancy at the time of treatment versus 138 days for the larger heifers.

Stage of pregnancy plays a significant role in efficacy of these compounds. Although there is no definite stage in gestation wherein efficacy abruptly decreases, as pregnancy advances beyond 120 days there is a reduction seen in efficacy of these compounds (Jackson and Cooper, 1976; Schultz and Copeland, 1979, 1981).

Lindell et al. (1980/81), using cloprostenol, found a difference in the post-abortion $PGF_{2\alpha}$ release as well as differences in the physical nature of the abortion when pregnancies were below 80 days rather than over 100 days. In animals pregnant less than 80 days the aborted foetuses were delivered within the intact foetal membranes with little or no bleeding while in those between 100 and 150 days of gestation the foetuses were delivered prior to the membranes which were retained for more than 24 hours. The degree of foetal membrane attachment was proposed as the reason for the differences in the nature of the induced abortions.

Another factor affecting the susceptibility of the pregnancy to prostaglandin treatment are progestagen implants or feed additives. Melengestrol acetate (MGA) has been shown to reduce the efficacy of cloprostenol (Youngquist et al., 1977; Schultz and Copeland, 1979) and $PGF_{2\alpha}$ (McAllister and Lauderdale, 1979). The role of concomitant treatments and previous treatments such as dewormers, insecticides, antibiotics, vitamins and vaccinations have been suspected of affecting the efficacy but this has not been proven when subjected to testing (Schultz and Wilson, 1978; Schultz and Copeland, 1979). The stress of long distance hauling, new environment, new feed and handling associated with entry into the feedlot might well be suspected of inducing spontaneous abortion. However, it was found by the authors (Schultz and Copeland, 1979) that quite the opposite appears to be true. When the animals were treated with the abortifacient prostaglandin (cloprostenol) at the time of induction into the feedlot (period of very high stress) there appeared to be a reduction in the ability of the prostaglandin to induce abortion. In these same studies, pregnant, non-treated control animals failed to abort. Perhaps the stress stimulation of the adrenal resulted in protective levels of progesterone secretion. On the other hand, many feedlot managers report that they do see spontaneous abortions in animals when their rations are changed to the high energy finishing rations ("hot rations"). The physiology which might explain these findings has not been reported.

Abortion induction beyond 150 days of gestation

Although the purpose of this paper is to address susceptibility of early pregnancies it is of interest to consider the status of abortion induction in cattle beyond 150 days of gestation. For the feedlot operator the animals beyond 150 days pregnant present a significant problem in that many of these animals will calve in the feedlot. In a very high percentage of these animals dystocia, retained placenta and acute metritis occur. Many do not survive and those that do cause profit losses for the feedlot since they gain weight and convert feed to meat very poorly.

Day (1977) and Murray et al. (1981) obtained increased efficacy in late pregnancy by administering dexamethasone followed ten days to two weeks later by cloprostenol. This is not a practical approach for use in feedlots since each handling and treating of the animals affects the rate of gain and feed efficiency. Therefore, several studies evaluating simultaneous administration of dexamethasone and cloprostenol to heifers in the last half of pregnancy have been conducted by the workers at Saskatoon (Barth et al., 1981a, b; Johnson et al., 1981; Mapletoft, 1983). They demonstrated that dexamethasone can induce abortion in heifers in late pregnancy when the corpus luteum was removed either by ovariectomy or administration of cloprostenol. Further, they demonstrated that the extra-luteal source of progesterone that was eliminated by the dexamethasone was not of maternal adrenal origin. They concluded that dexamethasone causes abortion in cattle deprived of an ovarian source of progesterone by reducing placental synthesis of progesterone.

Field trials were also conducted to evaluate efficacy of concomitant administration of cloprostenol and dexamethasone to pregnant feedlot heifers. In the first trial Barth et al. (1981a) found that 500 μg cloprostenol given simultaneously with 25 mg dexamethasone was effective. Overall, 111 of 117 (95%) of the treated animals aborted while none of the 29 control animals aborted. Of the heifers pregnant between one to four months, 37 of 37 aborted, while in the groups pregnant between four to six and six to eight months, 37 of 40 aborted in each group. Subsequently, dose titration trials have been conducted which indicate that 375 μg cloprostenol given with 10-15 mg dexamethasone may be sufficient to obtain desired efficacy (Mapletoft, 1983).

It must be remembered that whenever abortion is induced the treated animals should be kept under close observation since complications can occur. The earlier in pregnancy abortion can be induced the less chance for problems to occur.

REFERENCES

Adams, W.M. and Wagner, W.C., 1970. The role of corticoids in parturition. Biol. Reprod., 3: 223-228.

Ball, L. and Carroll, E.J., 1963. Induction of fetal death in cattle by manual rupture of the amnionic vesicle. J. Am. Vet. Med. Ass., 142: 373-374.

Barth, A.D., Adams, W.M., Manns, J.G., Kennedy, K.D., Sydenham, R.G. and Mapletoft, R.J., 1981a. Induction of abortion in feedlot heifers with a combination of cloprostenol and dexamethasone. Can. Vet. J., 22: 62-64.

Barth, A.D., Johnson, W.H., Mapletoft, R.J. and Manns, J.G., 1981b. Induction of abortion in feedlot heifers with dexamethasone and prostaglandin $F_{2\alpha}$. Proc. Soc. for Theriogenology, 78-95.

Bennett, B., 1983. Induction of abortion in feedlot heifers. Unpublished data.

Bennett, B. and Rupp, G., 1983. Survey of Colorado feedlots and packers. Personnel Communication, Colorado State Univ.

Brand, A., DeBois, C.H.W., Kommerij, R. and DeJong, M.P., 1975. Induction of abortion in cattle with prostaglandin $F_{2\alpha}$ and oestradiol valerate. Tijdschr. Diergeneesk., 100: 432-435.

Copeland, D.D., 1978. Therapeutic and feedlot abortion application of prostaglandins. Bov. Pract., 13: 54-56.

Copeland, D.D., Schultz, R.H. and Kemtrup, M.E., 1978. Induction of abortion in feedlot heifers with cloprostenol (a synthetic analogue of prostaglandin $F_{2\alpha}$): A dose response study. Can. Vet. J., 19: 29-32.

Day, A.M., 1977. Cloprostenol for termination of pregnancy in cattle. N. Z. Vet. J., 25: 139-144.

Estergreen, V.L. Jr., Frost, O.L., Gomes, W.R., Erb, R.E. and Bullard, J.F., 1967. Effect of ovariectomy on pregnancy maintenance and parturition in dairy cows. J. Dairy Sci., 50: 1293.

Fields, M.J., Schultz, R.H., Bass, J. and Warnick, A.C., 1977. Abortion in beef heifers using cloprostenol (a $PGF_{2\alpha}$ analogue). Meet. of So. Div. Am. So. Anim. Sci., Abstr., Atlanta, Georgia, 45-46.

Hill, J.J. and Pearson, R.E., 1958. Repositol diethylstilbestrol as an abortifacient in feedlot heifers. J. Am. Vet. Med. Ass., 132: 507-512.

Jackson, P.S. and Cooper, M.J., 1976. The use of cloprostenol (ICI 80996) in the treatment of infertility in cattle. Presented at World Ass. for Buiatrics, Paris, 903.

Johnson, W.H., Manns, J.G., Adams, W.M. and Mapletoft, R.J., 1981. Termination of pregnancy with cloprostenol and dexamthasone in intact cows. Can. Vet. J., 22: 288-290.

Keay, L.E., Kiracofe, G., Riley, J. and Simon, M., 1983. Aborting feedlot heifers with alfaprostol. Cattleman's Day Proceedings, March 4, Kansas State Univ., 22-23.

Lamond, D.R., Tomlinson, R.V., Drost, M., Hendricks, D.M. and Jochle, W., 1973. Studies of prostaglandin $F_{2\alpha}$ in the cow. Theriogenology, 4: 269-284.

Lauderdale, J.M., 1972. Effects of $PGF_{2\alpha}$ on pregnancy and the estrous cycle of cattle. J. Anim. Sci., 35: 246 (Abstract).

Lindell, J.O., Kindahl, H. and Edqvist, L.-E., 1980/81. Prostaglandin induced early abortions in the bovine. Clinical outcome and endogenous release of prostaglandin $F_{2\alpha}$ and progesterone. Anim. Reprod. Sci., 3: 289-299.

Lindell, J.O., Edqvist, L.-E. and Gustafsson, B., 1981. Oophorectomy during different stages of pregnancy in the cow. Acta. vet. scand., 22: 553-565.

Mapletoft, R.J., 1983. Unpublished data.

McAllister, J.F. and Lauderdale, J.W., 1979. Lutalyse in pregnancy termination. Proc. Lutalyse Symposium, Upjohn, 65-74.

McDonald, L.E., McNutt, S.H. and Nichols, R.E., 1953. On the essentiality of the bovine corpus luteum of pregnancy. Am. J. Vet. Res., 14: 539-541.

Millar, P.G., 1974. Methods of early termination of pregnancy in the cow. Vet. Rec., 94: 626.

Murray, R.D., Smith, J.H. and Harker, D.B., 1981. Use of cloprostenol with dexamethasone in the termination of pregnancy in heifers. Vet. Rec., 108: 378-380.

Parmigiani, E., Ball, L., Lefever, D., Rupp, G. and Seidel, G. Jr., 1978. Elective termination of pregnancy in cattle by manual abortion. Theriogenology, 4: 283-290.

Refsal, K.R., Seguin, B.E., Mather, E.C. and Gustafsson, B.G., 1976. Estradiol and $PGF_{2\alpha}$ as luteolytic agents and abortifacients in cattle. 57th Conf. Research Workers in Anim. Diseases, Chicago.

Rupp, G.P. and Kimberling, C.L., 1982. A new approach for spaying heifers. VMSAC., 77: 561-565.

Seguin, B.E., Refsal, K.R., Goerke, T.P. and Phatak, A., 1978. Field trial experiences with prostaglandins in Minnesota and Wisconsin beef cattle. Minn. Vet., 18: 9-14.

Schultz, R.H. and Wilson, D.C., 1978. Factors affecting cloprostenol-induced abortion in beef heifers. Presented at 70th Ann. Meet. Am. Soc. Anim. Sci., Mich. State U., 431 (Abstract).

Schultz, R.H. and Copeland, D.D., 1979. Unpublished data.

Schultz, R.H. and Copeland, D.D., 1981. Induction of abortion using prostaglandins. Acta vet. scand., Suppl. 77: 353-361.

Simms, D. and Kuhl, G., 1982. Effectiveness of Lutalyse as an abortifacient in feedlot heifers. Kansas Focus on Feedlots. November.

Syntex Agribusiness, Inc., 1982. Freedom of Information - N.A.D.A., 2-4:128-549.

Tanabe, T.Y., 1966. Essentiality of the corpus luteum for maintenance of pregnancy in dairy cows. J. Dairy Sci., 49: 731.

Wilson, D., 1983. Unpublished data.

Youngquist, R.S., Bierschwal, C.J., Elmore, R.G., Jenkins, A.L., Schultz, R.H. and Widmer, J.H., 1977. Induction of abortion in feedlot heifers with cloprostenol (ICI 80996). Theriogenology, 7: 305-312.

Animal Reproduction Science, 7 (1984) 245—267
Elsevier Science Publishers B.V., Amsterdam — Printed in The Netherlands

PERSISTENCE OF LUTEAL ACTIVITY IN THE NON-PREGNANT EWE

L. Zarco[1], G.H. Stabenfeldt[1], H. Kindahl[2], J.F. Quirke[1,a] and E. Granström[3]

[1] Department of Reproduction, School of Veterinary Medicine, University of California, Davis, California 95616, (USA).

[2] Department of Obstetrics and Gynaecology, College of Veterinary Medicine, Swedish University of Agricultural Sciences, 750 07 Uppsala, (Sweden).

[3] Department of Physiological Chemistry, Karolinska Institutet, 104 01 Stockholm, (Sweden).

ABSTRACT

Zarco, L., Stabenfeldt, G.H., Kindahl, H., Quirke, J.F. and Granström, E., 1984. Persistence of luteal activity in the non-pregnant ewe. Anim. Reprod. Sci., 7: 245-267.

Four cases of spontaneous persistence of the corpus luteum (CL) in the ewe are presented along with data concerning normal luteolysis in two ewes. The persistence of the CL in the four ewes resulted in oestrous cycle lengths of 26, 56, 139 and 145 days. Daily plasma levels of progesterone, 15-keto-13,14-dihydro-$PGF_{2\alpha}$ and 11-ketotetranor PGF compounds are reported. In addition, blood samples were taken at two h intervals during the initial prolongation of the CL in two of the ewes. In these ewes, pulsatile release of $PGF_{2\alpha}$ occurred at the expected time, and the release of $PGF_{2\alpha}$ during each pulse, as well as total release from the first to the last pulse, were similar to those of ewes undergoing normal luteolysis. However, the interval between individual pulses was greatly increased in ewes with persistence of the CL, having means of 18.0 h and 18.5 h for two of the ewes sampled intensively. In contrast, two ewes undergoing normal luteolysis had interpulse intervals of 6.8 and 7.2 h, respectively.

The results suggest that spontaneous persistence of the CL in the ewe is not due to a primary inability of the uterus to produce $PGF_{2\alpha}$. A disturbance of the system which acutely controls individual pulses of $PGF_{2\alpha}$ (and thus their frequency) seems to be a more likely cause for the malfunction.

In all four cases, the prolonged luteal phase was finally terminated by active release of $PGF_{2\alpha}$.

---

[a] On leave from the Irish Agricultural Institute, Ballinrobe, (Ireland).

0378-4320/84/$03.00

INTRODUCTION

Persistence of the luteal phase occurs in several domesticated animal species. Persistence of the corpus luteum (CL) in the mare is relatively common with an incidence of 15-20% in animals with normal cyclic ovarian activity (Stabenfeldt, et al., 1974) while the incidence in the cow is lower, i.e., 1.5% (Bulman and Lamming, 1977).

In the ewe, cases of persistence of the CL have been reported after the first oestrus of puberal ewes (Quirke and Gosling, 1979), after the first post-partum ovulation (Quirke et al., 1983), and in ewes that had several oestrous cycles of normal length before the condition occurred (Yenikoye et al., 1982). None of the above studies reported profiles of prostaglandin $F_{2\alpha}$ ($PGF_{2\alpha}$) during initiation and termination of luteal persistence.

We have recently conducted studies on ovarian physiology of ewes as a function of breed and age (Zarco, Quirke, Bradford and Stabenfeldt, unpublished observations). During these studies, persistence of the luteal phase was noted in four ewes. The details of these findings are reported, including data involving $PGF_{2\alpha}$ synthesis and release, particularly in conjunction with the onset and termination of prolonged luteal activity.

MATERIALS AND METHODS

Animals

During a study of ovarian physiology of the ewe, four animals were identified as having persistent luteal activity. They were of four different breeds, i.e., Finnish Landrace (ewe No. 22), Rambouillet (No. 26), Suffolk (No. 60) and Finnish Landrace-Rambouillet cross (No. 216). Three of the ewes (Nos. 22, 26 and 60) were part of an experiment on luteolysis which involved an intensive bleeding regime (2 h intervals) beginning two days before the expected time of luteolysis and continuing through oestrus. Oestrous cycle lengths had been determined for these ewes previously during the autumn, and also in previous years. Two of the ewes (Nos. 22 and 26) had shown oestrus at the expected time prior to the start of the experiment and these were sampled intensively during the time of initial prolongation of the CL. The third ewe in this group (No. 60), also scheduled for the intensive bleeding experiment, had failed to show oestrus at the expected time, but was included in the experiment on the assumption that oestrus had been missed. It subsequently became apparent that she had prolonged luteal activity during the previous oestrous cycle when she again failed to show oestrus at the end of the intensive bleeding regime. All three ewes were kept under the intensive bleeding regime until the third day after the expected onset of oestrus. Thereafter they were bled daily until plasma progesterone analysis indicated the luteal phase had terminated.

The fourth ewe (No. 216), not involved in the intensive bleeding regime, was detected as having a possible persistent CL as a result of an experiment involving laparoscopic examinations of ovaries at monthly intervals and on the basis of early seasonal cessation of oestrous activity. Upon identification of the animal, blood samples were collected daily until plasma progesterone levels indicated the end of luteal activity.

All four ewes were teased daily with a vasectomized ram except for the period when 3 of the ewes were subjected to the intensive bleeding regime when they were teased every two h. The day of the onset of oestrus was considered to be the first day of the oestrous cycle and thus oestrous cycles are numbered with the first day of oestrus equal to 1.

Data from two ewes of different breds undergoing normal CL regression are included for comparative purposes, i.e., Finnish Landrace-Rambouillet cross (No. 400) and Suffolk (No. 52).

Assays

Analysis of plasma progesterone was done by an enzyme immunoassay (Munro and Stabenfeldt, 1982) in order to monitor luteal activity. The antibody cross-reacts with 11α-hydroxy-progesterone (21.4%), 5α-pregnane-3,20-dione (29.5%), and 20β-hydroxy-progesterone (2.4%) while the cross reaction is <0.5% for 17α-hydroxy-progesterone, 17β-oestradiol, cortisol, Δ^4-androstenedione and testosterone. The sensitivity of the assay is 2.5 pg/ml. The inter-assay coefficient of variation ranges from 1.2% to 4.3% at different concentrations of hormone and the intra-assay coefficient of variation from 2.7% to 8.4%.

The initial main circulating metabolite of $PGF_{2\alpha}$, 15-keto-13,14-dihydro-$PGF_{2\alpha}$, was determined in plasma in order to identify the time at which synthesis and release of $PGF_{2\alpha}$ occurred (Kindahl et al., 1976b). The radioimmunoassay used has been previously described (Granström and Kindahl, 1976a; Kindahl et al., 1976a). The antibody cross-reacts with 15-keto-$PGF_{2\alpha}$ (16.0%), 13,14-dihydro-$PGF_{2\alpha}$ (4.0%), and with 15-keto-13,14-dihydro-PGE_2 (1.7%). The detection limit of the assay is 10 pg/ml of plasma. The inter-assay coefficient of variation is 14% and the intra-assay coefficient of variation ranges between 6.6% and 11.7% for different ranges of the standard curve.

The later-appearing, longer-lived metabolites, the 11-ketotetranor PGFs, were analyzed by a method previously described (Granström and Kindahl, 1976b, 1982a, 1982b).

Data analysis

Estimates of the following parameters, length of the oestrous cycle, length of luteal phase, number of $PGF_{2\alpha}$ pulses, time of their occurrence, interval between pulses, interval from first to last pulse, amount of $PGF_{2\alpha}$ released during single pulses, and total production of $PGF_{2\alpha}$, were calculated for each of the persistent CL ewes whenever possible, and for the 2 normal ewes. Luteolysis was considered to be initiated at the time of the first significant release of $PGF_{2\alpha}$ that was followed by a continued decline in progesterone levels. Luteolysis was considered complete at the time that progesterone concentrations first reached, and remained at, 200 pg/ml or less.

In order to identify significant peaks of 15-keto-13,14-dihydro-$PGF_{2\alpha}$, we used an iterative approach. As a first step we calculated, for each ewe, the mean and standard deviation of all the samples taken during the intensive bleeding period and considered any value greater than three standard deviations above the mean as a significant increase. However, since the height of the pulses of 15-keto-13,14-dihydro-$PGF_{2\alpha}$ is highly variable, the values of the larger pulses disproportionately increased the mean value and the standard deviation, so that usually only one or two of the largest peaks could be detected in this first step. The calculations were thus repeated after eliminating the values corresponding to points identified as peaks during the first step. The rationale for this was that values already identified as significant peaks should not be considered in the calculation of the mean and standard deviation of the basal level because they clearly form part of a different sub-population. As a result of the elimination of peak values, the mean value decreased, the standard deviation was reduced, and new points were identified as significant elevations. These newly identified values were again eliminated for the next round of calculations and the process was repeated until no new significant elevations were detected (usually 4-6 cycles of analysis). We thus considered all the values of 15-keto-13,14-dihydro-PGF$_{2\alpha}$ identified as significant elevations by the sequential method described above as representing significant PGF$_{2\alpha}$ synthesis and release.

Another estimate of PGF$_{2\alpha}$ release was obtained by the determination of other PGF$_{2\alpha}$ metabolites, i.e., 11-ketotetranor PGFs, which provides additional information because of their longer half-life in plasma. The levels of 11-keto-tetranor PGF compounds remain elevated for several sampling periods (2 h intervals) after each episode of PGF$_{2\alpha}$ release, i.e., in a pulse-like form. Each individual pulse of 11-ketotetranor PGFs was considered to be formed by a peak value and by all of the consecutive above-baseline values preceding and following the peak. An estimate of PGF$_{2\alpha}$ release occurring during each pulse was obtained by measuring the area under the curve. The baseline of all samples, for a particular ewe, was determined by calculation of a basal mean using the Hampel method (Andrews et al., 1972). This involves calculation of a robust estimate of the mean of all samples from a ewe, an estimate which is not affected by extreme values such as those obtained during bursts of PGF$_{2\alpha}$ release. In the case where a new pulse started before the values from the preceding pulse had reached baseline, the point where a descending line intersected with an ascending line was considered to be the end of the preceding pulse and the initiation of the next one. The areas used to estimate PGF$_{2\alpha}$ release during individual pulses were calculated from plots done in a linear scale. Pulse areas for each ewe were then summed to obtain an estimate of the total release of PGF$_{2\alpha}$. Neither individual nor total release parameters, reported in mm^2, are intended to reflect actual amounts of PGF$_{2\alpha}$ released, but they are useful for comparative purposes among animals.

RESULTS

Normal ewes

Plasma levels of progesterone, 15-keto-13,14-dihydro-PGF$_{2\alpha}$ and 11-keto-tetranor PGF metabolites of two ewes undergoing normal luteolysis are shown in Figs. 1 and 2. For ewe 400 (Fig. 1), luteolysis began 46 h before the onset of oestrus and was completed within 22 h as indicated by progesterone levels (or 24 h before the onset of oestrus). This ewe showed 8 significant peaks of PGF$_{2\alpha}$

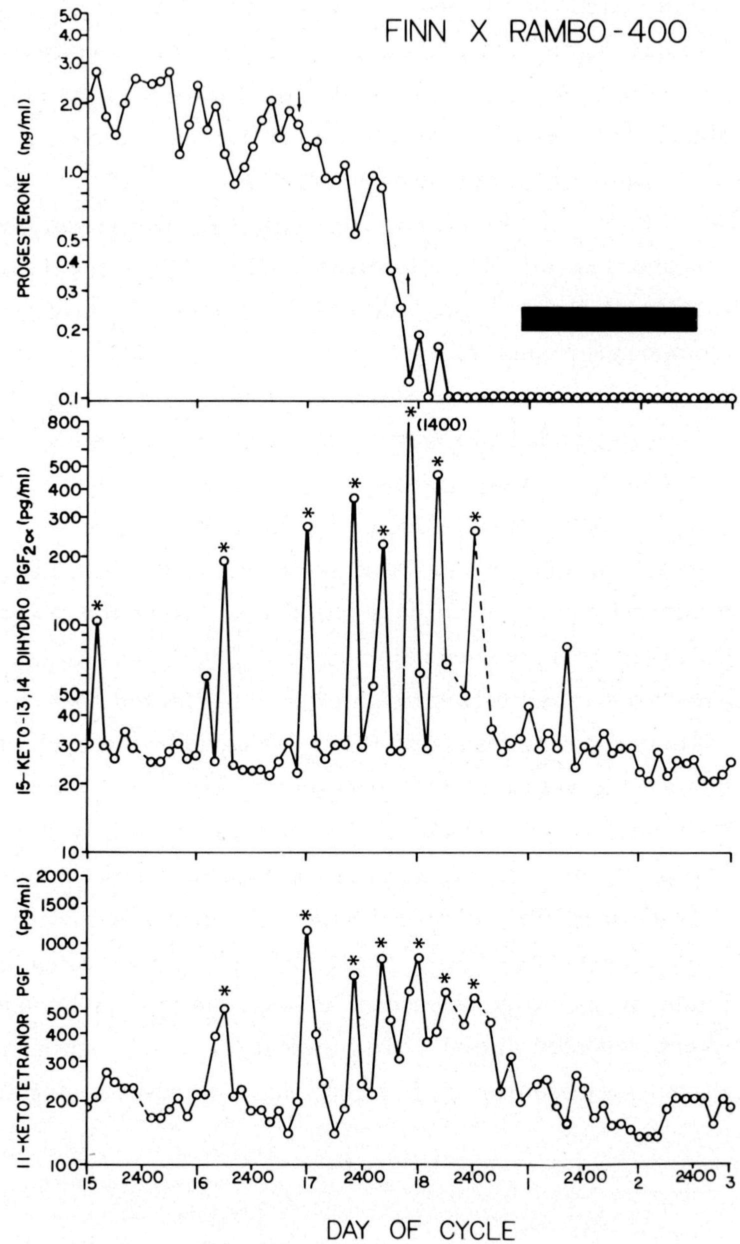

Fig. 1. Concentrations of progesterone, 15-keto-13,14-dihydro-$PGF_{2\alpha}$ and 11-ketotetranor PGF compounds in plasma samples (jugular vein) obtained at 2-h intervals during luteolysis in a normal ewe. Behavioral oestrus is indicated by a solid horizontal bar. Significant peaks of $PGF_{2\alpha}$ metabolites are indicated by asterisks. Arrows indicate the beginning (↓) and end (↑) of luteolysis. Note that the first pulse of $PGF_{2\alpha}$ was only detected by 15-keto-13,14-dihydro-$PGF_{2\alpha}$. Also note that logarithmic scales are used for concentrations.

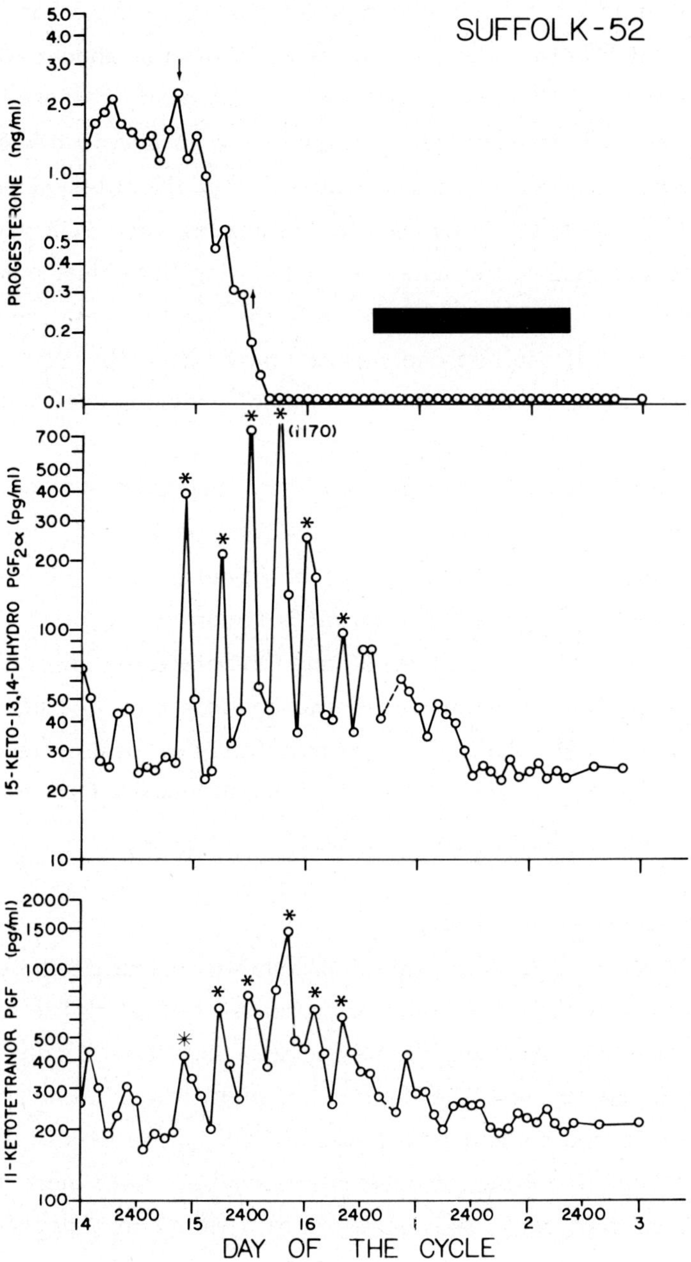

Fig. 2. Concentrations of progesterone, 15-keto-13,14-dihydro-$PGF_{2\alpha}$ and 11-ketotetranor PGF compounds in plasma samples (jugular vein) obtained at 2-h intervals during luteolysis in a normal ewe. Behavioral oestrus is indicated by a solid horizontal bar. Significant peaks of $PGF_{2\alpha}$ metabolites are indicated by asterisks. Arrows indicate the beginning (↓) and end (↑) of luteolysis.

with an interval of 11.7 ± 3.1 h between peaks. The first and second peaks, which occurred 92 and 64 h before the onset of oestrus, did not appear to result in the initiation of luteolysis. The following 6 peaks, apparently responsible for luteolysis, appeared at an increased frequency between 46 and 10 h before the onset of oestrus. If only the last 6 peaks are considered, the interval between peaks was 7.2 ± 0.8 h. The basal level of the 11-ketotetranor PGF compounds was estimated to be 211 pg/ml, the mean release during individual pulses was equivalent to a pulse area of 152 ± 27 mm^2, and the total release from start of the first pulse to the end of the last one was estimated at 1,216 mm^2. The length of the oestrous cycle was 19 days (mean length of previous cycles was 17.4 days, n=5).

The initiation of luteolysis in ewe 52 (Fig. 2) occurred 40 h before the onset of oestrus with completion occurring 16 h later (or 24 h before the onset of oestrus). The number of significant peaks of 15-keto-13,14-dihydro-$PGF_{2\alpha}$ associated with luteolysis was 6, the interval between peaks, 6.8 ± 0.5 h, and the occurrence of first and last peaks was 40 and 6 h before the onset of oestrus. The basal level of the 11-ketotetranor PGF metabolites was 272 pg/ml, the mean release during individual pulses, 170 ± 52 mm^2 and the total release, 1,020 mm^2. The length of the oestrous cycle was 16 days (mean length of previous cycles was 15.6 days, n=6).

Ewes with persistent luteal activity

Data from ewe No. 22, which failed to exhibit oestrus at the expected time during the intensive bleeding period, are presented in Fig. 3. The duration of the oestrous cycle was 26 days, considerably longer than the 18.7 day average (n=8) observed for previous cycles. The duration of significant luteal activity was 21 days. This ewe released 8 significant peaks of $PGF_{2\alpha}$, the first of which occurred 60 h before the expected onset of oestrus followed by three more peaks within a 38 h period. These four peaks thus occurred at a time consonant with the normal time for the onset of luteolysis in this ewe. However, luteolysis was not initiated and 4 additional peaks were observed at 22, 44, 80 and 88 h after the fourth one; none of the peaks was followed by decreased progesterone levels. The interval between peaks if all 8 peaks are considered was 18.0 ± 3.6 h; if only the first four peaks are considered the interval was 12.6 ± 1.7 h. The basal level of the

11-ketotetranor PGF compounds for this ewe was 252 pg/ml, the mean release during individual pulses was 341 ± 44 mm^2 (pulse area) and the total release was 2,016 mm^2.

The ewe was subjected to laparoscopic examination of the ovaries on days 12 and 24 of the cycle. One CL was present in the left ovary and two CLs in the right ovary at both days of examination.

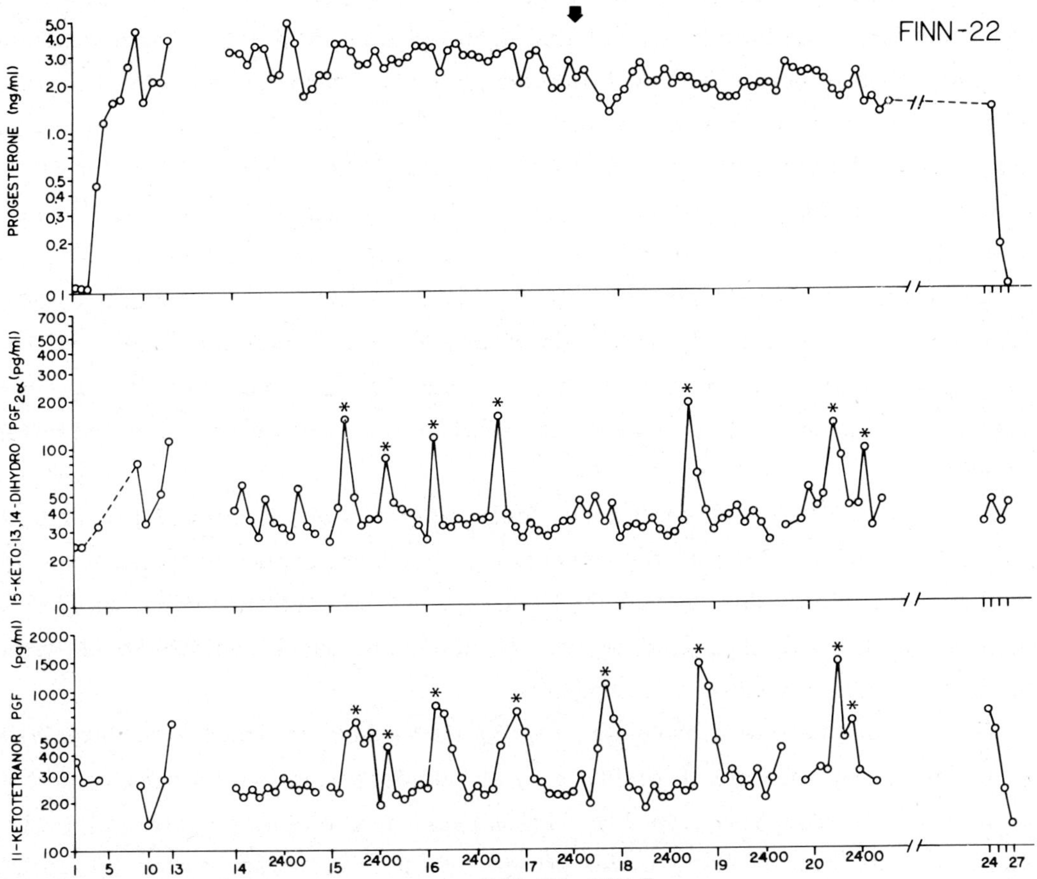

Fig. 3. Concentrations of progesterone, 15-keto-13,14-dihydro-$PGF_{2\alpha}$ and 11-ketotetranor PGF compounds in plasma samples (jugular vein) obtained daily during days 1 to 13 and 24 to 27 of the cycle and at 2-h intervals during the expected time of luteolysis in one ewe which developed a persistent CL. The expected onset of oestrus is indicated by an arrow. Significant pulses of $PGF_{2\alpha}$ metabolites are indicated by asterisks. Note that the fifth pulse was only detected by 11-ketotetranor PGF analysis.

The intensive bleeding schedule was discontinued on day 20 of the cycle of ewe No. 22 with samples taken daily thereafter. The precipitous decline in progesterone concentration observed between days 25 and 27, together with elevated levels of 11-ketotetranor PGFs on days 24 and 25, suggest that normal $PGF_{2\alpha}$ synthesis and release took place in conjunction with the termination of luteal activity. This ewe had four additional oestrous cycles prior to the termination of the reproductive season, all of them of normal length.

Fig. 4 presents data from the intensive bleeding period of another ewe (No. 26) which failed to show oestrus at the expected time. The data, obtained during the time when luteolysis and ovulation would have occurred in a cycle of normal length, show that progesterone declined on day 14 of the cycle from levels greater than 1 ng/ml, to less than 0.75 ng/ml, and then remained low for 60 h. During this period, progesterone reached baseline values in four occasions, three of which coincided with or immediately followed $PGF_{2\alpha}$ pulses; however, in none of these cases did progesterone remain at basal levels for more than one sampling period. After day 17, the levels of progesterone gradually increased, returning to values of more than 1 ng/ml by day 20. Five pulses of $PGF_{2\alpha}$ were detected in this ewe, all of which occurred within 90 h before the expected onset of oestrus.

In ewe No. 26 (Fig. 4), the interval between peaks of 15-keto-13,14-dihydro-$PGF_{2\alpha}$ was 18.5 ± 4.2 h with the interval from the first to the last pulse being 74 h. The mean release during individual pulses of 11-ketotetranor PGFs was 148 ± 32 mm^2 (peak area); total release was 740 mm^2. The basal level of the 11-ketotetranor PGF compounds was 235 pg/ml.

The luteal phase was prolonged to about 145 days at which time luteolysis was initiated by the release of $PGF_{2\alpha}$, as demonstrated by the high values of both metabolites observed on days 145-149 and by the abrupt decline in progesterone levels (Fig. 5). One CL was present on the right ovary during the entire luteal phase, as observed during repeated laparoscopic examinations.

Fig. 6 shows endocrine data from a ewe (No. 60) which appeared to have developed a persistent CL during the oestrous cycle previous to that involved with intensive bleeding. Although she failed to manifest oestrus at the onset of the cycle designated for study, she was placed in the experiment on the assumption of the occurrence of a silent ovulation. It subsequently became evident from

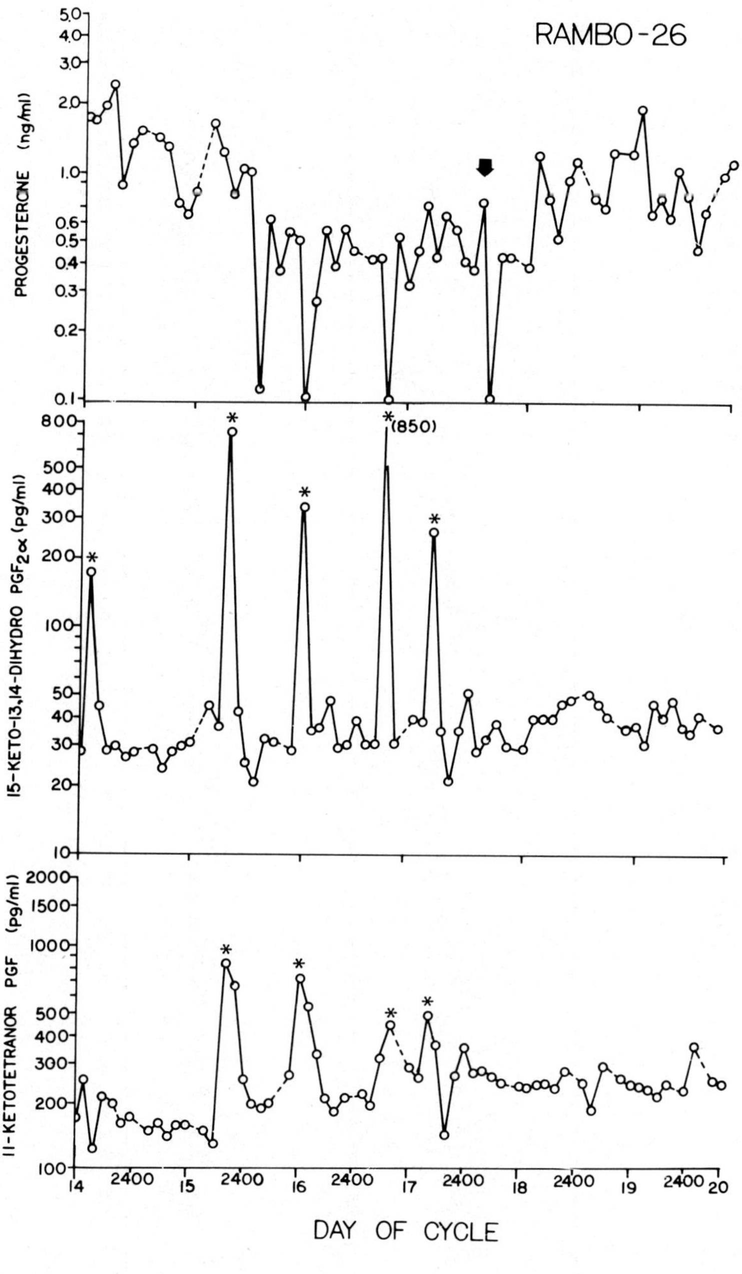

Fig. 4. Concentrations of progesterone, 15-keto-13,14-dihydro-PGF$_{2\alpha}$ and 11-ketotetranor PGF compounds in jugular vein plasma samples obtained at 2-h intervals during the expected time of luteolysis of one ewe (No. 26) which developed a persistent CL. The expected time of oestrus is indicated by an arrow. Significant pulses of PGF$_{2\alpha}$ metabolites are indicated by asterisks. Note the first pulse was only detected by 15-keto-13,14-dihydro-PGF$_{2\alpha}$ analysis.

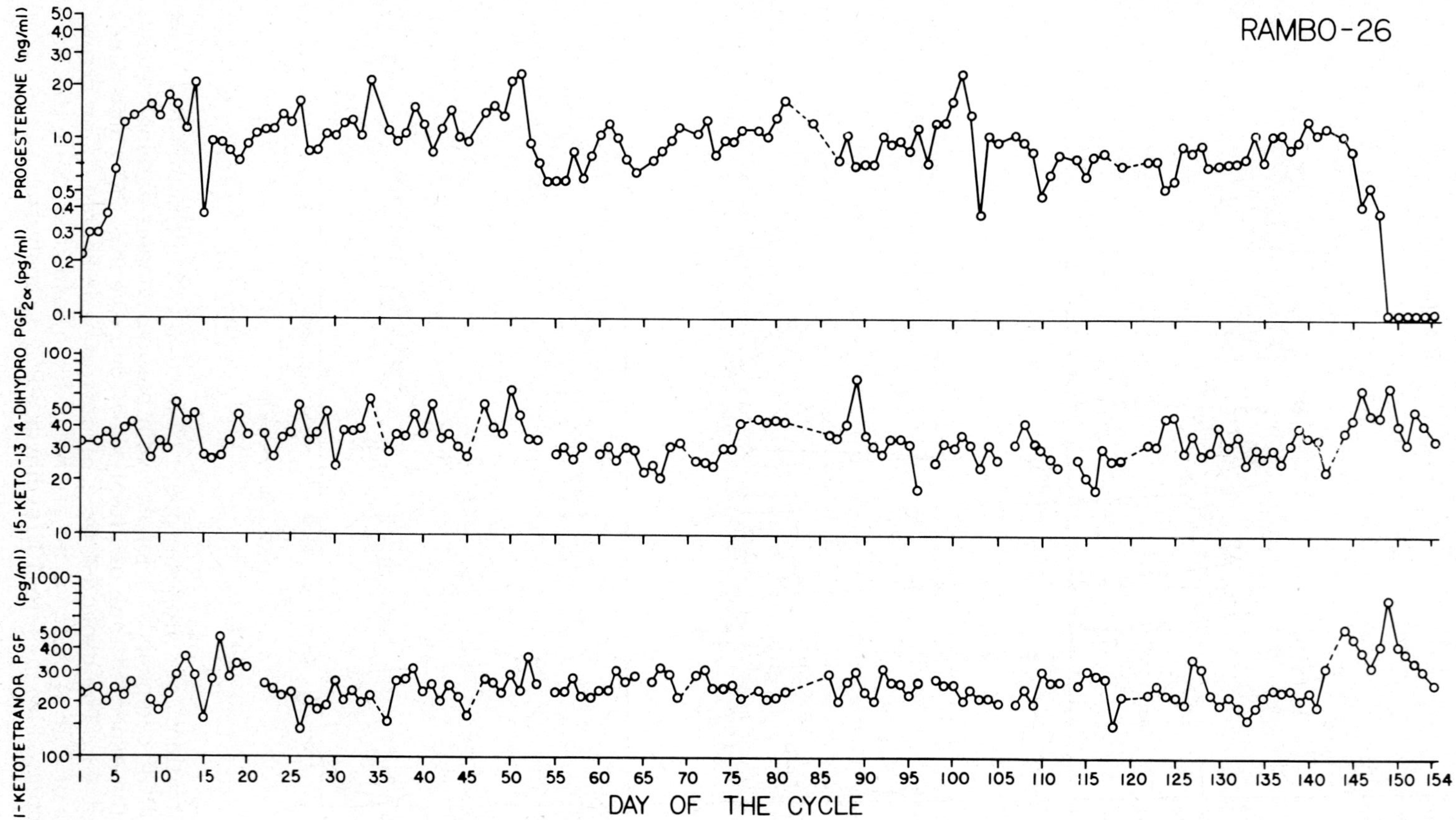

Fig. 5. Concentrations of progesterone, 15-keto-13,14-dihydro-$PGF_{2\alpha}$ and 11-ketotetranor PGF compounds in jugular vein plasma samples obtained daily during the luteal phase of one ewe (No. 26) with a persistent CL. Note the increase in the 11-ketotetranor PGF metabolites concentration associated with luteolysis.

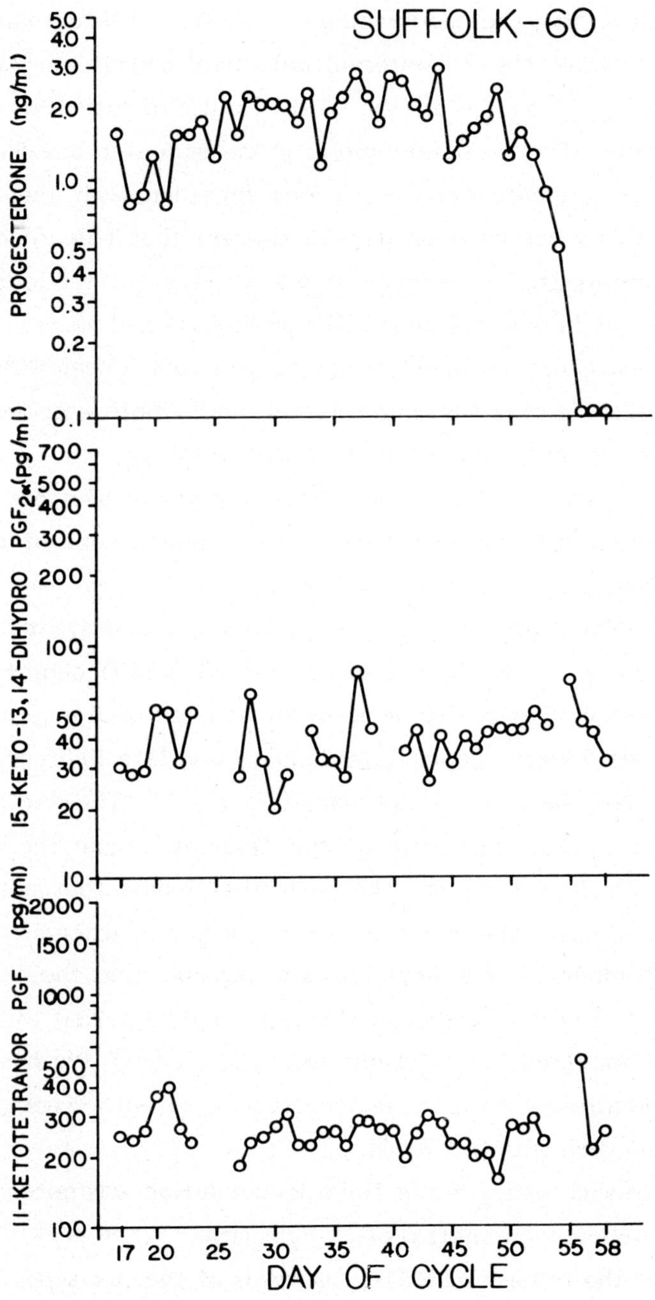

Fig. 6. Concentrations of progesterone, 15-keto-13,14-dihydro-$PGF_{2\alpha}$ and 11-ketotetranor PGF compounds in jugular vein plasma samples obtained daily during the luteal phase of one ewe with persistent CL. Note the increase in the 11-ketotetranor PGF metabolites concentration associated with luteolysis.

progesterone analysis that she had developed a persistent luteal phase during the previous oestrous cycle. The 2-h interval data (not shown) were thus collected beginning about 14 to 15 days after the initial failure of luteolysis with no significant changes noted. The persistent luteal phase ended on day 56. The precipitious decline in progesterone concentrations together with the high level of 11-ketotetranor PGFs observed on day 56 suggest that release of $PGF_{2\alpha}$ was responsible for luteolysis. The amount of plasma was not sufficient to analyze for concentrations of 11-ketotetranor PGFs on days 54 and 55.

Laparoscopic examinations done at regular intervals revealed the presence of one CL in each ovary during the entire luteal phase. Both uterine horns of ewe No. 60 appeared to be moderately distended with fluid.

Following luteolysis, ewe No. 60 had three subsequent oestrous cycles of normal length as indicated by progesterone levels although none of these oestrous cycles was accompanied by behavioral oestrus.

A fourth ewe (No. 216) was suspected of having a persistent luteal phase in late February because of the absence of oestrus after 26 December, and because of the presence of two CLs on the same ovary at each of 2 laparoscopic examinations made about 5 weeks apart (late January and late February). Blood sampling began on a daily basis on 28 February (see Fig. 7). The data are presented as beginning on day 65 of the cycle on the assumption that the oestrous cycle actually started on 26 December. The possibility exists that a sexually unexpressed ovulation(s) may have occurred during the period when no blood sampling was done (26 December-27 February). It is also possible that the persistent luteal phase lasted about 139 days although there is documentation for only 75 days. The luteal phase appeared to be terminated by the active synthesis and release of $PGF_{2\alpha}$ as indicated by elevated concentrations of both types of metabolites observed in conjunction with luteolysis.

An enlarged uterus suggestive of fluid accumulation was noted in this animal during the laparoscopic examination of 26 February.

A summary of the release of $PGF_{2\alpha}$ in terms of the two types of metabolites in conjunction with luteolysis in two ewes, and with the onset of luteal persistence in two ewes, is presented in Table I.

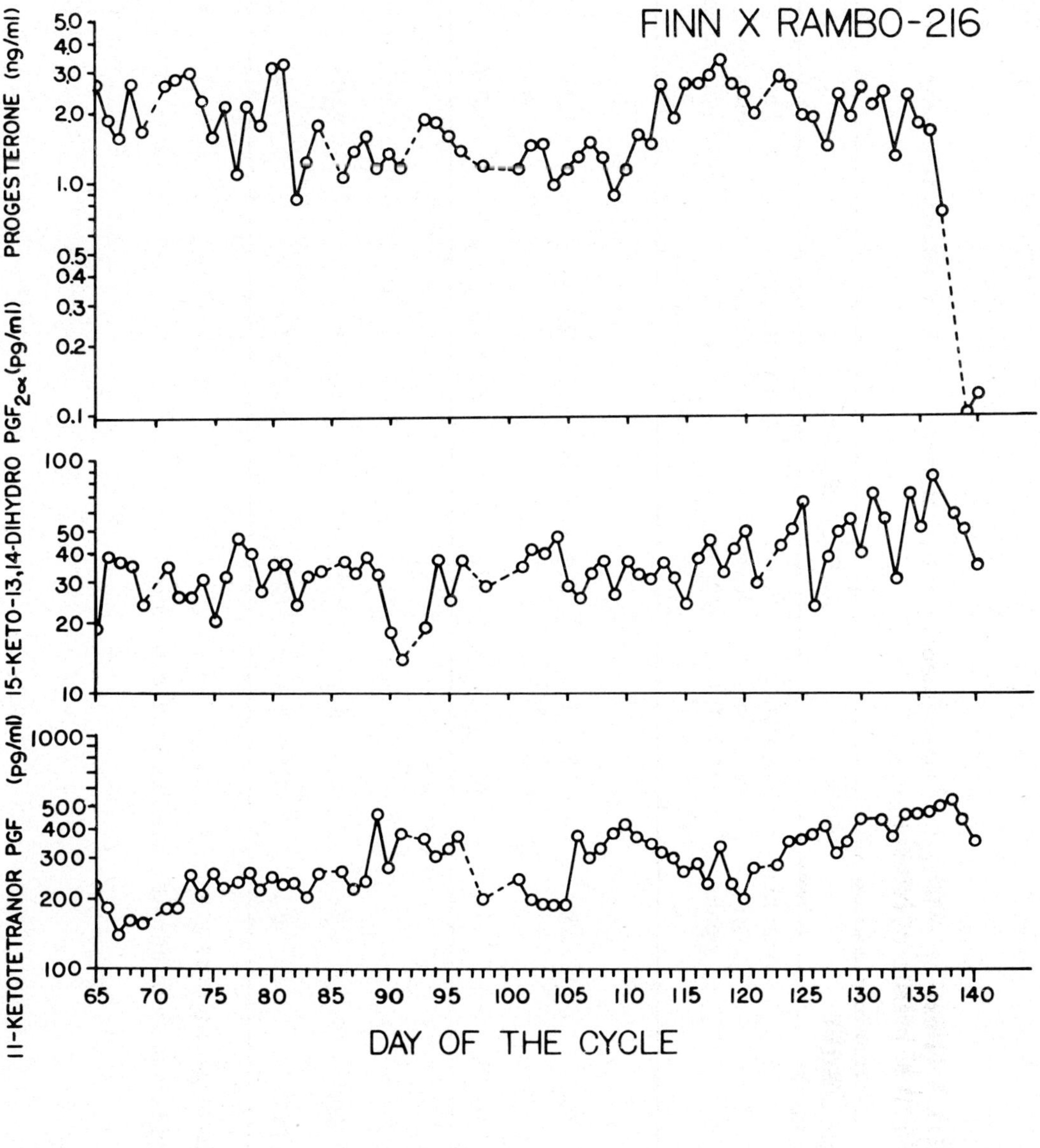

Fig. 7. Concentrations of progesterone, 15-keto-13,14-dihydro-$PGF_{2\alpha}$ and 11-ketotetranor PGF compounds in jugular vein plasma samples obtained daily during the luteal phase of one ewe with persistent CL.

Table I. Summary of the release of $PGF_{2\alpha}$ (as determined by metabolite analysis) in conjunction with normal luteolysis and development of persistent luteal phase

Animal	Cycle length (days)	$PGF_{2\alpha}$ parameters[1]				
		Number of peaks	Interval between peaks (h)	Basal level (pg/ml)	Mean release during single pulses (mm^2 peak area)	Total release (mm^2 peak area)
Normal ewes						
No. 400	19	8	11.7 ± 3.1 7.2 ± 0.8[a]	211	152 ± 27	1,216
No. 52	16	6	6.8 ± 0.5	272	170 ± 52	1,020
Ewes with persistent CL						
No. 22	26	8	18.0 ± 3.6	252	341 ± 44	2,016
No. 26	145	4	18.5 ± 4.2	235	148 ± 32	740

[1] See text for definition of parameters.

[a] Interval if only the 6 peaks associated with luteolysis are considered.

DISCUSSION

We found the analysis of two different types of $PGF_{2\alpha}$ metabolites useful in assessing the surge release of $PGF_{2\alpha}$ involved in luteolysis. In a series of 6 ewes which were bled every two h during the period of luteal regression (unpublished observations), only one of 32 significant increases of 11-ketotetranor PGF compounds was not accompanied by a significant increase in the levels of 15-keto-13,14-dihydro-$PGF_{2\alpha}$. Conversely, there was also only one instance of a significant increase in the levels of 15-keto-13,14-dihydro-$PGF_{2\alpha}$ not accompanied by an increase in 11-ketotetranor PGF levels. Hence, both types of metabolites appear to be equally useful for the detection of episodes of $PGF_{2\alpha}$ release in the ewe when plasma samples are obtained every two h. In fact, complementary information can be obtained by analyzing the patterns of both types of metabolites due to the different dynamics of formation and elimination. The 15-keto-13,14-dihydro-$PGF_{2\alpha}$ metabolite is the first one formed with levels reaching peak concentrations in the plasma before those of 11-ketotetranor PGFs (Granström and Kindahl, 1982a, 1982b). Thus the 15-keto-13,14-dihydro-$PGF_{2\alpha}$ metabolite indicates the time at which release takes place with more precision than the 11-ketotetranor PGF metabolites. For this reason we have considered that each episode of $PGF_{2\alpha}$ release takes place at the time when a significant peak of 15-keto-13,14-dihydro-$PGF_{2\alpha}$ is detected. On the other hand, the 11-ketotetranor PGF metabolites, due to their longer half-life, are less affected by the temporal relationship between the actual release of $PGF_{2\alpha}$ and the time of sampling, and thus are best suited for the determination of the amount of $PGF_{2\alpha}$ released during each pulse.

Failure of endometrial release of $PGF_{2\alpha}$ at the proper time, or in sufficient amounts, has been proposed to be responsible for spontaneous prolongation of luteal activity in the mare (Stabenfeldt et al., 1974; Neely et al., 1979). The data from the two ewes that were studied intensively during the period of initial prolongation of the luteal phase suggest that the prolongation was not due to a primary inability of the uterus to synthesize and release $PGF_{2\alpha}$. This is supported by the finding of similar values for pulses of 11-ketotetranor PGFs in normal and persistent luteal phase ewes as well as by the similar value for the estimate of total release of $PGF_{2\alpha}$ found in both groups of ewes. In fact, the values for individual pulses and total release for the persistent luteal phase ewe No.22 were about two-fold larger than those of normal ewes. Thus the enzymatic machinery for the production of $PGF_{2\alpha}$ in the uterus seems to be present for the synthesis of the hormone.

A reduced frequency of release of $PGF_{2\alpha}$, rather than insufficient amounts of this hormone, seems to be at the basis of the spontaneous prolongation of the luteal phase in the ewe. Both of the normal ewes released 6 surges of 15-keto-13,14-dihydro-$PGF_{2\alpha}$ in rapid succession at the time of luteolysis. In both cases, the 6 peaks were separated from each other by intervals of about 7 h. We have observed this pattern of frequent release of $PGF_{2\alpha}$ during luteal regression in all normal ewes we have studied (unpublished observations). In contrast, the interval between peaks for ewes with persistent CL (Nos. 22 and 26) had a mean of 18.0 h and 18.5 h, respectively.

The concept of persistent luteal activity due to inappropriate frequency of pulses of $PGF_{2\alpha}$ is consistent with the proposal of Thorburn et al. (1973) that a series of frequent peaks are required to cause complete luteolysis and to prevent functional recovery of the CL. Recent experimental evidence by McCracken and Schramm (1983) supports this view. They found that infusion of 5 pulses of $PGF_{2\alpha}$ into the ovarian circulation was more effective for the induction of luteolysis than 4 pulses were, even when the dose used in the 4-pulse treatment regimen was 10-fold greater than that used in the 5-pulse treatment. A reduction in the frequency of $PGF_{2\alpha}$ release has also been proposed to form part of the mechanism by which the CL is maintained at the initiation of pregnancy in the ewe (Nett et al., 1976; Lewis et al., 1978).

Since it appears that the uteri of ewes with persistent CL do not lack the ability to produce $PGF_{2\alpha}$, it seems reasonable to propose that the decreased frequency of $PGF_{2\alpha}$ release may be due to a disturbance in the system which regulates the surge release of $PGF_{2\alpha}$. It has been known for several years that the release of $PGF_{2\alpha}$ at the end of the luteal phase is initiated by increased levels of oestrogens of follicular origin acting on a progesterone-primed uterus (Barcikowski et al., 1974; for review see Horton and Poyser, 1976). In the two ewes with persistent CL that were intensively bled, secretion of $PGF_{2\alpha}$ started at a time appropriate for a cycle of normal length. This fact suggests that the factor(s) that initiates the release of $PGF_{2\alpha}$ was present. The disturbance thus appears to originate at the level of a system involved in the acute regulation of individual pulses of $PGF_{2\alpha}$. McCracken's group has proposed that oxytocin from the posterior pituitary is the acute regulator of $PGF_{2\alpha}$ release by the uterus, and that the effects of steroid hormones on $PGF_{2\alpha}$ release are mediated by changes in the concentration of receptors for oxytocin (McCracken et al., 1978; McCracken, 1980; McCracken and Schramm, 1983). The finding of Fairclough et al. (1980) that levels of neurophysin I and II (indicators of oxytocin release) were elevated coincidentally with the pulses of 15-keto-13,14-dihydro-$PGF_{2\alpha}$ observed at the time of luteolysis, supports the concept that the pulsatile nature of $PGF_{2\alpha}$ release at the time of luteolysis may be acutely regulated, or "fine tuned", by the pulsatile release of oxytocin from the posterior pituitary. Alternatively, it has been proposed that a positive feedback relationship exists between $PGF_{2\alpha}$ and oxytocin of luteal origin during each pulse of $PGF_{2\alpha}$ (Flint and Sheldrick, 1983; Sheldrick and Flint, 1983). In this model, $PGF_{2\alpha}$ is postulated to induce the secretion of oxytocin by the CL which, in turn, feeds back to the uterus to increase $PGF_{2\alpha}$ release, the system remaining operative during each pulse until depletion of oxytocin results in cessation of the $PGF_{2\alpha}$ pulse. In each of the above models (pituitary or luteal source of oxytocin), a disturbance of the mechanism regulating the release of oxytocin would be likely to result in the kind of $PGF_{2\alpha}$ patterns that we have found in our persistent CL ewes.

There is an alternative possibility that would explain the reduced frequency of $PGF_{2\alpha}$ associated with persistent CL, i.e., the reduced frequency may be the consequence, rather than the cause, of CL persistence. It has been observed in several species that the decline in progesterone concentrations that occurs after $PGF_{2\alpha}$ release is followed by an increase in the frequency and amplitude of $PGF_{2\alpha}$ pulses (Thorburn et al., 1973; for review see Rothchild, 1981). It is then possible that if the CL does not respond to the first few pulses of $PGF_{2\alpha}$ with decreased progesterone secretion, the frequency of the prostaglandin pulses will not increase, and luteolysis will not take place. The failure of the CL to respond to the initial $PGF_{2\alpha}$ release can, in theory, be caused by lack of receptors, failure of $PGF_{2\alpha}$ to reach the ovary, or the presence of competitive substances. The data from one of the normal ewes (No. 400, Fig. 1) supports this concept. In this animal, the first two pulses of $PGF_{2\alpha}$ did not result in progesterone decrease, and the intervals between the first and second pulse, and the second and third pulse were rather large, 28 and 18 h, respectively. Only after the third pulse of $PGF_{2\alpha}$ did progesterone levels decrease, and this was followed by the immediate establishment of a frequent release pattern which involved peaks Nos. 3-8. Against this model, however, are the results from persistent CL ewe No. 26 (Fig. 4) in which the observed decline in progesterone levels associated with three of the pulses of $PGF_{2\alpha}$ did not result in an increased frequency of $PGF_{2\alpha}$ release. The cause of the decreased frequency of $PGF_{2\alpha}$ release associated with the development of the persistent CL, as well as the significance of the reduced frequency as a cause or result of the persistence, thus remains at a speculative level.

The enlarged uterus suggestive of fluid accumulation (hydrometra or mucometra) noted for ewes No. 60 and 216 could have been either the result or cause of the prolonged luteal activity. The presence of uterine fluid, however, does not appear to be necessary for the development of a persistent luteal phase in that ewes Nos. 22 and 26 had normal uteri (determined by laparoscopy) at the time of prolongation of CL activity. It is suggested that the accumulation of fluid is the result of persistence of the luteal phase, not its cause. This observation is supported by the finding of Roberts and Fox (1968) who reported the presence of mucometra in a cow which developed the condition only after 3 months of exposure to a persistent CL.

We were interested to know if the prolonged luteal activity was terminated by an active lytic process, i.e., by the release of $PGF_{2\alpha}$, or by a passive gradual decline in CL function as we have observed previously in the hysterectomized mare (Stabenfeldt et al., 1974). As mentioned earlier, it seems clear that all 4 ewes terminated the persistent luteal phase in a precipitous manner involving the release of $PGF_{2\alpha}$. The season of the year apparently does not affect the ability of the uterus to synthesize $PGF_{2\alpha}$ in that both ewe No. 26 and No. 216 were in the anoestrous season (as judged by flock activity) at the time of luteal regression.

We have noted persistence of the luteal phase following the first post-partum ovulation in 3 of 17 autumn lambing ewes (Quirke et al., 1983). The persistence of the CL in the cow may have some association with a delay in the reestablishment of a normal endometrium in the post-partum period as observed in the data of Bulman and Lamming (1977). Thus some similarity may exist between the ewe and cow in this respect. However, all 4 ewes reported here were non-bred animals that had 4 to 7 oestrous cycles of normal length prior to luteal prolongation. The persistence of luteal function in ewes reported by Yenikoye et al. (1982) was also in animals that had undergone cyclic ovarian activity over a period of time. Furthermore, Quirke and Gosling (1979) reported persistence of the CL in puberal animals after their first oestrus. Thus the ewe appears to be similar to the mare (Stabenfeldt et al., 1974) in that it undergoes persistence of the luteal phase without regard to the post partum period.

ACKNOWLEDGEMENT

The authors thank Coralie Munro for doing the progesterone analysis and Samar Basu for the prostaglandin metabolite assays. We are indebted to Cindee Worland for typing the manuscript. Supported in part by the California Agricultural Experiment Station, Hatch project "Manipulation of Postpartum Period in Ruminants", by the Swedish Council for Forestry and Agricultural Research and by the Swedish Medical Research Council (proj. no. 03P - 5804 and 03X - 05915). Luis Zarco was supported by grant No. 33551 of CONACYT, Mexico.

REFERENCES

Andrews, D.F., Bickel, P.J., Hampel, F.R., Huber, P.J., Rogers, W.H. and Tukey, J.W., 1972. Robust estimates of location: Survey and Advances. Princeton University Press, Princeton.

Barcikowski, B., Carlson, J.C., Wilson, L. and McCracken, J.A., 1974. The effect of endogenous and exogenous estradiol-17β on the release of prostaglandin $F_{2\alpha}$ from the ovine uterus. Endocrinology, 95: 1340-1349.

Bulman, D.C. and Lamming, G.E., 1977. Cases of prolonged luteal activity in the non-pregnant dairy cow. Vet. Rec., 100: 550-552.

Fairclough, R.J., Moore, L.G., McGowan, L.T., Peterson, A.J., Smith, J.F., Tervit, H.R. and Watkins, W.B., 1980. Temporal relationship between plasma concentrations of 13,14-dihydro-15-keto-prostaglandin F and neurophysin I/II around luteolysis in sheep. Prostaglandins, 20: 199-208.

Flint, A.P.F. and Sheldrick, E.L., 1983. Evidence for a systemic role of ovarian oxytocin in luteal regression in sheep. J. Reprod. Fert., 67: 215-225.

Granström, E. and Kindahl, H., 1976a. Radioimmunoassays for prostaglandin metabolites. Advances in Prostaglandin and Thromboxane Research, 1: 81-92.

Granström, E. and Kindahl, H., 1976b. Radioimmunoassay for urinary metabolites of prostaglandin $F_{2\alpha}$. Prostaglandins, 12: 759-783.

Granström, E. and Kindahl, H., 1982a. Species differences in circulating prostaglandin metabolites - relevance for the assay of prostaglandin release. Biochem. Biophys. Acta, 713: 555-559.

Granström, E. and Kindahl, H., 1982b. Radioimmunoassay of 5α,7α-dihydroxy-11-ketotetranorprostane-1,16-dioic acid, a major prostaglandin F metabolite in blood and urine. Methods in Enzymology, 86: 339-353.

Horton, E.W. and Poyser, N.L., 1976. Uterine luteolytic hormone: A physiological role for prostaglandin $F_{2\alpha}$. Physiological Reviews, 56: 595-651.

Kindahl, H., Edqvist, L.-E., Granström, E. and Bane, A., 1976a. The release of prostaglandin $F_{2\alpha}$ in the peripheral circulation during normal luteolysis in heifers. Prostaglandins, 11: 871-878.

Kindahl, H., Granström, E., Edqvist, L.-E., Neely, D., Hughes, J. and Stabenfeldt, G., 1976b. The advantages of measuring a prostaglandin $F_{2\alpha}$ metabolite in peripheral blood in studies of the physiological role of prostaglandin release during luteolysis in domestic animals. Proceedings VIIIth Int. Congr. Anim. Reprod. and A.I., Krakow, 3: 145-148.

Lewis, G.S., Jenkins, R.L., Fogwell, R.L. and Inskeep, E.K., 1978. Concentrations of prostaglandins E_2 and $F_{2\alpha}$ and their relationship to luteal function in early pregnant ewes. J. Anim. Sci., 47: 1314-1323.

McCracken, J.A., 1980. Hormone receptor control of $PGF_{2\alpha}$ secretion by the ovine uterus. Advances in Prostaglandin and Thromboxane Research, 8: 1329-1344.

McCracken, J.A. and Schramm, W., 1983. Prostaglandins and corpus luteum regression. In: P.B. Curtis-Prior (Editor). A Handbook of Prostaglandins and Related Compounds, Churchill-Livingstone, Edinburgh. (in press).

McCracken, J.A., Gammal, L.M., Glew, M.E. and Underwood, L.F., 1978. Hormone receptor control of prostaglandin $F_{2\alpha}$ secretion from the ovine uterus. Endocrinology, 102: A440.

Munro, C. and Stabenfeldt, G.H., 1982. Development of a microtiter plate enzyme immunoassay (EIA) for the determination of progesterone. Endocrine Society Meeting, San Francisco, Abstract 641, p. 240.

Neely, D.P., Kindahl, H., Stabenfeldt, G.H., Edqvist, L.-E. and Hughes, J.P., 1979. Prostaglandin release patterns in the mare: Physiological, pathophysiological, and therapeutic responses. J. Reprod. Fert., Suppl. 27: 181-189.

Nett, T.M., Staigmiller, R.B., Akbar, A.M., Diekman, M.A., Ellinwood, W.E. and Niswender, G.D., 1976. Secretion of prostaglandin $F_{2\alpha}$ in cycling and pregnant ewes. J. Anim. Sci., 42: 876-880.

Quirke, J.F. and Gosling, J.P., 1979. Pre-puberal plasma luteinizing hormone concentrations during the oestrus cycle and early pregnancy in Galway and Fingalway female lambs. Anim. Prod., 28: 1-12.

Quirke, J.F., Stabenfeldt, G.H. and Bradford, G.E., 1983. Resumption of ovarian function in autumn lambing Dorset, Rambouillet and Finnish Landrace ewes. Theriogenology, 19: 243-248.

Roberts, S.J. and Fox, F.H., 1968. An unusual case of bovine mucometra associated with a persistent corpus luteum. Cornell Vet., 58: 116-120.

Rothchild, I., 1981. The regulation of the mammalian corpus luteum. Recent Progress in Hormone Research, 37: 183-298.

Sheldrick, E.L. and Flint, A.P.F., 1983. Regression of the corpora lutea in sheep in response to cloprostenol is not affected by loss of luteal oxytocin after hysterectomy. J. Reprod. Fert., 68: 155-160.

Stabenfeldt, G.H , Hughes, J.P., Wheat, J.D., Evans, J.W., Kennedy, P.C. and Cupps, P.T., 1974. The role of the uterus in ovarian control in the mare. J. Reprod. Fert., 37: 343-351.

Thorburn, G.D., Cox, R.I., Currie, W.B., Restall, B.J. and Schneider, W.S., 1973. Prostaglandin F and progesterone concentrations in the utero-ovarian venous plasma of the ewe during the oestrus cycle and early pregnancy. J. Reprod. Fert., Suppl. 18: 151-158.
Yenikoye, A., Pelletier, J., Andre, D. and Mariana, J.C., 1982. Anomalies in ovarian function of Peulh ewes. Theriogenology, 17: 355-364.

THE ROLE OF PROSTAGLANDINS IN THE MATERNAL RECOGNITION OF PREGNANCY IN MARES[1]

D.C. Sharp, M.T. Zavy[2], M.W. Vernon[3], F.W. Bazer, W.W. Thatcher and L.A. Berglund

Department of Animal Science, University of Florida, Gainesville, Florida 32611, (USA).

ABSTRACT

Sharp, D.C., Zavy, M.T., Vernon, M.W., Bazer, F.W., Thatcher, W.W. and Berglund, L.A., 1984. The role of prostaglandins in the maternal recognition of pregnancy in mares. Anim. Reprod. Sci., 7: 269-282.

Prostaglandin F (PGF) was quantified in intra-uterine luminal flushings, endometrium and in vitro incubations of endometrium alone and endometrium in the presence of conceptus membranes. There were significant day by status interactions in all cases, indicating that the patterns of PGF synthesis and/or secretion were affected differently according to the presence or absence of an embryo. Endometrial PGF content and in vitro production capability both reached peaks on or about day 14 post ovulation in cycling mares then declined following luteolysis. Likewise, uterine luminal PGF peaked abruptly on day 14, remained elevated, then declined to day 20. In contrast, endometrial content of PGF remained lower in pregnant mares until an elevation on day 20, while the in vitro production capability of endometria from pregnant mares continued to increase post day 14. Co-incubation of endometrial tissue and conceptus membranes in vitro, however, reduced PGF production suggesting that embryonic factors may be capable of blocking endometrial PGF synthesis or secretion. PGFM measurements suggested that reduced PGF in uterine luminal flushings or in in vitro incubations was not reflected by a higher rate of metabolism. Corpora luteal PGF receptors retained the ability to bind PGF during early pregnancy. These data suggest that luteal maintenance in pregnant mares may involve inhibition of PGF synthesis and/or release and therefore should be considered luteostasis.

---

[1] Supported by Morris Animal Foundation and NIH HD 10436.

[2] Present address: USDA/ARS, P.O. Box 1199, El Reno, Oklahoma 73036, (USA).

[3] Present address: Dept. Ob. Gyn., University of Kentucky, Medical Center, Lexington, Kentucky 40506, (USA).

INTRODUCTION

In non-bred mares, the corpus luteum begins to regress and circulating progesterone begins to decline about 14 days after ovulation (Stabenfeldt, Hughes and Evans, 1972; Sharp and Black, 1973). This decline in circulating progesterone is associated with changes in uterine secretions which likely make the uterus a hostile environment for embryonic development or survival. In our labs, we have demonstrated that at least two endometrial proteins produced by the mare are progesterone-dependent (Zavy et al., 1982; McDowell et al., 1982) and it seems reasonable to speculate, therefore, that the loss of these particular proteins following luteolysis reflects a general decline in the quality of histotrophe available for embryonic development.

The regression in corpus luteum function and decline in progesterone secretion is temporally associated with increased uterine prostaglandin $F2\alpha$ (PGF) synthesis and/or secretion (Zavy et al., 1978; Vernon et al., 1981) increased uterine venous PGF (Douglas and Ginther, 1976) and increased circulating 15-keto-13,14-dihydro-$PGF_{2\alpha}$ (PGFM) (Neely et al., 1979). Similarly, it has been recognized for some time that administration of PGF during the luteal phase of the cycle of the mare results in rapid regression of the corpus luteum (Douglas and Ginther, 1972; Noden et al., 1973; Sharp, 1980). Although definitive experiments which would clearly demonstrate PGF to be the natural luteolysin have not been done, current evidence strongly favors that hypothesis. Therefore, it seems clear that the maternal recognition of pregnancy involves primarily, a mechanism that prevents luteolysis allowing continued progesterone secretion and perhaps most importantly, continued production of histotrophe. There is little question that this unknown mechanism is initiated by the early embryo. Moor and Rowson (1966) first demonstrated that luteolysis could be averted in non-bred ewes if embryos were transferred into the uterus prior to day 12. In a similar experiment, Hershman and Douglas (1979) demonstrated that removal of equine embryos by transcervical lavage prior to day 15 resulted in luteolysis, but when the lavage was performed after day 15, corpus luteum function was maintained. This requirement for the presence of the embryo coincides with the time of PGF secretion and expected luteolysis and it may be correct, therefore, to refer to this time (day 14 to 16) as the critical period for the maternal recognition of pregnancy.

Although the nature of the embryonic signal for maternal recognition of pregnancy is not known, its action to prevent luteolysis must reasonably be one of the following: (1) prevent PGF from being synthesized and/or released, (2) prevent PGF from reaching the corpus luteum, (3) prevent PGF from binding to luteal PGF receptors, (4) stimulate luteal gonadotropin receptors. The results presented in this paper address and support the first option.

METHODS

Experiment 1 - PGF binding by corpus luteum

Corpora lutea were obtained from non-bred (n=20) and bred (n=20) pony mares at ovariohysterectomy on days 4, 8, 12, 14, 16, 18 and 20 post ovulation. Pregnancy was confirmed in the latter group by presence of an embryo in the uterus. Luteal membrane receptors for PGF were assessed in the particulate fraction following slicing, homogenization and centrifugation (1086g; 10 min; 100,000 g; 60 min) of luteal tissue (Vernon et al., 1979). Specific binding of PGF was calculated as the difference between total PGF binding, and the binding in presence of an excess unlabeled PGF. Dissociation constant and binding capacity of the PGF receptor were calculated by Scatchard analysis (Scatchard, 1949) and the binding of PGF was expressed as pg PGF specifically bound/mg protein. Protein content of the luteal membrane preparation was determined by the method of Lowry et al. (1951). Luteal binding of PGF in non-pregnant and pregnant mares was compared by least squares analysis of variance with all appropriate interactions.

Experiment 2 - Endometrial production of PGF

Endometria were obtained at hysterectomy from non-pregnant (n=20) and pregnant (n=21) mares on days 4, 8, 12, 14, 16, 18 and 20 post ovulation. Endometrial strips (approximately 300 mg) were incubated for 2 h at 37 C in a Krebs-Ringer carbonate-bicarbonate buffer (KRB) in a Dubnoff metabolic incubator. Comparable strips of endometria were obtained from the same mares, placed in KRB and frozen immediately. These latter tissues were used to estimate PGF content, whereas the tissues incubated for 2 h gave an estimate of <u>in vitro</u> production. Endometria were homogenized in the KRB and the homogenate was assayed for PGF using a validated radio-immunoassay (RIA)(Vernon et al., 1981).

In addition, endometria from the above mares, or endometria from ovariectomized, steroid-treated mares (progesterone 50 mg/day, n=4; oestrogen 50 mg/day, n=4; oestrogen 7 days then progesterone 14 days, n=4; or vehicle only, n=4) were incubated in vitro in the presence of oestrogen or progesterone to determine the influence of steroids on PGF production. Differences in in vitro PGF production were assessed by least squares analysis of variance.

Experiment 3 - Uterine luminal PGF and PGFM

Uterine secretions were collected from non-pregnant (n=30) and pregnant pony mares (n=30) by the transcervical lavage technique (non-pregnant mares) of Zavy et al. (1978) or at hysterectomy (pregnant mares) on days 4, 8, 12, 14, 16, 18 and 20 post ovulation. Prostaglandin F was analyzed in the uterine flushing by a validated RIA (Zavy, 1979; Zavy et al., 1982). Because of the different techniques used in collecting uterine flushings two additional experiments were performed. To determine the potential influence of cervical manipulation, uterine flushings were obtained from ovariectomized, steroid-treated mares (n=9) by transcervical lavage and again, after an appropriate rest period (2 weeks), at hysterectomy. The third experiment was collection of uterine flushings from non-pregnant (n=14) and pregnant pony mares (n=14) on day 14 only and using transcervical lavage technique only. PGF as well as PGFM was quantified in these flushings using validated RIA.

Experiment 4 - Effect of conceptus membranes on PGF in vitro production

Endometria were obtained from pregnant pony mares (n=20) on day 14, and were incubated in vitro in the presence or absence of conceptus membranes. After a 2 h incubation in KRB, the medium was changed to modified Eagles Minimum Essential Medium (MEM; Basha et al., 1979) and the tissues were incubated for an additional 12 h on a rocking platform in 95% O_2, 5% CO_2 atmosphere. At the end of incubation, tissues and medium were frozen for subsequent assay for PGF and PGFM. In this experiment, in contrast to experiment 2, only the medium was assayed.

RESULTS

Experiment 1 - PGF binding by corpus luteum

As reported by Vernon et al., (1979) the amount of PGF bound by luteal receptors increased during the luteal phase, reaching a peak on day 14. In non-bred mares, there was a sharp decline in PGF binding after luteolysis (see Fig. 1). In pregnant mares, however, the binding capability continued to increase after the expected time of luteolysis and then declined on day 20 (Fig. 1). There was a significant ($P < 0.01$) status (pregnant vs non-pregnant) by day interaction, indicating that the binding of PGF to luteal receptors differed over days of the oestrous cycle or pregnancy. These data further indicate the competence of luteal receptors for PGF during the critical period for maternal pregnancy recognition suggesting that luteal maintenance is not likely the result of altered PGF binding.

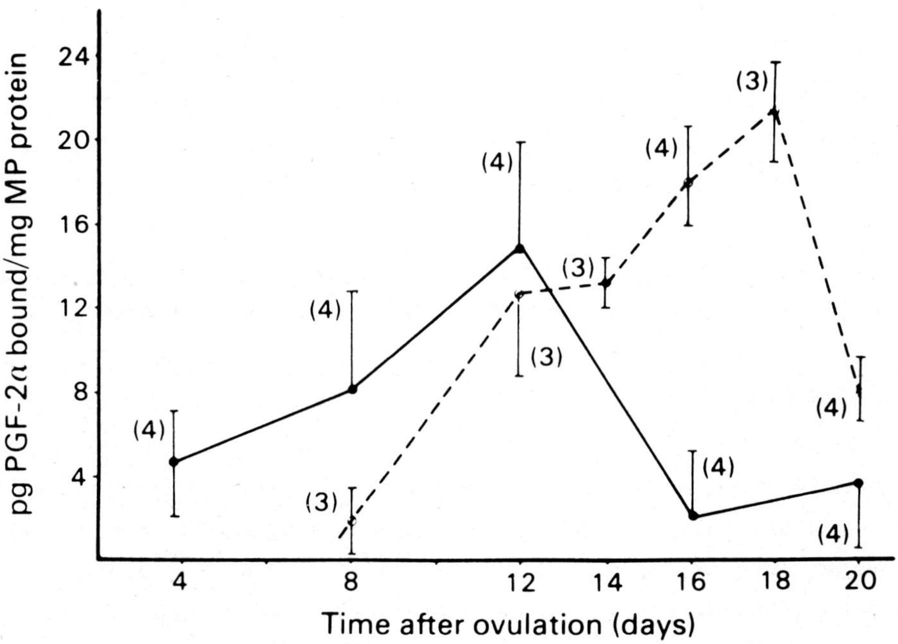

Fig. 1. Specific binding of PGF to the luteal cell membrane preparation of non-pregnant (●——●) and early pregnant (○- - -○) mare (least squares means ± SEM). (Vernon et al. 1979).

Experiment 2 - Endometrial production of PGF

Endometrial content and in vitro production of PGF both displayed clear time trends ($P < 0.001$) in non-pregnant mares as well as in pregnant mares (Vernon et al., 1981) (see Fig. 2). In non-pregnant mares, endometrial PGF content and in vitro production both increased 5 to 6 fold, reaching peaks about days 14-16. In pregnant mares, however, the in vitro production of PGF continued to increase after the equivalent day of luteolysis (day 14). In contrast, the content remained lower than in non-pregnant mares until day 20. The experiments in which endometrium from cycling or ovariectomized, steroid-treated mares was incubated with steroids in vitro showed that progesterone (1.0 µg) was ineffective in altering PGF production in vitro but that oestradiol (0.001 - 1 µg) was highly stimulatory when applied to endometrial tissue that had had prior exposure to progesterone in vivo.

Experiment 3 - Uterine luminal PGF and PGFM

Analysis of uterine flushings from non-pregnant mares revealed peak PGF levels about day 14 (Zavy et al., 1978) (see Fig. 3). However, there were no clear changes (no time trends) in intra-uterine luminal PGF concentrations in pregnant mares (Zavy, 1979). Because of the different procedures used (non-pregnant collected by transcervical lavage; pregnant collected at hysterectomy), the question of procedural differences was addressed in two additional experiments. Table I shows the mean (± SEM) PGF levels in the uterine flushings from ovariectomized, progesterone-treated mares. There was significantly ($P < 0.05$) more PGF obtained from transcervical lavage than when the flushings were obtained at hysterectomy (Berglund et al., 1982). These data suggested that cervical manipulation could indeed result in increased uterine luminal PGF. However, the differences between non-pregnancy and pregnancy could not be accounted for on the basis of cervical manipulation alone. When uterine flushings were collected from non-pregnant and pregnant mares using transcervical lavage in both statuses on day 14 (peak PGF), there was still a significant ($P < 0.05$) reduction in luminal PGF in the pregnant mares (Berglund et al., 1982) (see Table II). These data strongly suggested that in the presence of the embryo, PGF synthesis and/or release was inhibited. The converse question was raised, however, that the embryo might metabolize the PGF. If this were true, one might expect elevated

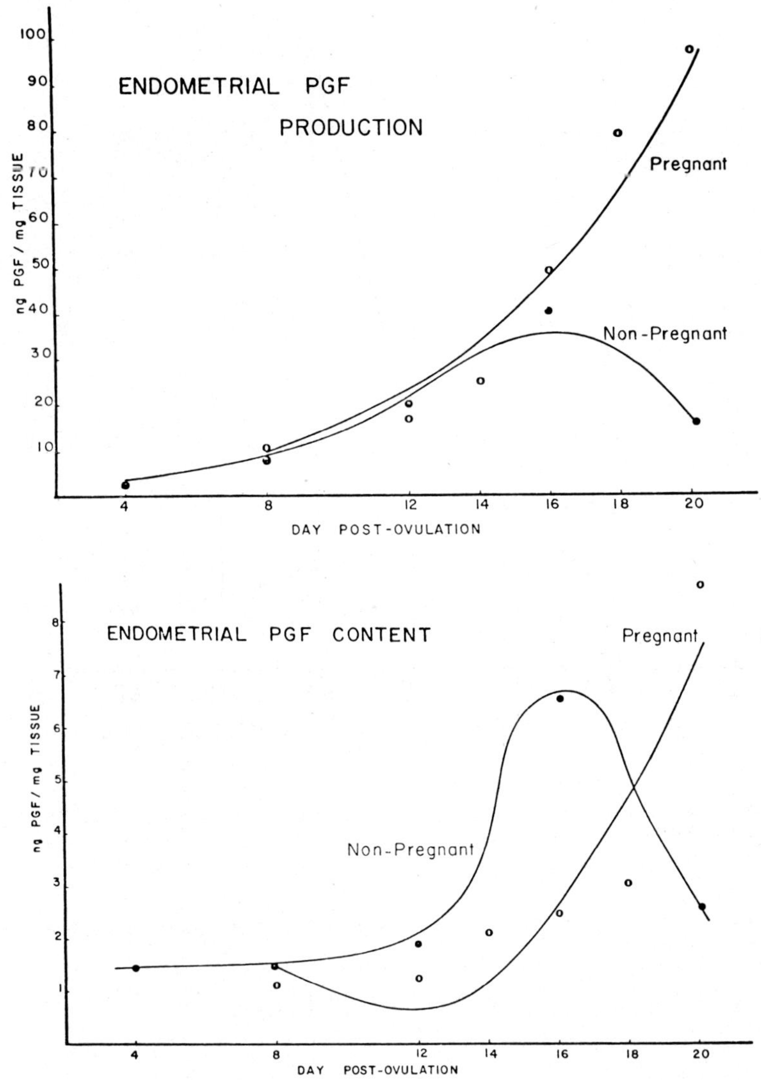

Fig. 2. Top panel: Regression curves with least squares means for production capacities of endometrial PGF during the oestrous cycle (●——●) and early pregnancy (○——○). Non-pregnant curve: $Y = -34.98 + 19.76X - 3.464X^2 + 0.2630X^3 - 0.00664X^4$ and pregnant curve: $Y = 48.04 - 9.623X + 0.5945X^2$. Bottom panel: Regression curves with least squares means for endometrial PGF content during the oestrous cycle (●——●) and early pregnancy (○——○). Non-pregnant curve: $Y = -21.25 + 11.43X - 1.921X^2 + 0.1297X^3 - 0.002965X^4$ and pregnant curve: $Y = 12.78 - 2.165X + 0.0956X^2$. (Adapted from Vernon et al. 1981).

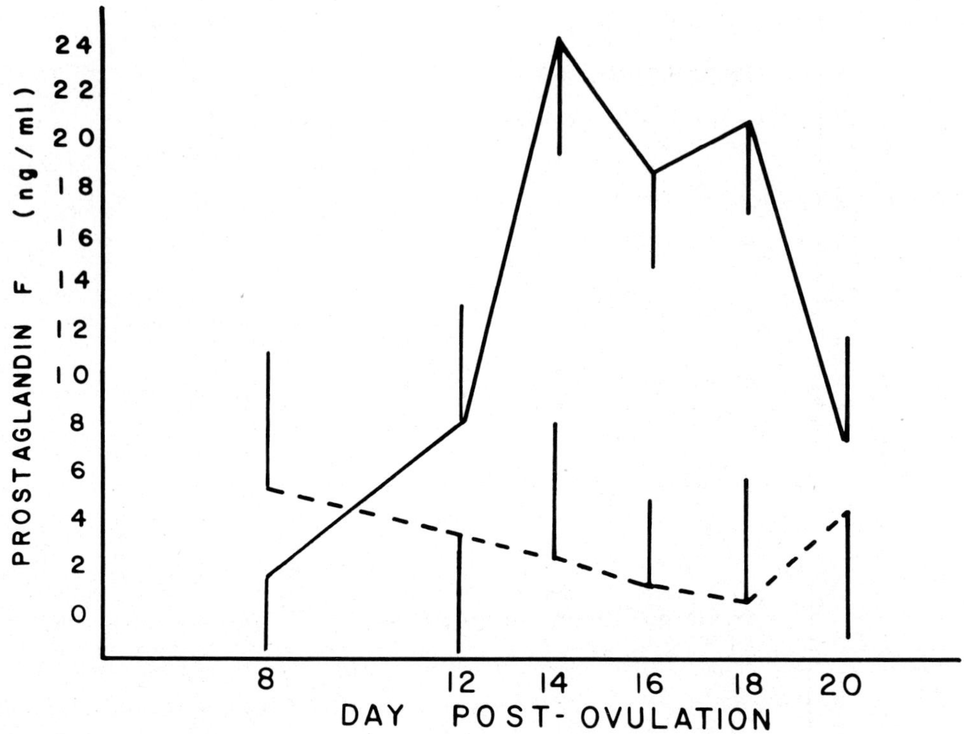

Fig. 3. Prostaglandin F concentrations (ng/ml; ± SEM) in uterine flushings on non-pregnant (solid line) and pregnant (dashed line) pony mares. (Zavy, 1979).

levels of PGFM in the flushings. There was no evidence of enhanced PGF metabolism using PGFM as an index (Berglund et al., 1983), suggesting that the former hypothesis was more accurate (see Tables I and II).

Experiment 4 - Effect of conceptus membranes on PGF in vitro production

When endometria were incubated in KRB for 2 h and then the medium was analyzed for PGF or PGFM, there were no clear trends according to incubation treatment. However, there were significant ($P < 0.05$) treatment effects during the 12 h MEM incubation. Specifically, endometrium from pregnant mares produced significantly less PGF when co-incubated with entire conceptus membranes or with membranes which comprised the capsule (see Fig. 4) (Berglund et al., 1982; 1983). Thus, these data indicated that PGF production in vitro was

Table I. Effects of collection technique on PGF and PGFM concentrations (ng/ml) in the uterine lumen of ovariectomized (OVX) pony mares on day 14 after ovulation

	PGF		PGFM	
	Transcervical	Hysterectomy	Transcervical	Hysterectomy
Treatment				
OVX-hormone treated*	12.6 ± 4.0^{a}	1.5 ± 3.8^{b}	2.1 ± 0.6	1.1 ± 0.6

a,b Significantly ($P < 0.05$) different.
All values are mean ± SEM.
*Oestradiol-17β, (50 mg) for 7 days; followed by progesterone, (150 mg) for 14 days.

Table II. Effects of pregnancy status on PGF and PGFM concentrations (ng/ml) in the uterine lumen of pony mares on day 14 after ovulation

	PGF	PGFM
Treatment		
Pregnant	0.7 ± 2.3^{a}	0.5 ± 0.3^{c}
Non-pregnant + PG-inhibitor*	1.9 ± 2.7^{a}	0.2 ± 0.4^{c}
Non-pregnant	12.7 ± 2.2^{b}	1.3 ± 0.3^{d}

a, b, c, d All values within a vertical column having the same superscript are not significantly ($P < 0.05$) different.
All values are mean ± SEM.
*Flunixen meglumine (Banamine; Schering Corporation) 1.1 mg/kg; 15 min prior to cervical penetration.

reduced in the presence of the conceptus membranes. PGFM concentrations, however, showed no treatment effects, were quite low, and generally reflected the direction of changes seen in PGF. This suggests that the reduced PGF in the presence of the conceptus membranes was not likely due to metabolism, but rather reflected reduced synthesis and/or secretion.

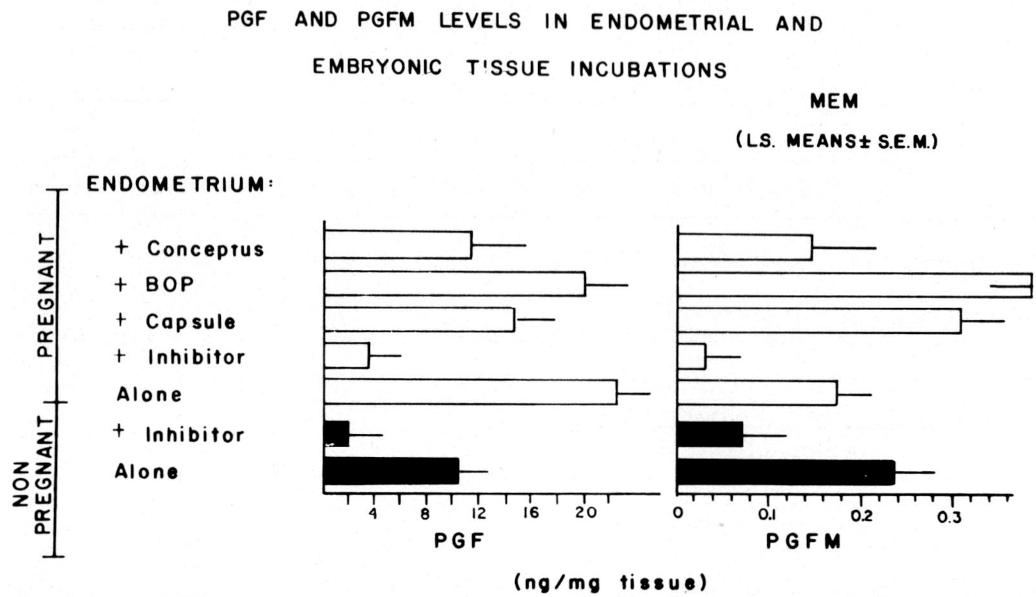

Fig. 4. PGF and PGFM levels and MEM after 12 h incubation. Values are least squares means ± SEM. Tissues from non-pregnant mares are shown in closed bars; tissues from pregnant mares are shown in open bars. Conceptus = all embryonic membranes minus the embryonic fluids. BOP = Bileminary omphalopleur. Inhibitor = Banamine (Schering Corp.). (Adapted from Berglund et al. 1983).

DISCUSSION

The role of the uterus in causing corpus luteum regression seems clear. It is well established that simple hysterectomy prior to day 12 to 14 results in prolonged luteal function (Ginther, 1967; Ginther and First, 1971). The explanation for this now seems apparent with a preponderance of evidence pointing to PGF release from the uterus at a critically timed stage. Thus, we have observed increased endometrial PGF content in vitro PGF production and intra-uterine luminal PGF content on or about 14 days after ovulation in non-bred mares (Vernon et al., 1981; Zavy et al., 1978; Berglund et al., 1982). These data are in good agreement with measurements of PGF in the uterine venous effluent (Douglas and Ginther, 1976) and with measurements of PGFM in the peripheral circulation (Neely et al., 1979; Berglund et al., 1983). Finally, this time of maximal PGF secretion coincides with the time of maximum PGF binding capability by luteal cell membrane receptors (Vernon et al., 1979).

The available evidence strongly indicates a precisely timed secretion of PGF by the non-pregnant uterus, leading to corpus luteum regression. A better understanding of the mechanism which controls this precisely timed luteolytic surge of PGF is requisite to both clinical management of maternal pregnancy recognition errors and to a clearer understanding of the embryonic signals which lead to pregnancy recognition and luteal maintenance.

The process by which the luteolytic surge is either altered or rendered ineffective by the embryo, however, is not so well understood. As mentioned in the introduction, it seems reasonable to speculate that the maternal recognition of pregnancy may be accomplished by one (or perhaps more) of the following: (1) prevent PGF from being synthesized and/or released, (2) prevent PGF from reaching the corpus luteum, (3) prevent PGF from binding to luteal PGF receptors, (4) stimulate luteal gonadotropin receptors.

The third option may not be appropriate in light of the report of Vernon et al. (1979) which indicated that luteal cell PGF receptors were present in corpora lutea of pregnant mares. The fourth possibility, that of an embryonic-induced luteotrophic factor is a popular notion, but currently has little supporting evidence. In a preliminary experiment, Garcia and Ginther (1974) administered HCG or equine pituitary extract to cycling pony mares from days 11 through 16 post ovulation. Progesterone levels were significantly reduced in control and HCG-treated mares on day 17, but were still elevated (> 1 ng/ml) in 3 of 4 mares treated daily with equine pituitary extract. Thus, these data are interesting and require further study, but do not conclusively indicate that the embryo might employ a similar mechanism.

The second possibility, that of preventing the luteolysin from reaching the corpus luteum, has been proposed as an explanation for the maternal recognition of pregnancy in the pig (Bazer and Thatcher, 1977; Bazer et al., 1984) These workers reported that uterine venous PGF was significantly reduced and that intra-uterine luminal PGF was significantly increased in pregnant gilts and they proposed that PGF is released in an endocrine fashion during the oestrous cycle, but is released in an exocrine fashion in the presence of an embryo. Thus, this mechanism results in corpus luteum maintenance by virtue of sequestering PGF within the uterus. The low luminal PGF in pregnant mares, however, suggests that this mechanism may not be at work in mares (Zavy, 1979; Berglund et al.,

1982). The possibility that the low luminal PGF in pregnant mares is a result of embryonic metabolism can also be discounted on the basis of the similarly low luminal PGFM levels (Berglund et al., 1983). A general reduction in PGF leaving the uterus of mares has been reported. Douglas and Ginther (1976) reported significantly reduced PGF levels in the uterine venous effluent of pregnant mares on day 14. Similarly, Kindahl et al. (1982) reported that PGFM levels in the peripheral circulation of pregnant mares were significantly reduced (i.e. no peaks). In mares with prolonged luteal function (i.e. pseudopregnant) Neely et al. (1979) reported similar findings. That is, no peaks of PGFM were detected in the peripheral plasma of non-bred mares at the expected time of luteolysis. These data do not indicate whether the reduced PGF was a result of sequestration or synthesis blockade, but strongly suggest that luteal maintenance can result from a failure of the presumed luteolysin to reach the corpus luteum. Finally, the reduction in in vitro PGF production in the presence of both endometrium and conceptus membranes suggests the possibility that embryonic factors, as yet unknown, may be inhibiting PGF synthesis and/or release (Berglund et al., 1982). This observation may explain the apparent paradox of increased in vitro PGF production capability in endometrium from pregnant mares (Vernon et al., 1981). The endometrium of pregnant mares is optimally primed for PGF secretion by virtue of its long exposure to progesterone (Vernon et al., 1981) and, in retrospect, it is not surprising it produces PGF. The reduction in PGF production in the presence of the conceptus, in the face of optimal PGF secretory preparation, suggests that the embryonic factor(s) may be very potent indeed.

These data, when viewed together, provide support for the notion that luteal maintenance with the establishment of pregnancy reflects a failure of PGF, the presumed luteolysin, to reach the corpus luteum. The lack of sequestered PGF in the uterine lumen of mares coupled with the inhibition of in vitro endometrial PGF production by conceptus membranes permits development of the hypothesis that maternal pregnancy recognition in mares involves an embryo-mediated inhibition of PGF synthesis and/or release. This hypothesis requires further testing. Meanwhile, it may be instructive to think of pregnancy recognition in the mare as more a process of luteostasis (luteal maintenance) than of luteotropism (luteal stimulation).

REFERENCES

Basha, S.M.M., Bazer, F.W. and Roberts. L.M., 1979. The secretion of a uterine specific, purple phosphatase by cultured explants of porcine endometrium. Dependency upon the state of pregnancy of the donor animal. Biol. Reprod., 20: 431-441.

Bazer, F.W. and Thatcher, W.W., 1977. Theory of maternal recognition of pregnancy in swine based on estrogen controlled endocrine versus exocrine secretion of prostaglandin $F_{2\alpha}$ by the uterine endometrium. Prostaglandins, 14: 397-402.

Bazer, F.W., Marengo, S.R., Geisert, R.D. and Thatcher, W.W., 1984. Exocrine vs endocrine secretion of prostaglandin $F_{2\alpha}$ in the control of pregnancy in swine. Anim. Reprod. Sci. 7: 115-132. (This symposium).

Berglund, L.A., Sharp, D.C., Vernon. M.W. and Thatcher, W.W., 1982. Effect of pregnancy and collection technique on prostaglandin F in the uterine lumen of pony mares. J. Reprod. Fert., Suppl. 32: 335-341.

Berglund, L.A., Sharp, D.C. and Thatcher, W.W., 1983. Mechanisms for maternal recognition of pregnancy in horses. Proc. 8th Equine Nutrition and Physiology Symposium,Lexington, Kentucky.

Douglas, R.H. and Ginther, O.J., 1972. Effect of prostaglandin on length of diestrus in mares. Prostaglandins, 2: 265-268.

Douglas, R.H. and Ginther, O.J., 1976. Concentration of prostaglandin F in uterine venous plasma of anesthetized mares during the estrous cycle and early pregnancy. Prostaglandins, 11: 251-260.

Garcia, M.L. and Ginther, O.J., 1974. Unpublished observations, University of Wisconsin, Madison, cited in Reproductive Biology of the Mare, 1979. O.J. Ginther, editor, McNaughton and Gunn, Ann Arbor, Michigan pp 194.

Ginther, O.J., 1967. Local utero-ovarian relationships. J. Anim. Sci., 26: 578-585.

Ginther, O.J. and First, N.L., 1971. Maintenance of the corpus luteum in hysterectomized mares. Am. J. Vet. Res., 32: 1687-1691.

Hershman, L. and Douglas, R.H., 1979. The critical period for the maternal recognition of pregnancy in pony mares. J. Reprod. Fert., Suppl. 27: 395-401.

Kindahl, H., Knudsen, O., Madej, A. and Edqvist, L.-E., 1982. Progesterone, prostaglandin $F_{2\alpha}$, PMSG and oestrone sulphate during early pregnancy in the mare. J. Reprod. Fert., Suppl. 32: 353-359.

Lowry, O.H., Rosebrough, N.J., Farr, A.L. and Randall, R.J., 1951. Protein measurement with the folin phenol reagent J. Biol. Chem., 193: 265-275.

McDowell, K.J., Sharp, D.C., Fazleabas, A., Roberts, R.M. and Bazer, F.W., 1982. Partial characterization of the equine uteroferrin-like protein. J. Reprod. Fert., Suppl. 32: 329-334.

Moor, R.M. and Rowson, L.E.A., 1966. The corpus luteum of the sheep: effect of removal of embryos on luteal function. J. Endocr., 34: 497-502.

Neely, D.P., Kindahl, H., Stabenfeldt, G.H., Edqvist, L.-E. and Hughes, J.P., 1979. Prostaglandin release patterns in the mare: physiological, pathophysiological, and therapeutic responses. J. Reprod. Fert., Suppl. 27: 181-189.

Noden, P.A., Hafs, H.D. and Oxender, W.D., 1973. Progesterone, estrus and ovulation after prostaglandin $F_{2\alpha}$ in horses. Proc. Soc. Exp. Biol. Med., 145; 145-149.

Scatchard, G., 1949. The attraction of proteins for small molecules and ions. Ann. N.Y. Acad. Sci., 51: 660-672.

Sharp, D.C., 1980. Factors associated with the maternal recognition of pregnancy in mares. Vet. Clinics of North America: Large Animal Practice, 2: 277-290.

Sharp, D.C. and Black, D.L., 1973. Changes in peripheral plasma progesterone throughout the oestrous cycle of the pony mare. J. Reprod. Fert., 33: 335-358.

Stabenfeldt, G.H., Hughes, J.P. and Evans, J.W., 1972. Ovarian activity during the estrous cycle of the mare. Endocrinology, 90: 1379-1384.

Vernon, M.W., Strauss, S.S., Simonelli, M., Zavy, M.T. and Sharp, D.C., 1979. Specific $PGF_{2\alpha}$ binding by the corpus luteum of the pregnant and nonpregnant mare. J. Reprod. Fert., Suppl. 27: 421-429.

Vernon, M.W., Zavy, M.T., Asquith, R.L. and Sharp, D.C., 1981. Prostaglandin $F_{2\alpha}$ in the equine endometrium: Steroid modulation and production capacities during the estrous cycle and early pregnancy. Biol. Reprod., 25: 581-589.

Zavy, M.T., Bazer, F.W., Sharp, D.C., Frank, M. and Thatcher, W.W., 1978. Uterine luminal prostaglandin F in cycling mares. Prostaglandins, 16: 643-649.

Zavy, M.T., 1979. A comparison of the nonpregnant and pregnant uterine luminal environments in the porcine and equine. Ph.D. Dissertation, University of Florida, Gainesville, Fl.

Zavy, M.T., Sharp, D.C., Bazer, F.W., Fazleabas, A., Sessions, F. and Roberts, R.M., 1982. Identification of stage specific and hormonally induced polypeptides in the uterine protein secretions of the mare during the oestrous cycle and pregnancy. J. Reprod. Fert., 54: 199-204.

Animal Reproduction Science, 7 (1984) 283—304
Elsevier Science Publishers B.V., Amsterdam — Printed in The Netherlands

HORMONAL CONTROL OF EARLY PREGNANCY IN THE MARE

W.R. Allen

TBA Equine Fertility Unit, Animal Research Station, 307 Huntingdon Road, Cambridge CB3 OJQ, (England).

ABSTRACT

Allen, W.R., 1984. Hormonal control of early pregnancy in the mare. Anim. Reprod. Sci., 7: 283-304.

Progesterone secretion during early pregnancy in equids is supplemented by the development of accessory corpora lutea between 40 and 150 days after ovulation. This coincides with the development and regression of the endometrial cups and their secretion of large quantities of chorionic gonadotrophin (eCG) into the maternal bloodstream. The dual gonadotrophic properties of eCG and its temporal relationship to secondary luteal activity would seem to suggest that the former stimulates the latter. However, from the results of interspecies matings between horse and donkey and the successful transfer of donkey embryos to horses and vice versa, the argument is raised that secondary luteal development in pregnancy is controlled more by maternal pituitary than chorionic gonadotrophin release. Furthermore, from an endocrinological standpoint, eCG is not essential to the maintenance of pregnancy.

INTRODUCTION

With the notable exception of the African elephant (Smith et al. 1969), an adequate and continuous supply of progesterone is considered fundamental to the maintenance of the pregnancy state in mammals. In man, the lower primates and most laboratory and domestic animals, the corpus luteum (CL) or corpora lutea which develop from the ovulations that give rise to pregnancy persist in the ovaries throughout gestation and continue to actively secrete progesterone until term. In many of these species the continued production of progesterone by the ovaries is superfluous beyond a certain stage of gestation and bilateral ovariectomy performed after this time is not followed by abortion; women, some primates and the ewe provide good examples of this type of pregnancy in which the

0378-4320/84/$03.00

placenta becomes the major source of progesterone during at least the second half of gestation. The goat and sow, on the other hand, are equally good examples of those species in which the maintenance of pregnancy depends upon the continued secretion of ovarian progesterone throughout gestation (see Short, 1969).

As in so many aspects of reproductive physiology, the mare is a striking exception to the majority rule. Her primary corpus luteum of pregnancy (1° CL) begins to decline in secretory activity from as early as day 14-16 after ovulation and thus at about the same time that complete luteolysis would normally occur in the cycling animal. This causes a slow and steady fall in peripheral plasma progesterone concentrations during the next 20-25 days until a pronounced secondary rise commences (see Fig. 1), usually at around day 35-45 and thus coincidentally with the onset of secretion of equine Chorionic Gonadotrophin (eCG, formerly known as Pregnant Mare Serum Gonadotrophin, PMSG; Cole and Hart, 1930; Allen, 1969a; Stewart and Allen, 1981). The secondary progesterone rise results from the formation of the first of what eventually becomes a whole crop of secondary corpora lutea (2° CL) that develop in the mare's ovaries between days 40 and 150. These accessory luteal structures arise singly and sporadically by both normal ovulation and luteinization of unruptured follicles and they may number from as few as 3 to as many as 30 by day 120 (Amoroso et al., 1948; Allen, 1970). After about day 150, all the corpora lutea, both primary and secondary, regress to leave the ovaries small and completely inactive for the remainder of gestation. Horse placental tissue contains appreciable quantities of progesterone (Short, 1958) and it has been generally assumed that the placenta is secreting sufficient progesterone to maintain pregnancy from about day 100. This hypothesis is supported by the findings of Holtan et al. (1979) that pregnancy continues in all mares ovariectomized on day 140 and will even do so in about half those ovariectomized as early as days 50-70.

The question arises as to what gonadotrophic or other stimuli are responsible for such pronounced changes in ovarian activity during pregnancy in equids and what controls the process which can, in many senses, be considered as a biphasic maternal recognition of pregnancy. That is, prolongation of the life of the cyclical corpus luteum at around day 14-16 in the first instance followed by renewed follicular growth leading to secondary luteal development in the second. Another

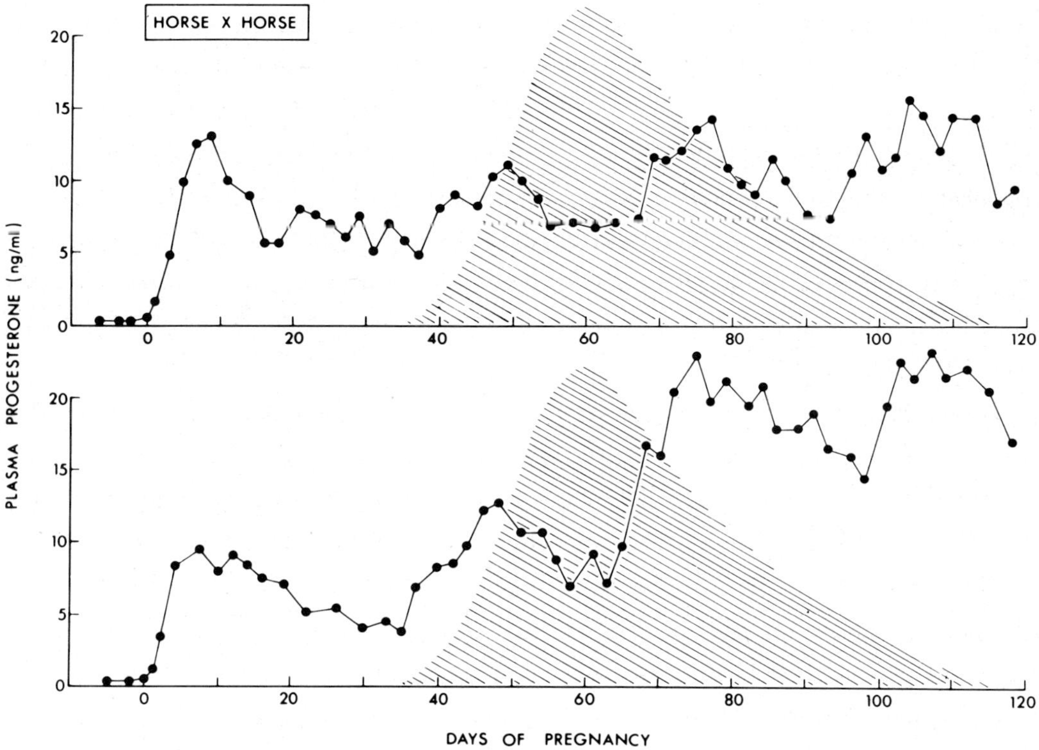

Fig. 1. Peripheral plasma progesterone concentrations measured during early pregnancy in 2 Welsh Pony mares. The hatched area is a diagramatic representation of serum eCG levels during this period.

speaker at this Symposium will be discussing the first of these essential cross-roads of pregnancy in the mare (see D.C. Sharp, this volume) and the present paper will therefore concentrate upon the second important milestone. Namely, the gonadotrophic control and physiological significance of accessory luteal development.

The equine endometrial cup reaction

One feature of equine pregnancy which is unique to this genus is the development and subsequent regression of the so-called endometrial cups. These comprise a series of small, ulcer-like, endometrial outgrowths which form in a circle around the conceptus in the gravid uterine horn. They first appear as pale, slightly raised plaques in the endometrium at day 38-40 after ovulation and enlarge steadily during the next 20-30 days, becoming distinctly concave or saucer-shaped on the surface due to a combination of overgrowth at the peripheral and commencing regression in the central regions of the structure. After about day 70-80 the degenerating cups become increasingly pale and cheesy in appearance and their concave surface becomes filled with a sticky, honey-coloured exocrine secretion. Eventually, between days 100-140, each necrotic cup and its attached exocrine secretion is sloughed off the surface of the endometrium to lie free in the uterine lumen (Schauder, 1912; Clegg et al., 1954; Amoroso, 1955; Allen, 1975).

Histologically, each cup is composed of a discrete and densely packed mass of large, binucleated, epithelioid cells. Very few blood vessels remain within the cup itself but a complex network of large lymph sinuses develops in the endometrial stroma beneath. The outlets and apical regions of most of the endometrial glands within the cups are destroyed during early development but their basal portions remain intact and become increasingly distended due to accumulation of secretory products. It is the release of this material, occasioned by the commencing degeneration of the cells at the lumenal surface of the cup after day 70, which gives rise to the glutinous secretion that adheres to the surface of the cup and overlying allantochorionic membrane.

There is no direct physical connection between the surface of the cups and the overlying allantochorion and it was widely assumed for many years that the cups represented some form of localized maternal decidual response to the presence of the developing conceptus in her uterus (Amoroso, 1955; Allen,1970). However, more recent studies demonstrated that the large endometrial cup cells are, in fact, foetal in origin and reach the maternal endometrium by an invasion of the specialized trophoblast cells of the chorionic girdle region of the foetal membranes between days 36 and 38 after ovulation (Allen and Moor, 1972; Allen et al., 1973). Throughout their 80-100 day lifespan, increasing numbers of lymphocytes, plasma cells and eosinophils accumulate in the endometrial stroma at the periphery of the cups. This band of leukocytes appears to "wall off" the foetal cup cells from the surrounding maternal tissue and, coincidentally with the commencing degeneration of the cells at the surface region of the cup beyond day 70, the leukocytes begin to actively invade the cup tissue and destroy the cells in the basal region. It has been proposed that the leukocyte reaction represents a maternal cell-mediated immune response to foreign antigens expressed by the foetal cup cells (Allen, 1975). This hypothesis is supported by the finding that >90% of primigravid mares develop high titres of cytotoxic antibody against the lymphocytes of the sire of their conceptuses (Allen, 1979; D.F. Antczak, personal communication). Specific alloantibody first appears in maternal serum between days 44 and 70 after ovulation and hence soon after invasion of the chorionic girdle cells to form the endometrial cups.

The endometrial cups are the source of the large quantities of chorionic gonadotrophin (eCG) which are present in the blood of mares and other female equids during the first half of pregnancy (Cole and Goss, 1943; King, 1965; Allen, 1969a). The hormone is first detected in maternal serum between 36 and 40 days after ovulation and concentrations rise rapidly to a well-defined peak between days 55 and 70. Thereafter they decline steadily until gonadotrophic activity completely disappears again from the blood at day 120-150 (Cole and Hart, 1930; Day and Rowlands, 1940; Allen, 1969a). It is the large, foetal endometrial cup cells which actually secrete the eCG (Allen and Moor, 1972) that reaches the maternal blood stream via the lymph (Allen, 1982). However, the exocrine cup secretion, consisting of necrotic cup cells and accumulated endometrial gland secretion, is also very rich in gonadotrophic activity and may contain as much as 1,000,000 i.u. eCG/gram wet weight (Rowlands, 1963).

eCG is a high molecular weight glycoprotein which is unique amongst the gonadotrophins in that it expresses both Follicle Stimulating Hormone (FSH) and Luteinizing Hormone (LH) biological activities within the same molecule (Papkoff, 1974; Stewart et al., 1976; Guillou and Combarnous, 1983). This property, together with the coincidental appearance and disappearance of eCG from the blood and the development and regression of the accessory corpora lutea, has led to the assumption that eCG provides both the gonadotrophic stimulus for secondary follicular development and the luteotrophic stimulus for their ovulation and/or luteinization to form the accessory corpora lutea.

Influence of foetal genotype on eCG secretion

Another interesting feature of reproduction in equids is the ability of the many different species to interbreed and produce viable, but infertile, offspring (see Gray, 1954). In recent years, foetal genotype has been shown to exert a profound influence on both the total amount of eCG secreted during pregnancy and its rate of disappearance from maternal blood after the peak concentration has been attained. In mares covered by a jack donkey so that they are carrying interspecies mule conceptuses, the peak serum eCG concentrations of only 8-30 i.u./ml reached by day 45-55 are some 5-10 fold lower than those commonly occurring in mares carrying normal intraspecies horse conceptuses. Furthermore, eCG activity disappears again from the blood as early as day 70-80 and hence much sooner in gestation than the day 120-140 period which is usual for normal horse pregnancy (Bielanski et al., 1956; Allen, 1969b). Conversely, the peak serum eCG concentrations of 250-350 i.u./ml measured in jenny donkeys covered by a horse stallion and so carrying hinny conceptuses, are some 5-8 times higher than the maximum levels of 30-50 i.u. found in donkeys carrying normal intraspecies donkey conceptuses.

The underlying cause of these marked differences in eCG production between the reciprocal hybrid pregnancies is found in the endometrial cups. In mares carrying mule foetuses, the cups are individually much smaller so that the total amount of cup tissue is appreciably less than that which normally develops in mares carrying horse foetuses. This results from the much narrower chorionic girdle which develops on the mule as compared the horse foetus (Allen, 1982). In addition to this reduced amount of cup tissue, however, the maternal leukocytic reaction against the cups is greatly increased in mares carrying mule foetuses. By as early as day 50 of gestation very large numbers of lymphocytes completely surround each cup and, instead of remaining clustered in the endometrial stroma, they actively invade the cup tissue and attack the large foetal cup cells. This results in the premature destruction and dehiscence of the cups by as early as day 60-70 with the resulting disappearance of eCG from maternal blood.

In donkeys carrying hinny pregnancies, on the other hand, the chorionic girdle which develops on the hinny conceptus is very much broader and more voluminous than its equivalent on the donkey conceptus. This gives rise to much larger and more active endometrial cups in the hinny as compared the normal donkey pregnancy and so results in the very elevated maternal serum eCG levels in the former (Allen, 1982). As in the mare carrying the mule foetus, the maternal leukocytic reaction mounted against the endometrial cups is considerably increased in the donkey when she is carrying the hinny foetus. However, in marked contrast to the mule situation, the accumulated leukocytes seem much less able to penetrate the hinny cups and kill the large cup cells. Instead they remain clustered in the peripheral endometrial stroma until a much later stage of pregnancy and so permit the longer lifespan of the cups and the resulting prolongation of eCG secretion (Allen, 1975).

In spite of the very large differences in eCG production between mares when they are carrying interspecies mule as compared normal intraspecies horse conceptuses, the amount of secondary luteal development, as reflected by peripheral plasma progesterone concentrations between 40 and 150 days of gestation, is similar in the two types of pregnancy (Fig. 2). Thus, plasma progesterone levels in a mare carrying twin horse conceptuses which has two complete sets of endometrial cups (Day and Rowlands, 1947) and may therefore exhibit a peak serum eCG concentration as high as 250-300 i.u./ml around day 60, will generally not be any higher than those in a mare carrying a single mule conceptus with a maximum level of <10 i.u./ml at the same stage of pregnancy. In the donkey, on the other hand, it would appear on first sight that a quantitative correlation does exist between eCG and the degree of secondary ovarian stimulation. In the presence of the elevated levels of eCG secreted by hinny endometrial cups plasma progesterone concentrations are also greatly increased, often reaching as high as 300-800 ng/ml between days 50 and 100 (Allen, 1975). These relatively enormous quantities of progesterone are secreted entirely by the maternal ovaries (Sheldrick et al., 1977) which become greatly enlarged due to the formation of numerous 2° CL and luteinized follicles (Allen, 1979).

Experiments using specific radioreceptor assays to measure the FSH-like and LH-like biological activities of eCG have indicated that it is the quality, not the quantity, of the eCG which is likely to be the cause of the striking difference between the horse and donkey in the apparent relationship of eCG and secondary luteal development. Whereas horse CG has an FSH:LH ratio of 1.23, this figure falls to only 0.2 for donkey CG. Mule and hinny CG, on the other hand, have FSH:LH ratios which are approximately midway between those of the two parental species gonadotrophins (0.7 and 0.5 respectively) (Stewart et al., 1977). Receptor binding studies have further shown that the affinity of horse CG for equine gonadal receptors is only 1-3% of that shown by equine pituitary FSH and LH and the pituitary gonadotrophins of other species (Stewart and Allen, 1979).

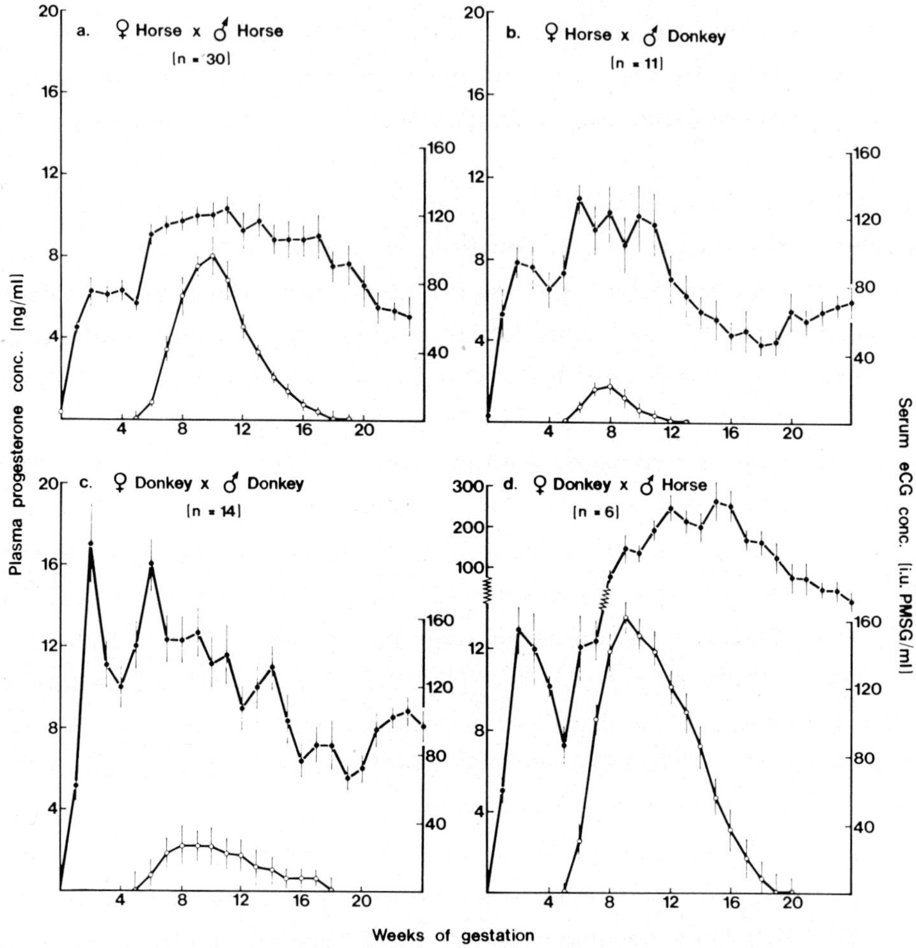

Fig. 2. Mean peripheral plasma progesterone and serum eCG concentrations measured during the first half of pregnancy in 30 mares carrying normal horse pregnancies, 14 donkeys carrying normal donkey pregnancies, 11 mares carrying interspecies mule pregnancies and 6 donkeys carrying hinny pregnancies. Note the change in the progesterone scale in the donkeys carrying hinny conceptuses.

The mechanism which permits equine gonadal receptors, but not the equivalent receptors in other species, to differentiate between pituitary and chorionic gonadotrophins of equine origin, remains obscure. However, its occurrrence proposes the hypothesis that, from necessity of avoiding ovarian hyperstimulation during normal intraspecies pregnancy, female equids have evolved an inherent ovarian resistance to at lease the FSH-like activity of their own chorionic gonadotrophin. Then, when the ovaries of the jenny donkey, which is "genetically organized" to be confronted in pregnancy with donkey CG that is so low in FSH content, are faced with larger quantities of hinny CG with a much higher FSH component, the protective barrier is overcome and excessive follicular development followed by ovulation and/or luteinization results in multiple 2° CL.

Xenogeneic pregnancy in horses and donkeys

Another pregnancy model which sheds light on the luteotrophic significance of eCG in pregnancy maintenance has been provided by the successful transfer of horse and donkey embryos between these two species (see Allen, 1982). In donkeys carrying extraspecies horse x horse embryos, the relatively broad and active (with respect to donkey) horse chorionic girdle readily invades the endometrium of the donkey host mother to give rise to large and active endometrial cups which secrete high quantities of horse CG into the maternal bloodstream. Just as in interspecies hinny pregnancy, the maternal donkey ovaries become hyperstimulated by the horse CG and in the presence of these very high blood levels of both eCG and progesterone, implantation occurs normally and the extraspecies foetus is carried safely to term.

The picture is very different in the horse carrying the transferred donkey x donkey conceptus, however. Here, the relatively narrow and underdeveloped donkey chorionic girdle completely fails to invade the horse endometrium at around day 36 with the result that endometrial cups do not develop and no eCG is secreted. Nonetheless, the donkey conceptus continues to grow and develop normally to around day 65, although normal implantation, involving the interdigitation of chorionic villi and endometrial crypts, fails to occur. Beyond this stage, increasing numbers of lymphocytes and other leukocytes begin to accumulate in the area of maternal endometrium that is in contact with the donkey allantochorion. This aggregation of immune cells is very reminiscent of that which occurs around the endometrial cups in all other types of equine pregnancy and from day 80 onwards the lymphocytes begin to pass through the endometrial epithelium and attack the donkey trophoblast. This results eventually in the death of the conceptus and its abortion or resorption between days 85 and 100 (Allen, 1982).

During the past 4 years, 22 extraspecies donkey-in-horse pregnancies have been established at the author's laboratory. Eleven of these were allowed to proceed uninterrupted, with or without various hormonal treatments, of which 10 failed between days 83 and 103; the remaining donkey foetus was carried to term and born live on day 346, despite a complete absence of eCG in the blood of the host horse mother and no indication of any secondary luteal development after day 40 (see Fig. 3). Two other donkey-in-horse pregnancies were surgically removed from the uterus at hysterotomy performed on days 43 and 55 respectively and both appeared to be developing normally at this early stage. The remaining 9 conceptuses were similarily removed at hysterotomy, but at the later stages of days 65-87 after ovulation. In 8 of these animals, gross degenerative changes in the foetus and placenta combined with the histological picture of a lack of implantation and a dense leukocyte accumulation made it clearly evident that the pregnancy would have failed had it been left in situ. In the 8th mare, however, the conceptus, removed on day 83, had obviously achieved stable implantation

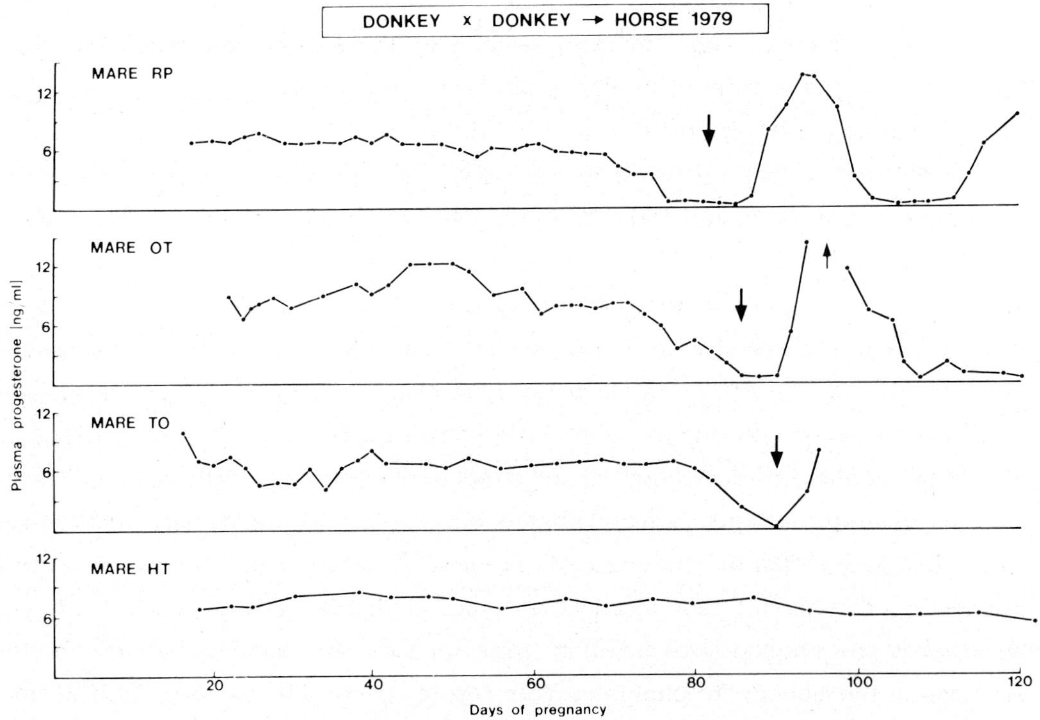

Fig. 3. Peripheral plasma progesterone concentrations in 4 Welsh Pony mares carrying transferred donkey conceptuses. Mare HT carried her donkey foetus to term and the vertical arrow indicates when spontaneous abortion occurred in the other 3 mares.

and it appeared viable and healthy. This was confirmed histologically and it was therefore concluded that this particular mare would also have proceeded to term had the pregnancy been left undisturbed. Thus, of the 20 mares allowed to carry extraspecies donkey conceptuses past day 60, only 2 (10%) did, or probably would have, gone to term. All the remainder were killed at around day 80-90 by a failure of implantation combined with what appears to be a generalized maternal cell-mediated attack against the foreign trophoblast.

Apart from the immunological aspects of the failure of the donkey-in-horse pregnancies, the lack of development of endometrial cups and resulting complete absence of eCG from the blood of these animals provides a valuable contrast to the other types of intra-, inter- and extraspecific equine pregnancies examined to date, all of which secrete varying amounts of eCG during the period when 2° CL are developing in the maternal ovaries. Fig. 3 shows the peripheral plasma progesterone concentrations measured in 4 of the mares in which the transferred donkey conceptus was not uninterrupted and to which no exogenous hormonal therapy was given. It is evident from these profiles that no 2° CL developed in 3 of these animals, including Mare HT which carried her foal to term. A definite, although relatively short-lived, rise in progesterone levels occurred in the fourth mare (OT) between days 41 and 52 after ovulation, suggesting that one secondary ovulation may have occurred at this time. Nevertheless, progesterone levels in this animal and in the other 2 mares which lost their pregnancies began to decline steadily from around day 75, reaching baseline values coincidentally with the expulsion of the degenerating conceptus and with the mare's return to oestrus with accompanying normal ovulation.

Fig. 4 shows the progesterone profiles measured in 4 more mares carrying donkey conceptuses. These pregnancies were established in the breeding season after those depicted in Fig. 3 and, in the mistaken belief that a lack of endogenous progesterone production resulting from the lack of 2° luteal development could have caused the loss of the 3 pregnancies in the previous year, 30 mg of the orally-active synthetic progestagen, allyl trenbolone (Regumate; Roussel Uclaf, Paris) was administered daily to each mare, commencing 33-53 days after ovulation; Shideler et al. (1982) showed that lower doses of allyl trenbolone than this will maintain pregnancy in mares bilaterally ovariectomized on day 35. Despite this exogenous progestagen therapy, however, all 4 mares lost their preg-

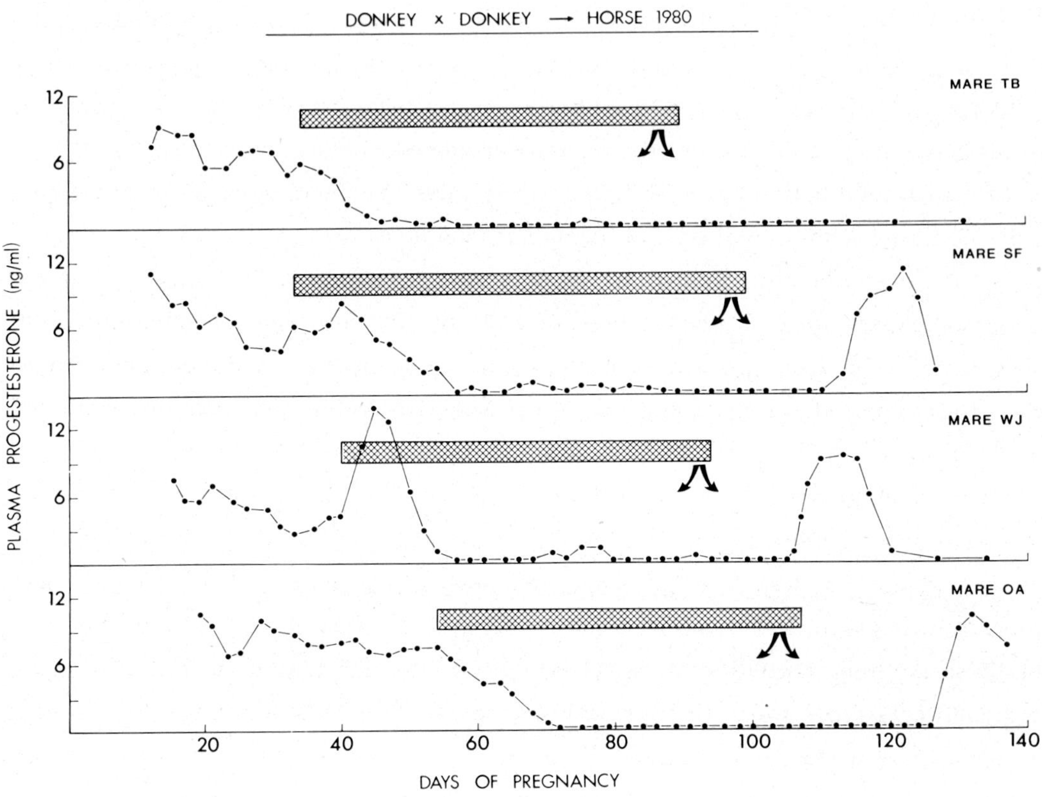

Fig. 4. Peripheral plasma progesterone concentration in 4 Welsh Pony mares carrying transferred donkey conceptuses. The horizontal hatched bars denote the period when 30 mg allyl trenbolone was administered daily to each animal and the splayed vertical areas show the period in which the conceptus was resorbed.

nancies at a similar stage of gestation, namely days 87-105 after ovulation. The progestagenic action of the allyl trenbolone kept the cervix closed in these mares which meant that the degenerating products of pregnancy were resorbed rather than aborted as they had been in the previous year. As before, eCG remained undetectable in the serum of all 4 mares during pregnancy and only one showed a pronounced rise in plasma progesterone levels indicative of a secondary ovulation (Mare WJ). This occurred 2 days after commencing allyl trenbolone therapy on day 38 and it was not followed by any further accessory luteal development.

Perhaps the most important difference between the progesterone profiles exhibited by the two groups of mares was the premature and more rapid decline of endogenous progesterone production in all 4 of the progestagen-treated animals. Whereas plasma progesterone concentrations remained >1 ng/ml until abortion occurred between days 83 and 89 in the 3 untreated mares, they fell to baseline as early as days 43-69 after ovulation (11-18 days after commencing allyl trenbolone) in all 4 of the treated mares, despite the continued presence of a viable conceptus in the uterus for some 20-30 days after complete luteolysis had occurred.

The hastened regression of the 1° CL in the progestagen-treated mares suggests the possibility that allyl trenbolone was either directly luteolytic or, more likely, it exerted a negative feedback action to suppress the secretion of an essential luteotrophin such as pituitary LH. While there is no firm evidence that the equine corpus luteum requires any form of luteotrophic stimulation to prolong its life beyond that of the normal oestrous cycle, experiments involving the passive immunization of cycling mares with an antiserum raised against a crude extract of equine pituitary glands have indicated that this is likely to be the case (Pineda et al., 1973). Administration of allyl trenbolone to seasonally anoestrous mares suppresses basal LH secretion rate (J.M. Kydd and W.R. Allen, unpublished observations) and indeed, measurement of serum FSH and LH concentrations in the 8 mares carrying transferred donkey conceptuses suggested some degree of suppression of LH release in the allyl trenbolone-treated animals. As in normal horse pregnancy (Evans and Irvine, 1975; Urwin and Allen, 1982) serum LH levels tended to remain low in both the treated and untreated groups from soon after ovulation until loss of the conceptus at day 85-95. However, whereas numerous daily spike-like increases in LH occurred at irregular intervals throughout this period in the untreated animals, these were absent in the allyl trenbolone-treated mares and the LH profiles remained low and flattened. Serum FSH concentrations, on the other hand, appeared to be unaffected by allyl trenbolone and they continued to show regular surge-like increases throughout gestation in both the treated and untreated animals (Urwin and Allen, 1982).

In a continuing attempt to determine the underlying cause of the loss of the donkey-in-horse pregnancies at day 90, a further 12 such pregnancies were established in the 1981 breeding season. Eight of these recipient mares were then given intravenous injections of large doses of eCG dissolved in phosphate buffered saline at varying times in gestation between days 35 and 83. The eCG was extracted from serum collected from mares carrying normal intraspecies horse pregnancies and it was purified to a specific activity of 1200-1500 i.u./mg by a modification of the method described originally by Goss and Cole (1940) (C.R. Brown, unpublished). The dose rates and frequency of treatment of the 8 mares ranged from 20,000 i.u. eCG given at weekly intervals to sporadic injections of as much as 300,000 i.u. eCG. Representative peripheral plasma progesterone profiles exhibited by 4 of the mares, selected to show the considerable variation in response to the exogenous eCG, are shown in Fig. 5a, b. In two of the total of

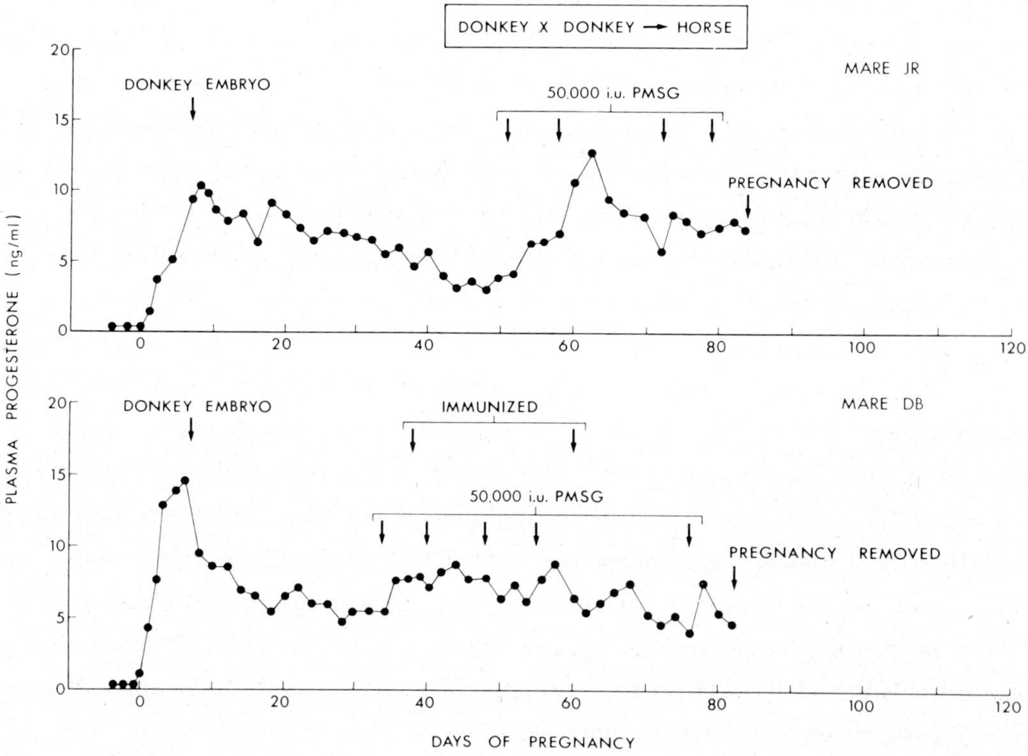

Fig. 5 a. Peripheral plasma progesterone concentrations in 2 Welsh Pony mares carrying transferred donkey conceptuses in 1981. The vertical arrows show the days on which the mares were given intravenous injections of varying doses of partially purified eCG (PMSG) dissolved in saline. Mare DB was also given intravenous injections of 2 x 10^8 washed lymphocytes from the genetic sire and dam of the donkey foetus on the days shown as Immunized.

8 treated mares (JR and SD) pronounced increases in plasma progesterone levels, indicative of secondary ovulation, occurred soon after 3 of the injections of eCG. In the other 6 mares, however, including Mare PM which had doses as high as 60,000, 150,000 and 300,000 i.u. eCG on days 38, 46 and 59 respectively, progesterone levels were completely unaffected by the treatment and it was clear that no secondary ovulations occurred.

As mentioned previously, the conceptus was removed surgically at hysterotomy at selected times between days 65 and 87 from 7 of the eCG-treated mares and it was resorbed spontaneously at around day 90 in the 8th. Gross and histological examination of the tissues indicated that 6 of the 7 removed conceptuses

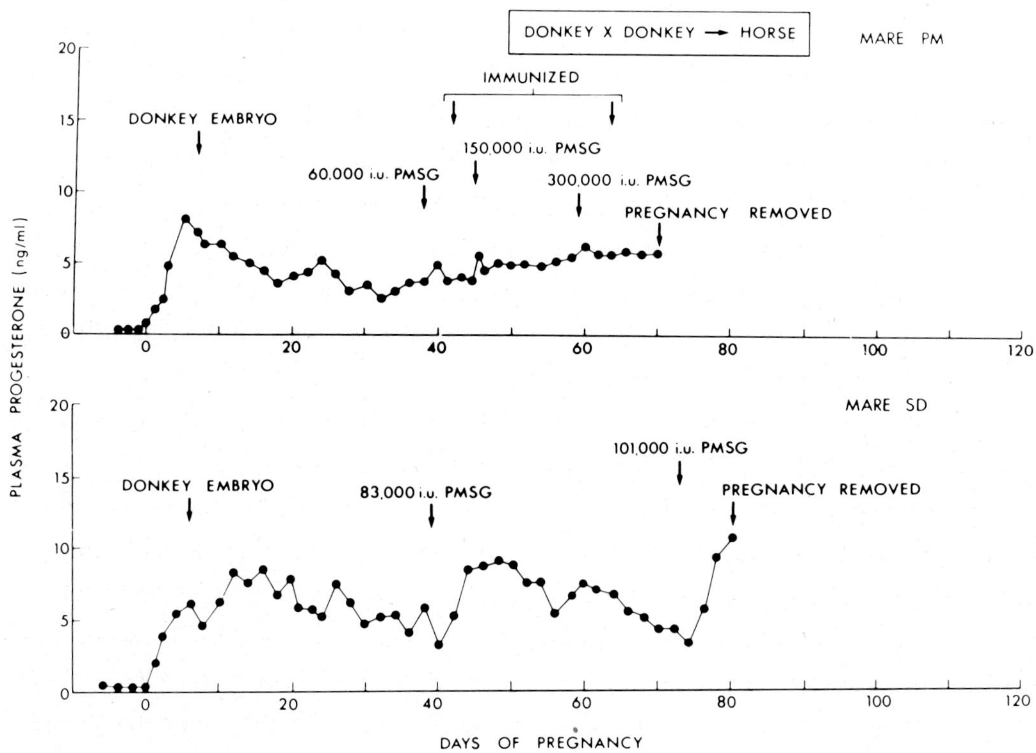

Fig. 5 b. Peripheral plasma progesterone concentrations in 2 Welsh Pony mares carrying transferred donkey conceptuses in 1981. The vertical arrows show the days on which the mares were given intravenous injections of varying doses of partially purified eCG (PMSG) dissolved in saline. Mare PM was also given intravenous injections of 2 x 10^8 washed lymphocytes from the genetic sire and dam of the donkey foetus on the days shown as Immunized.

would have failed at the expected time of day 85-95 had they been left in situ. In the remaining mare (JR), however, implantation was firmly established, the foetus was normal in size and appearance for the stage of pregnancy and the degree of leukocyte accumulation in the maternal endometrium was much less than in the other animals. From these findings it was concluded that the pregnancy would almost certainly have been carried to term had it remained in utero. Although Mare JR had been treated with exogenous eCG, the total dose of 200,000 i.u. given in 4 equal injections between days 50 and 78 was appreciably less than that given to some of the other mares in which pregnancy failed. Similarily, it is unlikely that the increase in progesterone levels which occurred in Mare JR as a result of a secondary ovulation induced by one of the eCG injections played a significant part in maintaining her pregnancy, since two such 2° CL developed in Mare SD following the eCG injections and, despite the relatively high progesterone concentrations in her plasma at the time of removal of her conceptus on day 80, the foetus was nonetheless in an advanced state of degeneration when recovered. From these observations it seems reasonable to conclude that eCG per se, functioning either directly to provide immunoprotection of the xenogeneic trophoblast or indirectly as a luteotrophin to stimulate ovarian progesterone secretion, is not the essential factor whose absence causes most of the donkey-in-horse pregnancies to fail at a very precise stage of gestation. Such a conclusion is supported by Mare HT in 1979 (see Fig. 3) which carried her donkey foetus to term in the absence of endometrial cup development, eCG secretion, 2° CL and exogenous hormonal therapy.

The finding in the present experiments that exogenous eCG was unable to stimulate the formation of 2° CL in most animals, is of some significance. Although the amounts given were, of necessity, less than that which could be expected to be present in the majority of mares carrying normal horse conceptuses, estimation of the blood volume of Mare PM, which failed to respond to any of the 3 large doses of eCG given to her, indicated that she could have had a serum level of around 5-10 i.u. eCG/ml after the second and third injections of 150,000 and 300,000 i.u. eCG respectively. This concentration is as high as the peak serum eCG levels attained by some mares carrying interspecies mule conceptuses which, as discussed previously, undergo the same degree of secondary luteal development as mares carrying normal horse foetuses (see Fig. 2).

The fact that secondary ovulations were apparently induced by the eCG injections on 3 occasions (Mares JR and SD, Fig. 5 a, b) seems to support the hypothesis, proposed originally by Evans and Irvine (1975) and confirmed by Urwin and Allen (1982), that it is the continuing rhythmical release of pituitary FSH during pregnancy, not eCG, which stimulates secondary follicular growth; eCG acts merely to induce ovulation or luteinization of the mature follicles to form the accessory CL. Thus, in the mares carrying donkey conceptuses which were given eCG, it may be argued that those which failed to respond did so because the injections were fortuitously given at times during the pituitary FSH-controlled waves of follicular development and regression when follicles of sufficient maturity to be able to respond to the LH-like activity of the eCG were not present in the ovaries. Conversely, a mature follicle did happen to be present in the ovaries of Mares JR and SD when they were treated and so secondary ovulations occurred.

Is eCG the luteotrophin of equine pregnancy?

In addition to the findings in the transferred donkey-in-horse pregnancies discussed above, many other observations in recent years have together painted a most confusing picture of the possible endocrinological functions and significance in pregnancy of the relatively enormous quantity of eCG that is secreted into the maternal circulation during such a short period of gestation. Does this unique hormone exert gonadotrophic and/or luteotrophic properties which are essential to maintain pregnancy, or is it perhaps nothing more than an irrelevant by-product of the endometrial cups which themselves have a more important and hitherto unrealized function concerning the establishment of implantation and the provision of immunoprotection for the allograft foetus?

As discussed previously, there is now little doubt that it is pituitary FSH, rather than eCG, which stimulates the necessary secondary follicular growth for accessory luteal development. Mature follicles are already present in the mare's ovaries by day 19-23 (Bain, 1967) and further waves of follicular growth occur at approximately 10-day intervals thereafter (van Rensburg and van Niekerk, 1968). Pituitary FSH is released in rhythmical wave-like surges at intervals of approximately 10-12 days throughout the breeding season in cycling mares (Evans and Irvine, 1975). This release pattern is basically controlled by photoperiod and it is not influenced by negative feedback from progesterone or other steroid hormones (Snyder et al., 1979). Thus, the pattern continues unchanged during early pregnancy (Evans and Irvine, 1975; Urwin and Allen, 1982) provided this period falls within the spring and summer months of the equine breeding season. The surges of FSH stimulate the observed waves of follicular growth and eCG, by virtue of its LH-like activity, will induce ovulation of any follicle that matures sufficiently to respond. From this it may be concluded that photoperiod, by virtue of its underlying control of pituitary FSH release, is the most important factor controlling secondary luteal development in pregnancy. Support for this hypothesis is provided by the observations of W.E. Allen that mares which conceive late in the breeding season and so are between days 40 and 120 of pregnancy during the winter anoestrous season have similar serum eCG profiles to, but fewer secondary ovulations as detected by rectal palpation of the ovaries than, equivalent mares mated earlier in the year so that the period of eCG secretion and secondary ovarian activity occurs during the long-day summer months (Allen, W.E., 1975).

Another question of interest is whether of not eCG is truly luteotrophic in the pregnant mare. That is, having provided the LH-like gonadotrophic activity to induce ovulation and/or luteinization of the secondary follicles, does it then act alone to prolong the life of the secondary corpora lutea and stimulate their secretion of progesterone? eCG almost certainly has some luteotrophic function during normal intraspecies pregnancy since a) the steady decline in secretory activity of the 1° CL which occurs during the early period of day 15-40 after ovulation is apparently halted when eCG secretion commences. Thereafter the 1° CL persists in the ovaries and continues to secrete progesterone for as long as the 2° CL (Squires and Ginther, 1975), b) both the primary and secondary CL remain functional while eCG secretion continues but regress soon after it disappears from the blood; the coincidence of these two phenomena in normal pregnancy is too great to be ignored, c) addition of eCG to the culture medium stimulates progesterone secretion by equine luteal slices incubated in vitro (Squires et al., 1979). Nevertheless, there is now strong evidence that eCG is not capable of maintaining luteal function alone and another luteotrophin, possibly of foetal origin, may also be involved. Spontaneous abortion or surgical removal of the conceptus from the uterus occurring after the invasion of chorionic girdle cells on day 36-38 does not interfere with the lifespan and secretory activity of the endometrial cups. They develop and regress in the normal manner and continue to secrete large quantities of eCG into the maternal bloodstream just as if the mare was still pregnant (Allen, 1970; Mitchell and Allen, 1975). Luteal function persists while eCG secretion continues in some of these animals but in others all the corpora lutea regress within 15-20 days after loss of the pregnancy despite the continued presence of high concentrations of eCG in the serum. Even more surprising, oestrus and ovulation does not normally recur in mares which have lost their pregnancy in this manner and which have no functional CL remaining in their ovaries until the endometrial cups have regressed completely and eCG has disappeared from the blood (Urwin, V.E. and Allen, W.R., unpublished data).

Further evidence in support of the existence of a luteotropin in pregnancy other than eCG is provided by the persistence of the 1° CL in the mares carrying the transferred donkey conceptuses. Examination of the plasma progesterone profiles shown in Fig. 3 of the 4 mares which were not given any exogenous hormones shows clearly that the 1° CL remained present and functional in the complete absence of eCG in all 4 animals while the conceptus remained alive. It continued beyond day 120 in Mare HT which carried her pregnancy to term and luteal regression only commenced in the other 3 animals as the foetus was dying. This would seem to indicate that the proposed non-eCG luteotrophin is also likely to be foetal in origin. However, the premature regression of the 1° CL that occurred in the mares carrying donkey conceptuses when they were given allyl trenbolone argues just as strongly in favour of pituitary LH as the missing factor.

Finally, the question arises as to whether 2° CL are, after all, really necessary to maintain pregnancy in equids. And the answer, in the light of Mare HT which carried her xenogeneic extraspecies donkey conceptus to term without the aid of any secondary luteal formation, together with other isolated mares carrying normal intraspecies conceptuses which similarily remain pregnant without accessory luteal support (see Allen, 1982), must be no. One is therefore left puzzling why nature saw fit to impose a minority system on the mare with regard to the progestagenic maintenance of pregnancy. Whatever advantage was gained for equids by the evolutionary adaptation which gave secondary luteal development in early pregnancy remains, as yet, unclear.

REFERENCES

Allen, W.E., 1975. Ovarian changes during early pregnancy in pony mares in relation to PMSG production. J. Reprod. Fert., Suppl. 23: 425-428.

Allen, W.R., 1969a. The immunological measurement of Pregnant Mare Serum Gonadotrophin. J. Endocr., 43: 593-598.

Allen, W.R., 1969b. Factors influencing Pregnant Mare Serum Gonadotrophin production. Nature, Lond. 223: 64-66.

Allen, W.R., 1970. Equine gonadotrophins. PhD thesis, University of Cambridge.

Allen, W.R., 1975. Immunological aspects of the equine endometrial cup reaction. In: Immunobiology of Trophoblast, pp. 217-253. Editors R.G. Edwards, C.W.S. Howe and M.H. Johnson. Cambridge University Press.

Allen, W.R., 1979. Maternal recognition of pregnancy and immunological implications of trophoblast-endometrium interactions in equids. In: Maternal Recognition of Pregnancy (Ciba Fdn. Symp. No. 64), pp. 323-352. Excerpta Medica, Amsterdam.

Allen, W.R., 1982. Immunological aspects of the endometrial cup reaction and the effect of xenogeneic pregnancy in horses and donkeys. J. Reprod. Fert., Suppl. 31: 57-94.

Allen, W.R. and Moor, R.M., 1972. The origin of the equine endometrial cups. I. Production of PMSG by fetal trophoblast cells. J. Reprod. Fert., 29: 313-316.

Allen, W.R., Hamilton, D.W. and Moor, R.M., 1973. The origin of the equine endometrial cups. II. Invasion of the endometrium by trophoblast. Anat. Rec., 177: 485-501.

Amoroso, E.C., 1955. Endocrinology of pregnancy. Br. med. Bull., 11, 117-125.

Amoroso, E.C., Hancock, J.L. and Rowlands, I.W., 1948. Ovarian activity in the pregnant mare. Nature, Lond., 161: 355-356.

Bain, A.M., 1967. The ovaries of the mare during early pregnancy. Vet. Rec., 80: 229-231.

Bielanski, W., Ewy, Z. and Pigoniowa, H., 1956. Differences in the level of gonadotrophin in the serum of pregnant mares. Proc. 3rd. Int. Congr. Anim. Reprod., Cambridge, pp. 110-111.

Clegg, M.T., Boda, J.M. and Cole, H.H., 1954. The endometrial cups and allantochorionic pouches in the mare with emphasis on the source of equine gonadotropin. Endocrinology, 54: 448-463.

Cole, H.H. and Goss, H., 1943. The source of equine gonadotropin. In: Essays in Honor of Herbert M. Evans, pp. 107-119. University of California Press, Berkeley.

Cole, H.H. and Hart, G.H., 1930. The potency of blood serum of mares in progressive stages of pregnancy in effecting sexual maturity of the immature rat. Am. J. Physiol., 93: 57-68.

Day, F.T. and Rowlands, I.W., 1940. The time and rate of appearance of gonadotrophin in the serum of pregnant mares. J. Endocr., 2: 255-261.

Day, F.T. and Rowlands, I.W., 1947. Serum gonadotropohin in Welsh and Shetland ponies. J. Endocr., 5: 1-8.

Evans, M.J. and Irvine, C.H.G., 1975. The serum concentrations of FSH, LH and progesterone in the oestrous cycle and early pregnancy in the mare. J. Reprod. Fert., Suppl. 23: 193-200.

Goss, H. and Cole, H.H., 1940. Further studies on the purification of mare gonadotropic hormone. Endocrinololy, 26: 244-249.

Gray, A.P., 1954. Mammalian Hybrids, pp. 45-49. Commonwealth Agricultural Bureaux, Farnham Royal, Bucks.

Guillou, F. and Combarnous, Y., 1983. Purification of equine gonadotropins and comparative study of their acid-dissociation and receptor-binding specificity. Biochim. Biophys. Acta, 755: 229-236.

Holtan, D.W., Squires, E.L., Lapin, D.R. and Ginther, O.J., 1979. Effect of ovariectomy on pregnancy in mares. J. Reprod. Fert., Suppl. 27: 457-463.

King, J.M., 1965. Comparative aspects of reproduction in Equidae. PhD thesis. University of Cambridge.

Mitchell, D. and Allen, W.R., 1975. Observations on reproductive performance in the yearling mare. J. Reprod. Fert., Suppl. 23: 531-536.

Papkoff, H., 1974. Chemical and biological properties of the subunits of Pregnant Mare Serum Gonadotropin. Biochem. Biophys. Res. Com., 58: 397-404.

Pineda, M.H., Garcia, M.C. and Ginther, O.J., 1973. Effect of an antiserum against an equine pituitary fraction on corpus luteum and follicles in mares during diestrus. Am. J. vet. Res. 34: 181-183.

Rowlands, I.W., 1963. Levels of gonadotrophins in tissues and fluids with emphasis on domestic animals. In: Gonadotropins: Their Chemical and Biological Properties and Secretory Control, pp. 75-107. Editor H.H. Cole. W.H. Freeman and Company, San Francisco.

Schauder, W., 1912. Untersuchungen über die Eihäute und Embryotrophe des Pferdes. Arch. Anal. Physiol. (1912), 259-302.

Sheldrick, E.L., Wright, P.J., Allen, W.R. and Heap, R.B., 1977. Metabolic clearance rate, production rate and source of progesterone in donkeys with fetuses of different genotypes. J. Reprod. Fert., 51: 473-476.

Shideler, R.K., Squires, E.L., Voss, J.L., Eikenberry, D.J. and Pickett, B.W., 1982. Exogenous progestagen therapy of ovariectomized pregnant mares. J. Reprod. Fert., Suppl. 32: 459-464.

Short, R.V., 1958. Progesterone in tissues and body fluids. PhD thesis. University of Cambridge.

Short, R.V., 1969. Implantation and the maternal recognition of pregnancy. In: Foetal Autonomy (Ciba Found. Symp.) pp. 2-31. Editors G.E.W. Wolstenholme and M. O'Connor. Churchill, London.

Smith, J.G., Hanks, J. and Short, R.V., 1969. Biochemical observations on the corpora lutea of the African elephant (Loxodonta Africana). J. Reprod. Fert., 20: 111-117.

Snyder, D.A., Turner, D.D., Miller, K.F., Garcia, M.C. and Ginther, O.J., 1979. Follicular and gonadotrophic changes during transition from ovulatory to anovulatory seasons. J. Reprod. Fert., Suppl. 27: 95-101.

Squires, E.L. and Ginther, O.J., 1975. Follicular and luteal development in pregnant mares. J. Reprod. Fert., Suppl. 23: 429-433.

Squires, E.L., Stevens, W.B., Pickett, B.W. and Nett, T.M., 1979. Role of Pregnant Mare Serum Gonadotropin in luteal function of pregnant mares. Am. J. Vet. Res., 40: 889-891.

Stewart, F. and Allen, W.R., 1979. The binding of FSH, LH and PMSG to equine gonadal tissues. J. Reprod. Fert., Suppl. 27: 431-440.

Stewart, F. and Allen, W.R., 1981. Biological functions and receptor binding activities of equine chorionic gonadotrophin. J. Reprod. Fert., 62: 527-536.

Stewart, F., Allen, W.R. and Moor, R.M., 1976. Pregnant Mare Serum Gonadotrophin: ratio of follicle stimulating hormone and luteinizing hormone activities measured by radioreceptor assay. J. Endocr., 71: 371-382.

Stewart, F., Allen, W.R. and Moor, R.M., 1977. Influence of foetal genotype on the follicle-stimulating hormone: luteinizing hormone ratio of Pregnant Mare Serum Gonadotrophin. J. Endocr., 73: 419-425.

Urwin, V.E. and Allen, W.R., 1982. The pituitary and chorionic gonadotrophic control of ovarian function during early pregnancy in equids. J. Reprod. Fert., Suppl. 32: 371-381.

van Rensburg, S.J. and van Niekerk, C.H., 1968. Ovarian function, follicular oestradiol-17β and luteal progesterone and 20α-hydroxy-pregn-4-ene-3-one in cycling and pregnant equines. Onderstepoort J. Vet. Res., 35: 301-318.